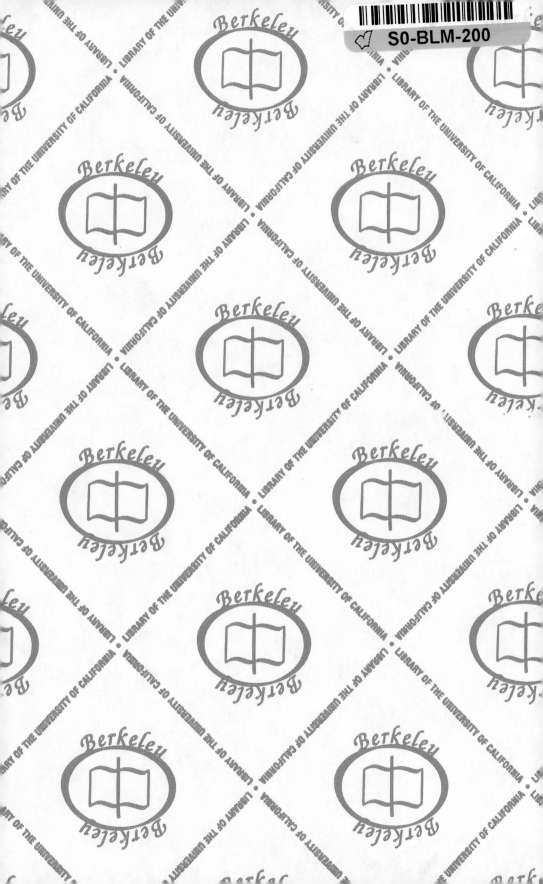

# Optics in Astrophysics

# NATO Science Series

*A Series presenting the results of scientific meetings supported under the NATO Science Programme.*

The Series is published by IOS Press, Amsterdam, and Springer (formerly Kluwer Academic Publishers) in conjunction with the NATO Public Diplomacy Division.

*Sub-Series*

| | |
|---|---|
| I.   **Life and Behavioural Sciences** | IOS Press |
| II.  **Mathematics, Physics and Chemistry** | Springer (formerly Kluwer Academic Publishers) |
| III. **Computer and Systems Science** | IOS Press |
| IV.  **Earth and Environmental Sciences** | Springer (formerly Kluwer Academic Publishers) |

The NATO Science Series continues the series of books published formerly as the NATO ASI Series.

The NATO Science Programme offers support for collaboration in civil science between scientists of countries of the Euro-Atlantic Partnership Council. The types of scientific meeting generally supported are "Advanced Study Institutes" and "Advanced Research Workshops", and the NATO Science Series collects together the results of these meetings. The meetings are co-organized by scientists from NATO countries and scientists from NATO's Partner countries — countries of the CIS and Central and Eastern Europe.

**Advanced Study Institutes** are high-level tutorial courses offering in-depth study of latest advances in a field.
**Advanced Research Workshops** are expert meetings aimed at critical assessment of a field, and identification of directions for future action.

As a consequence of the restructuring of the NATO Science Programme in 1999, the NATO Science Series was re-organized to the four sub-series noted above. Please consult the following web sites for information on previous volumes published in the Series.

http://www.nato.int/science
http://www.springeronline.com
http://www.iospress.nl

**Series II: Mathematics, Physics and Chemistry – Vol. 198**

# Optics in Astrophysics

edited by

## Renaud Foy

Observatoire de Lyon / CRAL,
France

and

## Françoise-Claude Foy

Observatoire de Lyon / CRAL,
France

Springer

Published in cooperation with NATO Public Diplomacy Division

phys

0 1439136 x

Proceedings of the NATO Advanced Study Institute on
Optics in Astrophysics
Cargèse, France
16–28 September 2002

A C.I.P. Catalogue record for this book is available from the Library of Congress.

ISBN-10 1-4020-3436-9 (PB)
ISBN-13 978-1-4020-3436-0 (PB)
ISBN-10 1-4020-3435-0 (HB)
ISBN-10 1-4020-3437-7 (e-book)
ISBN-13 978-1-4020-3435-7 (HB)
ISBN-13 978-1-4020-3437-4 (e-book)

Published by Springer,
P.O. Box 17, 3300 AA Dordrecht, The Netherlands.

*www.springeronline.com*

*Printed on acid-free paper*

MJT
2/15/06

# Contents

# Preface

Astronomy and Optics have a long tradition of cooperation, streching back at least to the invention of the first refractor by Galileo. Significant breakthroughs in Astronomy and later optical Astrophysics –from the ultraviolet to the far infrared– have been linked to some major advance in Optics. Indeed, theoretical astrophysics has to eventually rely on or be confronted with observations. For a long time this strong link has been based mostly on the design of (optical) telescopes. With the advent of photographic plates and, more recently, modern detectors, the field of focal instrumentation has grown enormously, thanks in particular to the emergence of new concepts in optics. Very often there has also been a feedback from astrophysics to optics.

Astrophysics is now facing a quantum leap in the understanding of major problems including the evolution of the very early Universe, the presence of exoplanets and exobiology. The tools of astrophysicists in Optics will move from the 8-10 m class telescopes to decametric optical telescopes (DOTs: 30 - 60 m) or even hectometric optical telescopes (HOTs: $\gtrsim$100m). The design of these telescopes and their focal plane instruments represents a fantastic challenge. With the exception of the 10m Keck telescope and the similar Gran Telescopio Canarias, 8 m telescopes are more or less extrapolated from the generation of 4-6 m telescopes. It will simply not work to extrapolate again towards DOTs. The same fact holds for the instrumentation: classical concepts sometimes leads to monstrous instruments at 8 m telescopes. This may be alleviated by adaptive optics, which can provide diffraction limited images at the focus, sharper by one to two magnitudes than images limited by the atmospheric turbulence. Adaptive optics relies on critical components such as wavefront sensors and deformable mirrors. However, adaptive optics at DOTs or HOTs is itself a big challenge.

Very large diameters, beyond 100 m, now correspond exclusively to diluted apertures and amplitude interferometry. Another big challenge for the next two decades to develop kilometric arrays of telescopes, having optical coherent recombination, presumably with fiber optics, focal instrumentation with a significant field of view, etc.

High angular resolution is a new-comer to astrophysics, arriving on the scene at the end of the last century. Speckle interferometry, adaptive optics and interferometry have shed new light on many fields of astrophysics, and their future is even more promising. Only the most recent advances in optics have allowed the frontier set by the atmospheric turbulence to be broken. However, there are two other major actors: the huge progress in detectors and a strong effort to develop mathematical algorithms to restore wavefronts and images with very complex point spread functions. Because of this complexity, design and system studies for the next generation of DOTS and HOTS and of their optical instruments will have to include these algorithms -and not only data processing softwares- from the very beginning.

The goal of this School is to address these new domains with an interdisciplinary approach, in particular bringing together young scientists from astrophysics, optics and image processing. These young scientists will play an important role in forthcoming instruments and facilities. We also tried to open the School to new approaches in the handling of photons, at least in Astrophysics. If they are feasible, then they will provide astrophysicists with diagnoses of physical processes which are barely or simply not accessible with current technologies e.g. gravitational waves and the degree of coherence of astrophysical sources.

The youth of the attendants and their very active participation in the discussions we had after each lecture and also during at least one hour each evening make us confident that the School will leave a lasting scientific impression on their minds [1]

*Renaud Foy*
*Françoise-Claude Foy*

---

[1]Most of the lessons given at the Institut d'Études Scientifiques de Cargèse, as well as a few posters and additional documents to this book, can be downloaded from *http://www-obs.univ-lyon1.fr/nato-corse/*.

# Acknowledgments

The Editors wish to sincerely thank the Lecturers of the Summer School for the excellent quality of their lectures, and the Contributors to this book for the time they have spent on their manuscripts.

We also thank the co-chairman, Y.Y. Balega, and the members of the Science Organizing Committee (P. Chavel, Ch. Dainty, C. Froehly, J. Nelson, M. Tallon) for their help in the preparation of the programme of the School, and in searching for lecturers. Our thanks also go to the organizations which have sponsored this Advanced Study Institute: the Scientific Affairs Division of NATO, the High-Level Scientific Conferences programme of the European Commission and the Formation Permanente programme of the French CNRS, and to their staff, in particular Drs F. Pedrazzini and Ph. Froissard, and Mme P. Landais.

Special thanks are due to the staff of the Institut d'Études Scientifiques de Cargèse for the warm hospitality and for the continuous and patient assistance, in particular Mme B. Cassegrain, ... and not forgetting the kitchen staff!

We thank the participants who kindly provided us with the photos included in this book. Finally it is a pleasure to thank all the participants for the many fruitful discussions and the very friendly atmosphere of the School.

# List of Participants

ALLINGTON-SMITH Jeremy
Durham Univ. (UK)
j.r.allington-smith@durham.ac.uk

AMORES Eduardo Brescansin de
Sao Paulo (Brazil)
amores@astro.iag.usp.br

AREZKI Brahim
LAOG/Grenoble (France)
arezki@obs.ujf-grenoble.fr

ASSEMAT François
GEPI/Meudon (France)
francois.assemat@obspm.fr

BABA AISSA Djounai
CRAAG - Alger (Algeria)
bdjounai@yahoo.fr

BALEGA Yura
SAO (Russia)
balega@sao.ru

BAYAZITOV Ural
Bashkortostan (Russia)
bayazit@mail.ru

BELETIC James
Rockwell Scientific Company
jbeletic@rwsc.com

BIANCO Andrea
Brera (Italy)
andrea.bianco@polimi.it

BOMBARDELLI Claudio
SAO (USA-Ma)
cbombardelli@cfa.harvard.edu

BUSONI Simone
OAA - Arcetri (Italy)
busoni@arcetri.astro.it

CAKIRLI Omur
Ege univ/Izmir (Turkey)

cakirli@astronomy.sci.ege.edu.tr

CARDWELL Andrew
IAC - Tenerife (Spain)
cardwell@ll.iac.es

CARVALHO Maria Ines Barbosa de
Porto (Portugal)
mines@fe.up.pt

COFFEY Deirdre
DIAS - Dublin (Ireland)
dac@mir.cp.dias.ie

COUDREAU Thomas
LKB - Paris (France)
Thomas.Coudreau@spectro.jussieu.fr

DEVANEY Nicholas
IAC/GRANTECAN - Tenerife (Spain)
ndevaney@iac.es

DIERICKX Philippe
ESO - Graching (Germany)
pdierick@eso.org

EL IDRISSI Moha
Marrakech (Marocco)
idrissi@pleiades.unice.fr

FERNANDEZ Enrique J.
Murcia (Spain)
enriquej@um.es

FERRUZZI CARUSO Debora
OAA - Arcetri (Italy)
ferruzzi@arcetri.astro.it

FLICKER Ralf
Lund (Sweden)
ralf@astro.lu.se

FOY Françoise-Claude
CRAL - Lyon (France)
fcfoy@obs.univ-lyon1.fr

FOY Renaud
CRAL - Lyon (France)
foy@obs.univ-lyon1.fr

FURESZ Gabor
Szeged (Hungary)
fureszg@neptun.physx.u-szeged.hu

GIRARD Julien
CRAL - Lyon (France)
girard@obs.univ-lyon1.fr

GONCHAROV Alexander
Lund (Sweden)
alexander.goncharov@nuigalway.ie

HASANOVA Leyla
SAO (Russia)
layla@sao.ru

HINNACHI Riadh
Tunis (Tunisia)
riadhzah@yahoo.fr

ISERLOHE Christof
M.P.G. - Garching(Germany)
iserlohe@mpe.mpg.de

KABLAK Natalia
Uzhgorod (Ukraine)
space@univ.uzhgorod.ua

KAMENICKY Milan
Tatranska Lom.(Slovaquia)
mkamen@nextra.sk

KAYDASH Vadym G.
Kharkov (Ukraine)
kvg@online.kharkiv.com

KUZ'KOV Vladimir
M.A.O. - Kyev (Ukraine)
kuzkov@mao.kiev.ua

KUZNETCHOV Maxim
Rostov (Russia)
max_kuznecov@mail.ru

KUZNETZOVA Juliana
M.A.O.- Kyev (Ukraine)

juliana@mao.kiev.ua

LAGE Armindo Luis Vilar Soares
Porto (Portugal)
alage@fe.up.pt

LANDRAGIN Arnaud
LPTF - Paris (France)
Arnaud.Landragin@obspm.fr

LANE Richard
Applied Research Associates NZ Ltd
email@aranz.com

LAURENT-PROST Florence
CRAL - Lyon (France)
prost@obs.univ-lyon1.fr

LEHUREAU Jean-Claude
Thalès/LCR - Orsay (France)
jean-claude.lehureau@thalesgroup.com

LIOTARD Arnaud
LAM - Marseille (France)
a.liotard@ic.ac.uk

LOMBARDI Marco
IAEF - Bonn (Germany)
lombardi@astro.uni-bonn.de

LORIETTE Vincent
E.S.P.C.I. - Paris (France)
loriette@optique.espci.fr

LÜHE Oskar von der
KIS - Freiburg (Germany)
ovdluhe@kis.uni-freiburg.de

MACKOWSKI Jean-Marie
IPN - Lyon (France)
j-m.mackowski@lma.in2p3.fr

MAGAGNA Carlo
Padova (Italy)
magagna@pd.astro.it

MANESCAU Antonio
IAC - La Laguna (Spain)
amh@ll.iac.es

MANZANERA Roman Silvestre
Murcia (Spain)
silmanro@um.es

MARTIN Fleitas Juan Manuel
IAC/Grantecan - La Laguna (Spain)
jmartin@ll.iac.es

MATOS Anibal Castilho Coimbra
Porto (Portugal)
anibal@fe.up.pt

MAWET Dimitri
Liège (Belgium)
Dimitri.Mawet@student.ulg.ac.be

MONNET Guy
E.S.O. - Garching (Germany)
gmonnet@eso.org

MOREAU Julien
ESPCI - Paris (France)
moreau@optique.espci.fr

MORRIS Timothy James
Durham Univ. (UK)
T.J.Morris@durham.ac.uk

MURRAY Larry
Sac Peak (USA-Az)
mrlarrypatrick@hotmail.com

NELSON Jerry
UC - Santa Cruz (USA-Ca)
jnelson@ucolick.org

NIRSKI Jaroslaw
Torun (Poland)
nirski@astri.uni.torun.pl

OREIRO Rey Raquel
I.A.C. - La Laguna (Spain)
ror@ll.iac.es

OTI Gonzalez Jose E.
Santander (Spain)
otije@unican.es

OZISIK Tuncay
Istanbul (Turkey)

tuncay@astroa.physics.metu.edu.trk

PARKS Robert E.
Optical Pers. (USA-Az)
reparks@optiper.com

PASANEN Mikko
Turun Univ.(Finland)
mianpa@utu.fi

PATERSON Carl
ICSTM - London (UK)
carl@ic.ac.uk

PENNINGTON Deanna
LLNL - Livermore (USA-Ca)
pennington1@llnl.gov

PERRIN Marshall D.
Berkeley (USA-Ca)
mperrin@astro.berkeley.edu

PICHON Christophe
Strasbourg (France)
pichon@pleiades.u-strasbg.fr

PIQUE Jean-Paul
LSP - Grenoble (France)
pique@spectro.ujf-grenoble.fr

POTANIN Sergey A.
Sternberg - Moskow (Russia)
potanin@sai.msu.ru

PRIBULLA Theodor
Astro. Inst. (Slovaquia)
pribulla@ta3.sk

QUIROS-PACHECO Fernando
ICSTM - London (UK)
f.quiros-pacheco@ic.ac.uk

RATAJCZAK Roman
Poznan (Poland)
wnuk@amu.edu.pl

REYNAUD François
IRCOM - Limoges (France)
f.reynaud@ircom.unilim.fr

RIAHI Riadh
Tunis (Tunisie)
riadhzah@yahoo.fr

ROMANYUK Yaroslav O.
M.A.O. - Kyev (Ukraine)
romaniuk@mao.kiev.ua

SAAD Somaya Mohamed
NRIAG - Cairo (Egypt)
somaya111@yahoo.com

SAILER Markus
Goettingen (Germany)
msailer@uni-sw.gwdg.de

SCHUMACHER Achim
IÁC - La Laguna (Spain)
Achim.Schumacher@Cern.Ch

SEVERSON Scott
CfAO/UCSC - Santa Cruz (USA-Ca)
severson@ucolick.org

SHEINIS Andrew
CfAO/UCSC - Santa Cruz (USA-Ca)
sheinis@ucolick.org

SILVA Diogo G.T.O.
Porto Univ. (Portugal)
ee01157@fe.up.pt

SKODA Petr
Ondrejov (Czech Rep.)
skoda@sunstel.asu.cas.cz

SOLOMOS Nikolaos H.
Appl.Opt.L./NO - Athens (Greece)
nsolom@mail.ariadne-t.gr

STÖSZ Jeffrey A.
DAO - Victoria (Canada)
Jeff.Stoesz@nrc.ca

STUPKA Michal
Ondrejov (Czech Rep.)
stupka@asu.cas.cz

STÜTZ Christian
Wien Univ. (Austria)
stuetz@tycho.astro.univie.ac.at

TATULLI Eric
LAOG - Grenoble (France)
Eric.Tatulli@obs.ujf-grenoble.fr

TAYLOR Luke Richard
LLNL - Liverore (USA-Ca)
taylor101@llnl.gov

THIÉBAUT Eric
CRAL - Lyon (France)
thiebaut@obs.univ-lyon1.fr

THOMAS Sandrine
AURA - La Serena (Chili)
ts_sandrine@yahoo.fr

THORSTEINSSON Hrobjartur
Cavendish - Cambridge (UK)
ht228@mrao.cam.ac.uk

TOZZI Andrea
OAA - Arcetri (Italy)
atozzi@arcetri.astro.it

VOLONTÉ Sergio
ESA - Paris (France)
sergio.volonte@esa.int

WASHÜTTL Albert
Postdam (Germany)
wasi@astro.univie.ac.at

WHELAN Emma
DIAS - Dublin (Ireland)
ewhelan@cp.dias.ie

ZAMKOTSIAN Frédéric
L.A.M. - Marseille (France)
Frederic.Zamkotsian@oamp.fr

ZOUHAIR Benkhaldoun
Marrakech (Marocco)
zouhair@ucam.ac.ma

# CONSTRAINTS OF GROUND-BASED OBSERVATIONS: THE ATMOSPHERE

Carl Paterson

*Imperial College London*

carl@ic.ac.uk

**Abstract**    This is a tutorial about the main optical properties of the Earth atmosphere as it affects incoming radiation from astrophysical sources. Turbulence is a random process, of which statitical moments are described relying on the Kolmogorov model. The phase structure function and the Fried parameter $r_0$ are introduced. Analytical expressions of the degradation of the optical transfer function due to the turbulence, and the resulting Strehl ratio and anisoplanatism are derived.

**Keywords:**    atmosphere, turbulence, Kolmogorov model, phase structure function, transfer function, Strehl ratio, anisoplanatism

## Introduction

Here we present an overview of the impact that the Earth's atmosphere, or more precisely, the refraction of the atmosphere, has on astronomical observations made through it. There are two major sources: the first being the essentially fixed variation of the refractive index with height above the ground, as the atmosphere becomes more sparse, the second being the dynamic random fluctuations in the refractive index which arise in atmospheric turbulence. Our emphasis will be on the latter dynamic effects. In fact it is worth bearing in mind that the refractive effects are not the only constraints of the atmosphere: absorption and scattering, which vary strongly with wavelength, result in finite spectral observation windows and regions of the spectrum where observations are not possible, magnitude and spectral distortions, and non-zero sky background. These will not be discussed here.

## Atmospheric Refraction

The variation of refractive index with height above the surface gives rise to an apparent shift in the positions of astronomical objects. The angle depends

*R. Foy and F. C. Foy (eds.), Optics in Astrophysics, 1–10.*

on the air density at the telescope and approximately on the tan of the zenith angle of the object. Although the exact position may be important for some observations, it does not affect the resolution of imaging. More important is the effect of atmospheric dispersion, which gives rise to a variation of the positional shift with wavelength (typically a fraction of an arcsec to a few arcsecs.) It is important that this is corrected for high-resolution imaging, for example by use of a Risley Prism Pair—a pair of prisms which can be rotated to act as an adjustable dispersion element. More challenging is the dynamic effect of atmospheric turbulence.

## Atmospheric Turbulence

The origin of atmospheric turbulence is diurnal heating of the Earth's surface, which gives rise to the convection currents that ultimately drive weather. Differential velocities caused perhaps when the wind encounters an obstacle such as a mountain, result in turbulent flow. The strength of the turbulence depends on a number of factors, including geography: it is noted that the best observation sites tend to be the most windward mountaintops of a range—downwind sites experience more severe turbulence caused by the disturbance of those mountains upwind.

For an incompressible viscous fluid (such as the atmosphere) there are two types of flow behaviour: 1) Laminar, in which the flow is uniform and regular, and 2) Turbulent, which is characterized by dynamic mixing with random subflows referred to as turbulent eddies. Which of these two flow types occurs depends on the ratio of the strengths of two types of forces governing the motion: lossless inertial forces and dissipative viscous forces. The ratio is characterized by the dimensionless Reynolds number $Re$.

For planar flow in a viscous fluid, frictional forces between the flow planes give rise to shear stress across the planes of the flow. For flow along $\hat{x}$, the shear stress across an area A is

$$\frac{\vec{F}_{\text{visc}}}{A} = \eta \frac{\partial v_x}{\partial y} \hat{x}, \tag{1}$$

where $\eta$ is the viscosity. Gradients in the shear stress give rise to local acceleration on the fluid,

$$\frac{\mathrm{d}v_x}{\mathrm{d}t} = \nu \frac{\mathrm{d}^2 v_x}{\mathrm{d}y^2}, \tag{2}$$

where $\nu = \eta/\rho$ is the kinematic viscosity, $\rho$ being the density of the fluid.

If we consider a flow, characterized by a typical velocity of $V$ and a scale size $L$, the strength of the viscous forces on a unit mass of the fluid is $\sim \nu V L$. The strength of the of the inertial forces in the flow is $\sim V^2/L$ (i.e., the force required to change the flow velocity by $V$ in distance $L$.). The ratio is,

$$\frac{\text{inertial forces}}{\text{viscous forces}} \sim \frac{VL}{\nu} = Re, \tag{3}$$

which is the Reynolds number. At small Reynolds numbers,the dissipative viscous forces are strong enough to keep the flow stable. For large Reynolds numbers, the dissipative forces are weak and the flow is turbulent. The crossover between laminar flow and turbulent flow occurs at a critical Reynolds number $R_c$. $R_c$ depends on the geometry of the flow, but typical values are in the range of tens or hundreds.

Consider as an example, flow of air near ground level with velocity $V \sim 5\,\mathrm{ms}^{-1}$, and size $L \sim 2\,\mathrm{m}$. The kinematic velocity of the atmosphere near ground level is $\nu \approx 1.5 \times 10^{-5}\,\mathrm{m^2 s^{-1}}$, which gives for the Reynolds number $Re \sim 10^5 \gg R_c$ and thus the flow will be turbulent.

By its random nature, turbulence does not lend itself easily to modelling starting from the differential equations for fluid flow (Navier–Stokes). However, a remarkably successful statistical model due to Kolmogorov has proven very useful for modelling the optical effects of the atmosphere.

## Random Processes — a short primer

To describe single-point measurements of a random process, we use the first-order probability density function $p_f(f)$. Then $p_f(f)\,df$ is the probability that a measurement will return a result between $f$ and $f + df$. We can characterize a random process by its moments. The $n$th moment is the ensemble average of $f^n$, denoted $\langle f^n \rangle$. For example, the mean is given by the first moment of the probability density function,

$$\langle f \rangle = \int_{-\infty}^{\infty} p_f(f) f^1 \, df. \tag{4}$$

It is often useful when looking at higher order moments first to subtract the mean. This gives us the *central moments*, the $n$th central moment is

$$\mu_n = \langle (f - \langle f \rangle)^n \rangle = \int_{-\infty}^{\infty} p_f(f)[f - \langle f \rangle]^n \, df. \tag{5}$$

An example is the variance, which is the second central moment $\mu_2$:

$$\mu_2 = \langle [f - \langle f \rangle]^2 \rangle = \langle f^2 \rangle - \langle f \rangle^2. \tag{6}$$

For normal statistics, the mean and the variance are completely sufficient to characterize the process: all the other moments are zero. For standard normal or Gaussian statistics (i.e., normal statistics with zero mean), the variance $\mu_2$

alone is sufficient. We shall consider turbulence to obey standard Gaussian statistics.

So far, we have been discussing so called first-order statistics, since we have only been describing the results of measurements at a single point. If we wish to describe the relationship between two measurements (e.g., values of the random variable measured at two different points in space, or at two different times), then we must use *second order statistics*. The *correlation* of two measurements at points $x_1$ and $x_2$ is defined as

$$B(x_1, x_2) = \langle f(x_1)f(x_2) \rangle, \tag{7}$$

and as its name suggests, gives a measure of how much the measurements at the two points $x_1$ and $x_2$ are related—a correlation of zero implies the measurements are statistically independent.

If a random process is described as *statistically homogeneous*, then its statistical moments are translation invariant, i.e., they do not depend on the positions $x_1$ and $x_2$ only on their difference. Thus the correlation reduces to:

$$B(x_1, x_2) = B(x_2 - x_1) = B(\Delta x) = \langle f(x)f(x + \Delta x) \rangle. \tag{8}$$

Similarly, if a process is described as *statistically stationary*, then its statistical moments are time invariant.

As with the central moments in first-order statistics, we can first subtract the mean. We define the *co-variance* (for a statistically homogeneous process) as

$$C(\Delta x) = \langle [f(x) - \langle f \rangle] [f(x + \Delta x) - \langle f \rangle] \rangle . \tag{9}$$

Note that the co-variance reduces to the variance for $\Delta x = 0$.

It is often useful to deal with the statistics in Fourier space. The Fourier transform of the correlation is called the *power spectrum*

$$\Phi(\kappa) = \int_{-\infty}^{\infty} B(x') \exp\left(-2\pi i\kappa x'\right) \, dx'. \tag{10}$$

For Kolmogorov statistics, it turns out that the power spectrum is infinite at the origin, which means that the variance is infinite. The *structure function* can be used instead of the co-variance to overcome this problem. It is defined as

$$D(\Delta x) = \left\langle [f(x) - f(x + \Delta x)]^2 \right\rangle \tag{11}$$
$$= 1/2 \left[ C(0) - C(\Delta x) \right].$$

Finally, a random process is described as *isotropic* if its statistics do not depend on orientation, i.e., for the three dimensional structure function,

$$D(\vec{\Delta r}) = D(|\vec{\Delta r}|) = D(\Delta r). \tag{12}$$

## Kolmogorov model of turbulence

The Kolmogorov model for turbulence is based on the so-called energy cascade. In this model, kinetic energy is injected into the large scale motions, for example, by convection or wind hitting an obstacle. This energy is transferred, by turbulent mixing, to successively smaller and smaller eddies, until eventually at the smallest scales, the viscous forces are sufficient to dissipate the energy and it is removed from the system. Within this motion there is a range of flow scales $\ell_0 \ll L \ll L_0$ over which no kinetic energy is injected and viscous dissipation is negligible. This is the *inertial subscale*, and its limits, $\ell_0$ and $L_0$ are the *inner scale* and *outer scale* respectively. For motion within the inertial subscale, the statistical properties of the turbulence are homogeneous, isotropic and depend only on $\varepsilon$, the rate of energy flow through the cascade, from which dimensional arguments lead to the 2/3 power law for the velocity structure function,

$$D_{vv}(\Delta r) = C_v^2 \Delta r^{2/3}, \tag{13}$$

where $C_v^2$ is the velocity structure constant.

The Kolmogorov velocity field mixes "packets" of air with different passive scalars; a passive scalar being one which does not exchange energy with the turbulent velocity flow. (Potential) temperature is such a passive scalar and the temperature fluctuations also follow the Kolmogorov law with a different proportionality constant

$$D_T(\Delta r) = C_T^2 \Delta r^{2/3}. \tag{14}$$

Using the ideal gas law and the relationship $(n-1) \propto \rho$ between refractive index $n$ and density $\rho$ leads us to the refractive index structure function,

$$D_n(\Delta R) = C_n^2 \Delta r^{2/3}. \tag{15}$$

$C_n^2$ is the refractive index structure constant and characterises the strength of the refractive index fluctuations. Taking the (3-D) Fourier transform leads to the Kolmogorov power spectrum of the refractive index fluctuations,

$$\Phi_n(\vec{\kappa}) = 0.033 C_n^2(z) |\vec{\kappa}|^{-11/3}. \tag{16}$$

Note that the Kolmogorov power spectrum is unphysical at low frequencies—the variance is infinite at $\kappa = 0$. In fact the turbulence is only homogeneous within a finite range—the inertial subrange. The modified von Karman spectral model includes effects of finite inner and outer scales,

$$\Phi_n(\vec{\kappa}) = \frac{0.033 C_n^2(z)}{(|\vec{\kappa}|^2 + \kappa_0^2)^{11/6}} \exp\left(-\frac{|\vec{\kappa}|^2}{\kappa_m^2}\right), \tag{17}$$

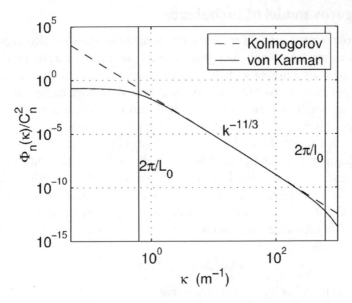

*Figure 1.* Kolmogorov and modified von Karman spectral models. ($L_0 = 10\,\mathrm{m}$ and $\ell_0 = 0.01\,\mathrm{m}$ for the von Karman plot.)

where $\kappa_0 = 2\pi/L_0$ and $\kappa_m = 5.92/\ell_0$. The outer scale beyond which the Kolmogorov spectrum is invalid depends on the turbulence causing mechanism but is typically many metres for astronomy. The inner scale takes is typically a few millimetres; experimental evidence points to a slight bump in the spectrum near the inner scale.

## The phase structure function and Fried's parameter

In astronomy, we are interested in the optical effects of the turbulence. A wave with complex amplitude $U(\vec{x}) = \exp[i\phi(\vec{x})]$ propagating through a layer of turbulence will pick up random retardations across its wavefront from the fluctuations in the refractive index, resulting in a random phase structure by the time it reaches the telescope pupil. If the turbulence is weak enough, the effect of the aberrations can be approximated by summing their phase along a path (the weak phase screen approximation), then the covariance of the complex amplitude at the telescope can be shown to be

$$C(\vec{x}) = \langle (U(\vec{x}')U^*(\vec{x}' + \vec{x})) \rangle$$
$$= \exp\left[-\tfrac{1}{2}D_\phi(\vec{x})\right]. \tag{18}$$

$D_\phi$ is the *phase structure function* which in turn is given by

$$D_\phi(\vec{x}) = 6.88\,(|\vec{x}|/r_0)^{5/3}, \tag{19}$$

$r_0$ is the *Fried parameter* and is given by a simple integral of the $C_n^2(z)$ height profile,

$$r_0 = \left[ \frac{2.91}{6.88} k^2 (\cos \gamma)^{-1} \int_0^\infty C_n^2(z) \, dz \right]^{-3/5}, \tag{20}$$

where $\gamma$ is the zenith angle, $k = 2\pi/\lambda$.[1] The Fried parameter characterises the strength of the optical effects of the turbulence. Its usefulness is illustrated with a number of related interpretations:

- It gives the crossover between diffraction-limited and turbulence-limited resolution. For aperture diameters smaller than $r_0$, close to diffraction limited imaging is possible without phase correction, for aperture diameters larger than $r_0$, the resolution is limited by the turbulence. For a circular aperture of diameter $D$, the phase variance over the aperture is

$$\sigma_\phi^2 = 1.03 \, (D/r_0)^{5/3}, \tag{21}$$

  which is approximately 1 for an aperture of diameter $r_0$.
- $r_0$ can be considered at the atmospheric spatial coherence length for the aberrations.
- The size of the turbulence seeing disk is roughly the size of the diffraction limited disk of a telescope with diameter $r_0$.

For adaptive optics $r_0$ is roughly the spacing of mirror actuators and of wavefront samples required to achieve close to diffraction-limited performance.

## The Strehl ratio

A useful measure of imaging quality is the Strehl ratio. It is defined as the ratio of the on-axis intensity in the image of a point source, to that given by the diffraction limited system with the same aperture. A phase aberration $\phi(\vec{x})$ will result in a Strehl ratio of

$$S = \frac{\left| \iint_\Omega e^{i\phi(\vec{x})} \, dx^2 \right|^2}{\left| \iint_\Omega \, dx^2 \right|^2}, \tag{22}$$

where $\Omega$ is the aperture. For strong phase aberrations ($\sigma_\phi > 1$), the Strehl ratio is small $S \ll 1$. For relatively small phase aberrations $\sigma_\phi < 2$, the Strehl ratio is given approximately by

$$S \approx e^{-\sigma_\phi^2}. \tag{23}$$

Imaging is considered to be diffraction-limited for $S \geq 0.8$— the Marechal criterion. (This corresponds to a wavefront error of $\lambda/14$.) The Strehl ratio decreases rapidly with increasing wavelength since $r_0 \propto \lambda^{6/5}$.

---

[1]Note that we have slipped in a height dependence of $C_n^2$ while still assuming homogeneity on smaller scales.

## Optical transfer functions

The performance of an optical imaging system is quantified by a *point-spread function* (PSF) or a *transfer function*. In astronomy we image spatially incoherent objects, so it is the *intensity* point-spread function that is used. The image $I(\xi, \eta)$ is given by a convolution of the object $\mathcal{O}(\xi, \eta)$ with the PSF $\mathcal{P}(\xi, \eta)$,

$$
\begin{aligned}
I(\xi, \eta) &= \mathcal{O}(\xi, \eta) \otimes \mathcal{P}(\xi, \eta) \\
&= \iint \mathcal{O}(\xi', \eta') \mathcal{P}(\xi' - \xi, \eta' - \eta) \, d\xi' \, d\eta'.
\end{aligned} \tag{24}
$$

Equivalently, in Fourier space, where tilde denotes the Fourier transform,

$$
\tilde{I}(u, v) = \tilde{\mathcal{O}}(u, v) \mathcal{H}(u, v). \tag{25}
$$

$\mathcal{H}$ is the optical transfer function (OTF), which is given by the *autocorrelation* of the *pupil function*

$$
\mathcal{H}(u, v) = P(\lambda f u, \lambda f v) \star P(\lambda f u, \lambda f v). \tag{26}
$$

[2] The intensity PSF is the Fourier transform of the OTF.

To model the temporal behaviour of the turbulence induced aberrations we assume that a single layer of turbulence can be considered as "frozen", but translated across the aperture by the wind. This is known as Taylor's frozen-flow hypothesis. The temporal behaviour can then be characterized by a time constant,

$$
\tau_0 = r_0/v, \tag{27}
$$

where $r_0$ is the Fried parameter, and $v$ the translation speed. It can be considered as a coherence time for the atmospheric aberrations. For exposure times $T \ll \tau_0$, we use the short exposure OTF. The resulting image will be characterized by high spatial frequency "speckle" right down to the diffraction limit of the telescope—this forms the basis for speckle imaging, where statistical processing is used to recover high spatial frequency information about the object from the speckle pattern.

For exposures much longer than the atmospheric coherence time ($T \gg \tau_0$), the image is formed by the average of many short exposure images,

$$
\langle I(x, y) \rangle = \langle \mathcal{O}(x, y) \otimes \mathcal{P}(x, y) \rangle = \mathcal{O}(x, y) \otimes \langle \mathcal{P}(x, y) \rangle, \tag{28}
$$

---

[2]Note that image spatial frequencies $(u, v)$ are related to pupil coordinates $(x, y)$ by $(x, y) = (\lambda f u, \lambda f v)$.

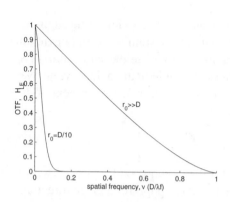

*Figure 2.* Long-exposure OTF for diffraction limited ($r_0 \gg D$) and atmosphere-limited ($r_0 = D/10$) cases.

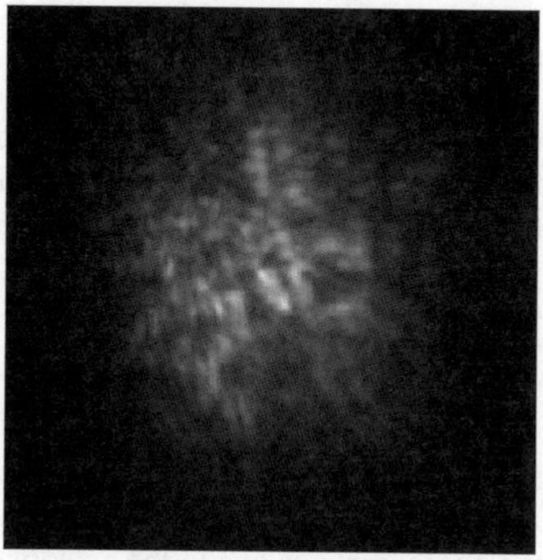

*Figure 3.* A typical short-exposure "speckle" image of a 0.2 arcsec binary star. The image width is about 2.2 arcsec. (courtesy N.J. Wooder)

where $\langle \cdot \rangle$ denotes the average, i.e., the long exposure PSF is the average of the short-exposure PSFs. The long exposure transfer function is given by

$$
\mathcal{H}_{\mathrm{LE}}(u, v) = \langle P(x, y) \star P(x, y) \rangle = \langle \psi(x, y)\Omega(x, y) \star \psi(x, y)\Omega(x, y) \rangle
$$
$$
= C_\psi(\lambda f u, \lambda f v)\mathcal{H}_{\mathrm{T}}(u, v).
$$

$$(29)$$

$C_\psi$ is the autocorrelation of the complex amplitude of the wave, given by equation 18 since we have zero-mean statistics, and $\mathcal{H}_{\mathrm{T}}$ is the telescope transfer function.

$\tau_0$ can be considered as a temporal analogue to the Fried parameter for adaptive optics systems. Another measure used in adaptive optics is the Greenwood frequency,

$$
f_G = \left[ 0.102 k^2 (\cos\gamma)^{-1} \int_0^\infty C_n^2(z)[v(z)]^{5/3} \, \mathrm{d}z \right]^{3/5}.
$$

$$(30)$$

For a single layer of turbulence this becomes

$$
f_G = 0.43|v|/r_0 = 0.43\tau^{-1}.
$$

$$(31)$$

The practical use of the Greenwood frequency is that in a first order loop compensator, the residual phase variance is

$$
\sigma^2_{\mathrm{closed\text{-}loop}} = (f_G/f_{\mathrm{3dB}}),
$$

$$(32)$$

where $f_{\mathrm{3dB}}$ is the bandwidth of the AO system.

## Anisoplanatism

The volume nature of the turbulence means that when viewed through different angles, the aberrations will be different. The system is anisoplanatic (an isoplanatic system being one in which the aberrations are the same across the field.) We define the *isoplanatic angle* $\Theta_0$ as the angle in the field of view over which we have approximate isoplanatism. It is given as a $5/3$ z-moment of the $C_n^2$ profile

$$\Theta_0 = \left[ 2.91 k^2 (\cos\gamma)^{-1} \int_0^\infty C_n^2 z^{5/3} \, \mathrm{d}z \right]^{-3/5}. \tag{33}$$

Note the similarity of this to the formula for the Greenwood frequency (with $v$ replaced by $\omega z$.) If the aberration is corrected at the pupil for a given direction, the phase variance due to the anisoplanatism at an angle $\Theta$ is

$$\sigma_\Theta^2 = (\Theta/\Theta_0)^{5/3}. \tag{34}$$

The isoplanatic angle can be quite small: a single layer of turbulence at height $5\,\mathrm{km}$ of strength such that $r_0 = 10\,\mathrm{cm}$ would give an the isoplanatic angle of about 1.3 arcsec. In addition to this angular anisoplanatism, when using laser guide stars for adaptive optics, we may also encounter focal anisoplanatism, a difference in the aberrations experienced by the light from the guide star in the atmosphere and that from a natural star at infinity.

## References

Andrews, L.C., Phillips, R.L., *Laser beam propagation through random media*, SPIE, Bellingham, 1998

Hardy, J.W., *Adaptive optics for astronomical telescopes*, Oxford University Press, 1998

Mandel, L., Wolf, E., *Optical coherence and quantum optics*, Cambridge University Press, Cambridge, 1995

Roy, A.E. and Clarke, D., *Astronomy*, volume Principles and Practice, Adam Hilger, Bristol, 2nd ed., 1982

Walker, G., *Astronomical observations*, Cambridge University Press, 1987

# INTERFERENCES

Carl Paterson

*Imperial College London*

carl@ic.ac.uk

**Abstract**    This tutorial shows how fundamental is the role plaid by interferences in many of the physical processes involved in astrophysical signal formating and consequently instrumentation. It is obvious in interferometry. Grating spectroscopy is explained within the same framework as Young experiment, and Fabry-Perot filters are explained as Michelson interferometers.Polarization interferences, used in Lyot filters, are discussed, emphasizing the analogy with échelle gratings.

**Keywords:**    interferences, amplitude interferometry, mutual intensity, grating spectroscopy, Fabry-Perot interferometers, polarization interference, birefringence, Lyot filters

## Introduction

The aim of this lecture is to give an all too brief overview of interference to illustrate its wide-ranging importance to astronomical observation and at the same time to show the underlying identity of seemingly different interference phenomena. At visible wavelengths, we do not observe the electromagnetic oscillations directly (to do so would require a sampling rate of order $10^{16}$ Hz.) We measure time-averaged intensity. In fact all our visible astronomical observations rely to some extent on interference followed by intensity measurements.

We shall begin with the principle of superposition, illustrated by Young's experiment (Fig. 1). If two pinholes separated by $a$ in an opaque screen are illuminated by a small source S of wavelength $\lambda$, under certain conditions it is observed that the intensity in front of the screen varies cosinusoidally with the angle $\theta$ according to

$$I(\theta) = I_0[1 + \cos(2\pi a\theta/\lambda)]. \tag{1}$$

11

*R. Foy and F. C. Foy (eds.), Optics in Astrophysics, 11–20.*

The total (complex) amplitude of the waves at a point $\mathbf{r}$, $V(\mathbf{r})$ is the superposition or sum of the amplitudes due to the waves from each pinhole,

$$V(\mathbf{r}) = K_1 V(\mathbf{r}_1) \, e^{\,i(2\pi s_1/\lambda - \omega t)} + K_2 V(\mathbf{r}_2) \, e^{\,i(2\pi s_2/\lambda - \omega t)}, \qquad (2)$$

where $V(\mathbf{r}_1)$ and $V(\mathbf{r}_2)$ are the amplitudes of the disturbance at the pinholes ($K_1$ and $K_2$ are *propagators*, factors proportional to $1/s$ and to the area of the pinhole). Taking the intensity to be $I = |V|^2$, gives

$$\begin{aligned} I(\mathbf{r}) &= |K_1 V(\mathbf{r}_1)|^2 + |K_2 V(\mathbf{r}_2)|^2 + 2 \, \mathrm{Re}\left[ K_1 V(\mathbf{r}_1) K_2 V(\mathbf{r}_2) \, e^{\,i2\pi(s_1 - s_2)/\lambda} \right] \\ &= I^{(1)} + I^{(2)} + 2\sqrt{I^{(1)}}\sqrt{I^{(2)}} \cos(2\pi(s_1 - s_2)/\lambda), \end{aligned}$$

$$(3)$$

where $I^{(1)}$ and $I^{(2)}$ are the intensities at $\mathbf{r}$ arising from the disturbances at each pinhole alone.

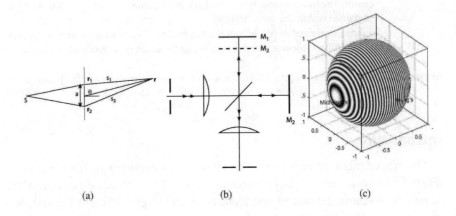

Figure 1.  Monochromatic two source interference: (a) Young's points, (b) Michelson interferometer, (c) 3D representation of far-field interference fringes over all viewing angles showing both Michelson fringes at the "poles" and Young's fringes at the "equator".

The superposition principle is illustrated further with the Michelson interferometer. Light is divided between two arms at a beamsplitter, recombined and the resulting intensity is observed. For a monochromatic source, the on-axis intensity is a superposition of the two recombined beams, and varies cosinusoidally with the difference in path lengths $\Delta z$

$$I = I_0[1 + \cos(2\pi\Delta z/\lambda)]. \qquad (4)$$

For a fixed $\Delta z$, the fringes vary cosinusoidally approximately with the square of the viewing angle. The Michelson and Young's interferences are in fact the same: with the Michelson, the effect of the two arms is to produce two virtual sources, separated by $\Delta z$ along the viewing axis, whereas with the Young's experiment we observe the interference of two pinhole sources separated by $a$ across the viewing axis.

## Mutual Coherence

A useful concept in understanding interference is that of coherence. Consider Young's experiment again (this time with an arbitrary source S which need not be monochromatic illuminating the pinholes). The optical disturbance at point $\mathbf{r}$ and time $t$ can be written alternatively as a sum of the disturbances at the pinholes, with propagation factors,

$$V(\mathbf{r}, t) = K_1 V(\mathbf{r}_1, t - t_1) + K_2 V(\mathbf{r}_2, t - t_2), \tag{5}$$

where $K_1$ and $K_2$ are propagators, $t_1 = s_1/c$ and $t_2 = s_2/c$ are the propagation delay times. The average intensity is then

$$
\begin{aligned}
\langle I(\mathbf{r}, t) \rangle &= \langle V(\mathbf{r}, t) V^*(\mathbf{r}, t) \rangle \\
&= |K_1|^2 \langle V_1(t - t_1) V_1^*(t - t_1) \rangle + |K_2|^2 \langle V_2(t - t_2) V_2^*(t - t_2) \rangle \\
&\quad + 2|K_1 K_2| \operatorname{Re} \langle V_1(t - t_1) V_2^*(t - t_2) \rangle.
\end{aligned} \tag{6}
$$

For stationary fields $\langle V_1(t - t_1) V_1^*(t - t_1) \rangle = I_1$, and we get

$$I = |K_1|^2 I_1 + |K_2|^2 I_2 + 2|K_1 K_2| \operatorname{Re} \langle V_1(t+\tau) V_2^*(t) \rangle, \qquad \tau = t_2 - t_1. \tag{7}$$

$\tau$ is the relative delay between the sources at the point of interference. The last term contains the correlation between $V_1$ and $V_2$ which is known as the *mutual coherence*,

$$\Gamma_{12}(\tau) = \langle V_1(t + \tau) V_2^*(t) \rangle. \tag{8}$$

The intensity can then be written in terms of the mutual coherence as

$$I = |K_1|^2 I_1 + |K_2|^2 I_2 + 2|K_1 K_2| \operatorname{Re} \Gamma_{12}(\tau). \tag{9}$$

Normalizing gives the general interference law for stationary fields

$$I = I^{(1)} + I^{(2)} + 2\sqrt{I^{(1)}}\sqrt{I^{(2)}} \gamma_{12}^{(r)}(\tau), \tag{10}$$

where $\gamma_{12}^{(r)}(\tau)$ is the real part of the *complex degree of coherence*,

$$\gamma_{12}(\tau) = \frac{\Gamma_{12}(\tau)}{\sqrt{I_1}\sqrt{I_2}}. \tag{11}$$

The modulus of $\gamma_{12}(\tau)$ gives a measure of the coherence ranging from incoherent (0) to coherent (1); its phase determines whether interference is constructive or destructive.

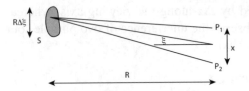

*Figure 2.*    Spatial coherence and the mutual intensity

When $r_1 = r_2$, we get the *self coherence* of the disturbance at $r_1$,

$$\Gamma_{11}(\tau) \equiv \langle V_1(t + \tau) V_1^*(t) \rangle. \tag{12}$$

This is the autocorrelation and by the Wiener–Khintchine theorem the *power spectrum* of the disturbance is given by its Fourier transform,

$$S(\nu) = \int_{-\infty}^{\infty} \Gamma_{11}(\tau)\, e^{2\pi i \nu \tau}\, d\tau. \tag{13}$$

For a unit intensity monochromatic source, $S(\nu) = \delta(\nu_0)$, which gives by the inverse Fourier transform $\Gamma_{11}(\tau) = e^{2\pi \nu_0 \tau}$, resulting in the familiar cos-squared interference fringes,

$$I = 2(I_1 + I_1 \operatorname{Re} e^{2\pi \nu_0 \tau}) = 2I_1(1 + \cos 2\pi \nu_0 \tau), \tag{14}$$

In Young's experiment the relative delay is $\tau = (s_2 - s_1)/c \approx a\theta/c$ and in the Michelson interferometer it is $\tau = \Delta x/c$

It is the self-coherence function that is measured in Fourier transform spectroscopy. Writing the measured on-axis intensity at the output of the Michelson interferometer as

$$\begin{aligned} I &= I^{(1)} + I^{(2)} + 2\operatorname{Re}\Gamma_{11}(\tau) \\ &= \langle I \rangle \left[ 1 + \operatorname{Re}\gamma_{11}(\Delta x/c) \right], \end{aligned} \tag{15}$$

it can be seen that by scanning $\Delta x$, the self coherence function of the source can be measured. The spectrum is then obtained via a Fourier transform,

$$S(\nu) = \int_0^{\infty} \operatorname{Re}\Gamma_{11}(\tau)\, e^{2\pi i \nu \tau}\, d\tau. \tag{16}$$

## Spatial coherence

Whereas temporal coherence is important for spectroscopy, spatial coherence is important for imaging. Consider the disturbance at two points $P_1$ and $P_2$ due to a finite sized source S (Fig. 2). If the source is small and distant

$(\Delta\xi \ll 1, \lambda R \gg x^2)$, for light of wavevector $k$ the optical disturbances at $P_1$ and $P_2$ are approximately

$$V_1 = C \iint V_s(\xi, \eta) \frac{e^{is_1 k}}{R} R \, d\xi R \, d\eta, \tag{17}$$

$$V_2 = C \iint V_s(\xi', \eta') \frac{e^{is_2 k}}{R} R \, d\xi' R \, d\eta' \tag{18}$$

The spatial coherence properties are described by the *mutual intensity* or equal-time coherence between these points (c.f., mutual coherence) which is defined as

$$J_{12} \equiv J_{P_1, P_2} \equiv \langle V_1 V_2^* \rangle. \tag{19}$$

But, for a spatially incoherent source, $\langle V_s(\xi, \eta) V_s^*(\xi', \eta') \rangle$ is non-zero only over a very small area of the object i.e., for $(\xi, \eta) \approx (\xi', \eta')$ and so substituting for $V_1$ and $V_2$ and simplifying gives

$$J_{12} = \langle V_1 V_2^* \rangle \propto \iint \langle V_s(\xi, \eta) V_s^*(\xi, \eta) \rangle e^{-i[s_2 - s_1]\bar{k}} \, d\xi \, d\eta$$
$$= \iint I_s(\xi, \eta) e^{-i\bar{k}(x\xi + y\eta)} \, d\xi \, d\eta. \tag{20}$$

Thus the mutual intensity at the observer is the Fourier transform of the source. This is a special case of the *van Cittert–Zernike theorem*. The mutual intensity is translation invariant or homogeneous, i.e., it depends only on the separation of $P_1$ and $P_2$. The intensity at the observer is simply $I = J_{11}$. Measuring the mutual intensity will give Fourier components of the object.

Now let us consider imaging by a lens. The disturbance at a point $(\xi, \eta)$ in the image plane is given by a superposition of the disturbances from points $(x, y)$ across the pupil $A$, with appropriate propagation delay,

$$V(\xi, \eta) = K \iint_A V_p(x, y) e^{-2\pi k(\xi x + \eta y)/f} \, dx \, dy$$
$$= K \mathcal{F}[A(x, y) V_p(x, y)]. \tag{21}$$

The measureable intensity is

$$I(\xi, \eta) = \langle V(\xi, \eta) V^*(\xi, \eta) \rangle. \tag{22}$$

After substituting for $V$ and with a little manipulation, we obtain

$$I(\xi, \eta) = \mathcal{F}[A(x, y) \star A(x, y) J(x, y)], \tag{23}$$

where $J(x, y)$ denotes the mutual intensity $J_{12}$ between points separated by $(x, y)$ in the pupil plane. This familiar result shows that the image is given by the object [whose Fourier transform is $J(x, y)$] filtered in the Fourier domain by the autocorrelation of the pupil function (i.e., the pupil transfer function.)

Information about the object can be obtained from direct measurements of the mutual intensity without forming a conventional image. In the Michelson stellar interferometer, light from two telescopes separated by $\Delta x$ is interfered: the fringe visibility gives the modulus of the degree of complex coherence and therefore a sample of the mutual intensity. Fitting measured data from a range of telescope separations to a model of the complex coherence for disk objects (which will in fact be given by the Airy function since that is being the Fourier transform of a disk) enables estimation of stellar diameters. The resolution is limited not by the diameters of the telescopes but by their maximum separation. More generally in long-baseline interferometry, the spatial frequencies of the object are sampled sparsely according to the available baselines between pairs of telescopes, for example the OHANA project aims to link up the large exisiting telescopes of Mauna Kea with optical fibre to enable them to be used as an interferometer. A further variation is the intensity, or Hanbury Brown and Twiss, interferometer (such as that at Narrabri in Australia) where the technical difficulties of interfering the light from two telescopes are avoided: the intensity is measured directly at each telescope and the coherence properties are inferred by correlating fluctuations in these intensities.

## Grating Spectroscopy

Thus far we have been considering—in Young's experiment, Michelson interference, and even measurements of mutual intensity—what is essentially two source interference. We now turn to multiple source interference. For example, a diffraction grating divides up a beam into an array of sources, which then interfere. By the principle of superposition, the disturbance in the far field is the sum over the sources

$$V \propto \sum_n e^{i2\pi s_n/\lambda} = \sum_n e^{i2\pi n\delta s/\lambda}, \tag{24}$$

where $\delta s = d(\sin\theta - \sin\theta_0)$ is the optical path difference between adjacent grating rulings (Fig 3). The condition for constructive interference is that $\delta s$ is a whole number of wavelengths,

$$\delta s = m\lambda \quad \Rightarrow \quad d(\sin\theta - \sin\theta_0) = m\lambda. \tag{25}$$

Grating spectroscopy makes use of the strong wavelength dependence (or dispersion) in the positions of the diffraction peaks. The width of the diffraction peak will give the resolution: for a grating with $N$ rulings, the closest minimum occurs at angle $\Delta\theta$ from a diffraction maximum,

$$\Delta\theta \approx (\lambda/2)/(Nd/2) = \lambda/Nd. \tag{26}$$

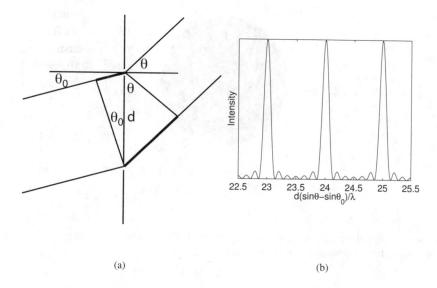

(a)

(b)

*Figure 3.* Grating diffraction: (a) geometry for the diffraction condition, (b) high-order diffraction for a grating with 7 rulings.

Using the Rayleigh criterion, the resolution of the $m$th diffraction order is then given by

$$R = \frac{\lambda}{\Delta\lambda} = \frac{\theta}{\Delta\theta} \approx \frac{m\lambda/d}{\lambda/Nd} = mN. \qquad (27)$$

which is proportional both to the number of rulings $N$ and to the diffraction order $m$. Echelle gratings, used in astronomical spectroscopy, are blazed gratings with large ruling period $d$, designed to give high diffraction efficiency for very high diffraction orders (e.g., $m \sim 100$) and have high dispersion. Their high dispersion also results in a low free spectral range ($\lambda/m$), however, diffraction orders can be separated by crossing the high resolution grating with one of lower resolution.

If a diffraction grating can be considered as a many-slit extension of Young's slits, then Fabry–Perot interference can be viewed as a many-source extension of Michelson interference. A Fabry–Perot étalon consists of two plane parallel surfaces with reflectivity $R$ and transmission $T$ (Fig 4). Multiple reflections between these surfaces give rise to interference effects. The optical path between the surfaces for propagation at a small angle $\theta$ to the optical axis is $n'D \cos\theta$, $D$ being the étalon spacing, $n'$ the refractive index inside it, and so the phase change after one pass is

$$\delta = 2\pi n'D \cos\theta/\lambda, \qquad (28)$$

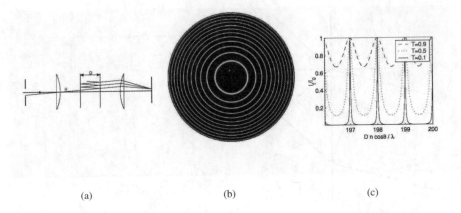

(a)                                    (b)                                    (c)

*Figure 4.*    Fabry–Perot interference: (a) Fabry–Perot étalon, (b) Fabry–Perot fringes, (c) Étalon transmission function for different plate transmissions.

and the complex amplitude of the light transmitted after one pass is

$$A_1 = A_0 t\, e^{i\delta} t = A_0 T\, e^{i\delta}, \tag{29}$$

where $t = \sqrt{T}$ is the amplitude transmission of a plate. After each successive double pass of the étalon the emerging amplitude is reduced by $r^2 = R$ and has extra phase delay $2\delta$. The total amplitude transmitted is the superposition of all these components,

$$A_t = \sum_{m=1,3,\ldots} A_m = TA_0\, e^{i\delta} \sum_{p=0}^{\infty} [R\, e^{2i\delta}]^p \frac{T}{1 - R\, e^{i\delta}} A_0. \tag{30}$$

Evaluating the sum, the intensity transmission of the étalon is

$$\frac{I_t}{I_0} = \frac{T^2}{|1 - R\, e^{i\delta}|^2} = \frac{(1 - R)^2}{|1 - R\, e^{i\delta}|^2}. \tag{31}$$

Note that for high reflectivities $R$, the transmission of the étalon varies from 1 to $T^2/4$. The higher the reflectivity $R$, the narrower the transmission peak and the higher the resolution of the étalon. Despite the finite reflectivities of the plates, interference means that the étalon is totally transparent at some combinations of wavelength and incident angle.

## Polarization interference and birefringence

So far we have only considered scalar interference, however, the vector nature of electromagnetic radiation can give rise to polarization interference

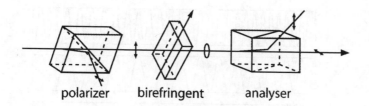

polarizer     birefringent     analyser

*Figure 5.* Polarization interference

effects. The refractive index experienced by light propagating through a bire-fringent crystal depends on the orientation of its electric field (i.e., the polarization) with respect to the fast and slow axes of the crystal. Birefringent crystals between polarizers can be used as half-wave plates, quarter-wave plates, adaptive optical elements (using birefringent liquid crystals,) dichroic mirrors etc. Birefringence allows us to construct polarization interference filters (Fig 5). Linearly polarized light (from the polarizer) incident on the birefringent plate is resolved into two polarization components, one parallel to the fast and the other to the slow axes of the crystal. The optical paths seen by these two components are different, since they experience different refractive indices. After traversing the crystal, the polarization of the two components are brought into the same plane by the analyser so that they interfere. The intensity of the output varies cosinusoidally with wavelength according to equation 4 for Michelson interference, but with $\Delta z$ now representing the optical path difference between the fast and slow axes. In fact, polarization interference is *identical* in principle to Michelson interference—propagation along the two arms of the Michelson being replaced by propagation with polarization along the fast or slow axes. As with the Michelson, the resolution of the filter is proportional to the path-length difference while the free-spectral range is inversely proportional to it.

By joining a series of polarization filters, whose birefringent plates vary in thickness in the geometrical progression 1:2:4:8:... we can construct a polarization filter which has both high resolution and free-spectral range—the resolution is determined by the longest plate, the free-spectral range by the shortest. This is a Lyot filter. Note that the transmission of a Lyot filter consisting of $N$ birefringent plates is identical to the diffraction pattern from an Echelle grating with $2^N$ rulings (Fig 6). If we look at the action of each single polarization filter, we can see that the optical path difference between fast and slow axis components has the effect of dividing the incident light into two virtual copies from sources shifted with respect to each other along the optical axis—just as the Michelson interferometer does. However, each successive plate in the Lyot filter repeats the division, until there are $2^N$ equally spaced virtual sources along the axis, the interference of which we measure at the output. It is therefore not surprising that the output should resemble that of a diffraction grating,

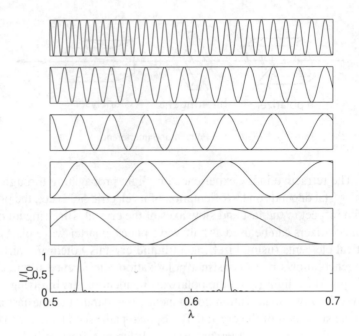

*Figure 6.* The transmission of a Lyot filter with five components (bottom plot) is given by the product of those of the five component filters (upper plots).

which also produces an array of equally spaced sources and so the Lyot filter is directly analogous to an Echelle grating.

## References

Born, M., Wolf, E., *Principles of optics*, Pergamon Press, Oxford, 6th ed., 1980

Jenkins, F.A., White, H.E., *Fundamentals of Optics*, McGraw-Hill, 4th ed., 1976

Mandel, L., Wolf, E., *Optical coherence and quantum optics*, Cambridge University Press, Cambridge, 1995

Roy, A.E., Clarke, D., *Astronomy*, volume Principles and Practice, Adam Hilger, Bristol, 2nd ed., 1982

Thorne, A, Litzén, U., and Johansson, S., *Spectrophysics: Principles and Applications*, Springer, Berlin, 1999

# A BRIEF HISTORY OF ASTRONOMICAL TELESCOPES

Philippe Dierickx

*European Southern Observatory*
*Karl-Schwarzschild str. 2*
*D-85748 Garching bei München2*
pdierick@eso.org

**Abstract**    A brief review of major milestones of the historical race to large telescopes is given, from the Galileo's refractor to the 4-6m class telescopes.

**Keywords:**    history of astonomy, refractors, optical telescopes

The invention of the telescope arguably constitutes an essential milestone in the history of science. Galileo's turning to the sky an instrument - a very modest one by today's standard- would not only change radically our understanding of the universe, it would eventually shatter the foundations of science, philosophy, and faith.

This brief summary is devoted to machines, not to the science they permitted. Yet, science relies critically on experimentation, the making of which may start in a laboratory, workshop, or in a factory. Astronomy is quite peculiar in respect of experimentation. It relies almost exclusively on contemplation: recording images of inaccessible objects and their spectral properties i.e., recording data cubes (two angular coordinates and a spectral one), without any capability to act on the parameters of the observed object. Few sciences have lesser means to experiment; yet none, perhaps, delivers so much with so little.

The scarcity of data and their invaluable content require that they be recorded in the most careful manner. The human eye is an incredibly powerful and versatile instrument, with an amazing dynamic range. But almost the entire universe is beyond its reach. Modern telescopes have a sensitivity 9 to 10 order of magnitude higher, yet they are barely able to provide a glimpse at the youngest galaxies, least to say stars.

*R. Foy and F. C. Foy (eds.), Optics in Astrophysics, 21–36.*

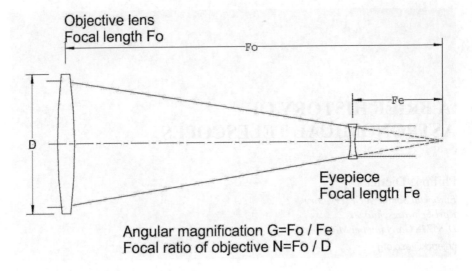

Angular magnification G=Fo / Fe
Focal ratio of objective N=Fo / D

*Figure 1.*    Principle of a Galilean refracting telescope.

Galileo's telescope, built in 1608, was made of a positive, converging objective and a negative, diverging eyepiece (Fig. 1). In spite of the limitations
of such design and of the rather poor fabrication quality allowed by the technology of the time, Galileo could make the sensational discoveries that almost
cost him his life. In his design, the positive lens forms of an object at infinite
distance an image located in its focal plane; the second, negative lens placed
in front of the first one's focus, re-images this image at infinite distance. An
observer will see the final image still at infinite distance, with its angular extent however magnified by the ratio of the focal distances $f$ of the two lenses
(Fig. 2). Modern sighting scopes still rely on the very same principles, with an
objective forming an image relayed by a sighting eyepiece, and a magnification
equal to the ratio of the two $f$s (see Fig. 1). All enhancements are essentially
related to compensating aberrations [1]. With conveniently small apertures and
small field angles, the aberrations will be minimal and the most primitive imaging system will deliver usable images.

The refracting telescope predated Galileo; devices able "to bring far objects
closer than they appear" were known well before his time, and their first origin

---

[1]For the purpose of this brief account we will provide only a notional definition of optical aberrations. In an
optical system, the angular coordinates of incident rays are transformed according to sequential applications
of Descarte's law from one optical surface to the next. Aberrations are essentially the non-linear terms of
the transformation, the angular coordinates of emerging rays not being strictly proportional to those of the
incident ones -thereby generating distorted and/or blurred images.

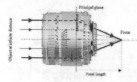

*Figure 2.* The focal length is the distance between the principal plane and the focus, where the principal plane is defined as the surface where the input and output light rays would intercept.

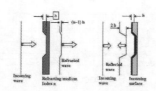

*Figure 3.* Wavefront error generated by a surface error of amplitude h.

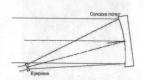

*Figure 4.* Zucchi's 1616 design of a reflecting telescope.

is difficult to trace. But it is most likely with the very intention to turn it to the sky that Galileo had one made.

Most glasses having an index of refraction close to 1.5, a reflective surface will have misfigure tolerances four times tighter than a refractive one. While Newton is generally credited with the invention of the reflecting telescope (1668), such credit is actually misplaced. Indeed Newton's telescope may have never delivered usable sky images or, at least, fared badly in comparison to Galileo's one. First, mirrors were made of speculum, with poor reflectivity. Second, for identical image quality, the tolerances applicable to reflecting surfaces are about four times tighter than with refractive ones (Fig. 3). A refractive medium of index $n$, with a surface misfigure $dh$, will generate distortion of the incident wave equal to $(n-1)dh$, being it assumed that the incident wave travels through vacuum. On a reflective surface, the wave distortion will be equal to $2dh$. Image contrast [2] degrades rapidly with increasing wavefront error, i.e. the departure of the wave from a perfectly spherical or flat one (Fig. 3).

The idea to produce a converging objective from a reflecting, concave surface, instead of a refractive lens, was recognised by Galileo himself, and tried by Zucchi in 1616 (Fig. 4). Such arrangement, however, does not deliver acceptable image quality as a spherical surface used off-axis produces strong aberrations, and Zucchi's attempt was doomed for reasons he could not know.

---

[2]Contrast, i.e. the rendering of spatially variable object brightness, is an essential performance characteristic of an optical system. It is also far more accurate and physically realistic than the traditional concept of angular resolution, which is meaningless in diffraction theory -as exemplified by the fact that the Hubble Space Telescope, with its initial flaw, could separate point sources nearly to diffraction limit, but failed at rendering acceptable images of extended, diffuse objects. In optical engineering, the term *Resolution* is reserved for the Strehl Ratio. An optical system is a low-pass spatial Fourier filter, and it can be shown that the Strehl Ratio is equal to the integral of the spatial Modulation Transfer Function of the system, i.e. an integrated measure of its ability to render contrast.

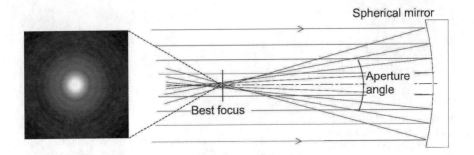

*Figure 5.*   Spherical aberration: rays corresponding to different aperture angles focus at different locations along the optical axis.

Newton's arrangement, which preserved the axisymmetry of the optical propagation and provided a more convenient observing position, did not suffer from this basic, and at the time not understood, flaw. Newton-type telescopes, however, would not compete with refractors until John Hadley's 1721's successful design and making of a 15-cm, 1500mm focal length ($f$/10) one. It is an unfortunate twist of history, or perhaps the result of imbalanced consideration for ideas rather than results, that John Hadley's work is little known, for he should be given credit for the successful advent of reflecting telescopes. His success is not only attributable to excellent craftsmanship, but also to his devising of an auto-collimation test at centre of curvature of the primary mirror. For crude such test may have been, it was the first step towards modern optical fabrication, where metrology plays a role equal to process. If you can measure it, you can make it [3].

Between 1634 and 1636, Descartes correctly interpreted spherical aberration (Fig. 5), and derived the aspheric shape lenses or mirrors should be given in order to prevent it. Technology, however, was not ready. When a surface is ground or polished by lapping, the tool and the workpiece naturally tend to converge towards spherical shapes, which will match in any position. This is no longer true with aspherical surfaces, and their making can be considerably more difficult. It requires tools smaller than the workpiece or flexible enough to adjust to the workpiece, or both. This difficulty, added to the fact that there anyway was no metrology able to control the convergence of the figuring process, explains the failure of early attempts at making aspherical lenses.

Even if the first attempts at making aspherical lenses had been successful, chromatism (Fig. 6, which was not properly understood, would have prevented such lenses to deliver acceptable images anyway. Solutions to the problem of chromatic aberrations had been delayed by Newton himself, who incorrectly

---

[3] J. Espiard, private communication.

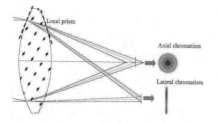

*Figure 6.* Chromatism is due to the wavelength dependence of the index of refraction and produces coloured, diffuse and/or elongated images. For simplicity reasons, the off-axis image in the figure above is shown with pure lateral chromatism; in reality both axial and lateral chromatisms would combine.

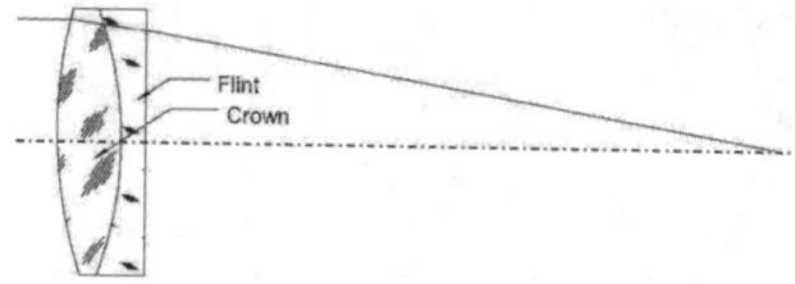

*Figure 7.* Schematic principle of an achromatic objective.

interpreted them. Gregory challenged Newton's conclusions in 1695, but it is not until 1729 that a solution, by Moor Hall, emerged. This solution is the achromatic doublet, whereby the objective is made of two lenses, a converging and a diverging ones (Fig. 7). The two lenses are made of different glasses, with different indices of refraction and different dispersions, chosen in such a way that the foci at two different wavelengths -usually the extreme ones of the visible spectrum, are superimposed. Chromatism is still present for intermediate wavelengths, but the overall effect is substantially reduced [4].

In Fig. 7, the ratio of the focal length to the objective diameter is exaggerated. In practice, achromatic objectives still require fairly long telescope tubes. The first achromatic objective was made by Dollond in 1758, and in 1761 Clairaut derived the modern form of the achromatic doublet, whereby the lenses are calculated to compensate each other's chromatic and spherical aberrations. Refracting telescopes would no longer need prohibitively long focal lengths, and their diameter would progressively grow up to the 1-m range by the end of the 19th century (Yerkes, 1897). Even today, casting large homogeneous slabs of glass is a daunting task; hopefully reflecting telescopes would eventually reach maturity around the mid-19th century, and successfully pick up the challenge. Perhaps the most famous refractors are those built by Fraunhofer.

The poor reflectivity of speculum and the tighter figure tolerance, combined with the difficulty to accurately measure aspherical surfaces have slowed the evolution of the reflecting telescope down. These issues are of a pure technological nature; optical solutions for reflecting telescopes were derived in the

---

[4] Using three lenses, exceptionally two with favourable glasses, it is possible to superimpose the foci at three wavelengths. Such systems are called apochromats

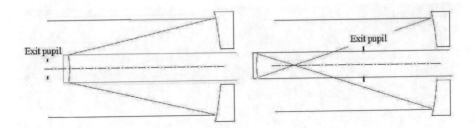

*Figure 8.*    Marin Mersenne's designs of all-reflecting telescopes, 1636.

## Entrance pupil

**Exit pupil**

*Figure 9.*    The entrance and exit pupils are the surfaces where the entrance and exit rays coming from the different field positions cross each other. In different terms, the entrance pupil is the aperture [5] of the optical system as seen by an observer located at the position of the object, or at the location of the image for the exit one.

17th century. In 1636, Mersenne proposed two original designs, where a secondary mirror plays the role of eyepiece (Fig. 8). Unfortunately, the exit pupil (see Fig. 9), where the observer shall be for his own pupil to be filled with the beams coming from the instrument, is inconveniently located. In the better known Gregory and Cassegrain solutions (Fig. 10), the secondary mirror is not used for the purpose of an eyepiece, but to relay the image given by the primary mirror at a location where a classical eyepiece could be conveniently located. In these designs the problem of the location of the exit pupil disappears, as the telescope pupil is re-imaged through the eyepiece, and the spherical aberration can be suppressed by combining a parabolic primary mirror with an elliptical, concave secondary (Gregory's solution), or hyperbolic, convex one (Cassegrain's solution).

---

[5]Entrance and exit pupils are conjugates i.e. images of each other through the optical system. The real physical aperture may be neither of them, but set by a physical diaphragm or component contour, located inside the optical system.

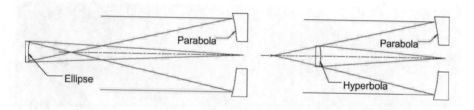

*Figure 10.* Gregorian (left) and Cassegrain (right) solutions.

A fundamental advantage of Gregory's and above all Cassegrain' solutions is that, contrarily to refractors, they are far shorter than their focal length and thereby allow more compact, lighter and stiffer opto-mechanical designs. The same principle is applied in camera's lens design, where a last group of diverging lenses extends the focal length of the front converging group. Hence the expression "telephoto effect" for this kind of design. All modern telescopes rely on it, with structures typically ten times shorter than the focal length by today's standard.

Because it is concave and elliptical, the secondary mirror of a Gregorian telescope allows for simpler fabrication test set-ups, as a point-like source placed at one of the foci of the ellipse should be re-imaged aberration-free at the other focus. Convex hyperbolic mirrors require more complex set-ups. This disadvantage is usually more than balanced by the more compact arrangement, the savings implied by the shorter telescope structure and more compact enclosure offsetting the additional cost and complexity of the secondary mirror test set-up. This trade-off, however, applies in the context of current technology.

Gregorian and Cassegrain solutions, proposed in 1663 and 1672, respectively, would take centuries before being put to practical use in major telescopes. In the absence of suitable figuring techniques and above all adequate fabrication metrology, the making of not one but two matching aspherical surfaces was a nearly impossible task. It is worth noting that the theory of reflecting telescopes was basically completed by the time Gregory and Cassegrain laid it down -indeed there would be no major progress until 1905, with Schwarzschild's milestone work. In contrast to this, the theory of refracting telescope would take about another century to come to maturity, as indicated before.

Reflecting telescopes would, however, still compete with refracting ones, mostly for their feasibility in much larger diameters. Herschel's telescopes,

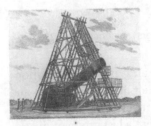

*Figure 11.* Herschel's f/10, 1.2-m telescope, 1789.

*Figure 12.* Lord Rosse's 1.8-m Newtonian telescope at Birr Castle (Ireland), 1845.

*Figure 13.* Nasmyth alt-azimuthal 50 cm telescope.

with 50 cm in 1784 and 1.2 m in 1789 (Fig. 11), were the true behemoths of the time; neither their quality nor their efficiency could rival that of the refractors, but their huge collecting power could offset their disadvantages.

In order to provide acceptably low aberrations and avoid aspherization of the primary mirror, Herschel's telescopes had to have fairly long focal lengths, hence long tubes and bulky structures supporting them. The mirrors being made of speculum, had poor reflectivity (65% after fresh polishing), and tarnish required them to be re-polished periodically. Casting the polishable speculum slabs was no simple task either; large castings having a tendency to crack open while cooling. In the evolutionary tree of telescopes, this species would culminate with Lord Rosse's 1.82-m Newtonian giant (Fig. 12), which accommodated original solutions such as the whiffle-tree support of its primary mirror.

Ideally, a mirror is supported by two decorrelated systems, an axial and a lateral ones, each overtaking gravity loads at any orientation in an independent manner and ensuring that the mirror is never subjected to stresses that would deform it. A reasonably small mirror, a couple of hundreds millimetres across at most, may rest axially on three pads, designed is such a way that the mirror will not be constrained by differential thermal expansion with its support system. Larger mirrors, however, require more than just three axial points as the deflection under gravity between adjacent support points would become unacceptably high. For this precise reason, larger mirrors are also thicker, classically in a ratio of 1:7 to 1:8 of their diameter. But this does not prevent the need of distributing support forces amongst a larger number of support points. In a

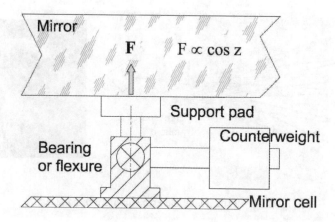

*Figure 14.* Principle of an astatic levers support system. The mirror rests on three fixed points but the total weight is distributed on a number of levers, which, by design, apply reaction forces proportional to the cosine of the zenithal angle.

whiffle-tree, these support are mechanically -or hydraulically- interconnected through successive stages of low-friction, force-equalizing balance systems, so as to provide astatic supporting through three virtual fixed points. Modern design always incorporate extensive modelling and analysis of support systems, as the performance of any optical part is contingent to that of its support. In essence, a telescope is not just the product of a set of optical equations, but a true opto-mechanical system, which requires extensive analysis and trade-offs to ensure adequate performance.

At the time Lord Rosse was building his leviathan, a smaller (50 cm) but innovative one was built by Nasmyth. It was probably the first successful Cassegrain combination. The telescope had an alt-azimuth mount and a folding mirror placed on the elevation axis allowed a convenient observing position under all orientations (Fig. 13). Because of its geometry, the azimuthal mount is ideal in terms of load transfers to the axes and foundations of the telescope. It requires, however, the combination of three rotations (azimuth, elevation, field de-rotation [6]) to compensate for earth' rotation, and this arrangement would come to maturity more than a century later, with computer-controlled motions and the Russian 6-m alt-azimuthal telescope (Nizhny Arkhyz, Russia) in 1976. Another major reflecting telescope of the 19th century is the 1.22-m Newtonian built by Lassel on Malta (1861). The innovative design included an equatorial mount allowing earth rotation to be compensated by a single rotation around the right ascension axis, a lightweight open tube eliminating internal turbulence, and a mirror support system based on astatic levers (Fig. 14).

---

[6]Field de-rotation is obviously not mandatory in case of visual observations.

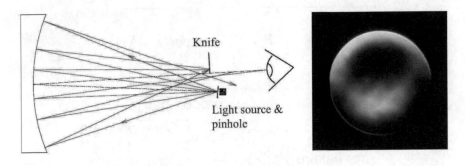

*Figure 15.*    Principle of the Foucault or knife-edge test.

Not only did Foucault inaugurate the era of glass mirrors, he also devised an accurate test that prefigured modern optical fabrication. Although essentially qualitative, the Foucault or knife-edge test (Fig. 15), devised in 1859, allowed, for the first time, to verify the convergence of a polishing process. In this test, a blade placed at the focus of the surface under test (or at its centre of curvature if it is illuminated by a point-like source located near the centre of curvature) intercept rays in a selective manner, revealing errors of shape as shadows on said surface. For primitive it may seem, this test is exquisitely sensitive to surface misfigure, and will immediately reveal errors that may escape the most modern metrology. Being it sensitive to slope errors, it may still today usefully cross-check elaborate interferometric tests, the latter being limited in spatial resolution by the detector sampling or by the interferometer's aperture.

The largest telescope made by Foucault was 80 cm in diameter and today is on display at the Observatoire de Marseille, France. A superb instrument, it incorporated even more innovation than meets the eye. First, Foucault ensured that the mirror was polished to a shape that would match the aberrations of the eyepiece, thereby delivering optimal visual images; second, he devised a rudimentary pneumatic support of the primary mirror, the shape of which being optimized by blowing or deflating supporting vessels until the mirror reached its optimal figure.

The failure of the Melbourne 1.22-m telescope (1869) temporarily shifted attention back to refractors, which is a plausible explanation for the giant refractors of the end of the 19th century (Nice, 76 cm, 1887; Lick, 91 cm, 1888; Yerkes, 1 m, 1897). Reflecting telescopes would however make a spectacular come-back, thanks to the productive work of Ritchey. By 1901, he produced a 60 cm with Newton and Cassegrain foci, and a primary mirror with an amazingly short focal ratio of f/3.9. The difficulty, or the minimal achievable misfigure, to generate an aspherical surface is, with conventional lapping technologies, proportional to the cube of the focal ratio. Hence Ritchey's 70 cm mirror was roughly 7 times more difficult to figure than the Melbourne one. Still, Ritchey's telescope delivered excellent photographic results, outperforming visual observations with much larger refractors. The next step became a 1.5-m reflector. Ritchey cautiously opted for a longer primary mirror focal ratio (f/5) than for the 60 cm, but incorporated novel design features such as the possibility to switch between Newton, Cassegrain and Coudé foci [7]. Still in operation today, after numerous upgrades, the telescope will see its 100th birthday in 2008.

Confidence in Ritchey's abilities was such that by 1906, Hale had ordered a 2.5 m blank, which eventually would become the primary of Mt Wilson's Hooker 2.5 m telescope. To put it mildly, the bubble content of the blank did not inspire confidence, but successive attempts at casting better ones failed. Ritchey finally agreed to work on it, and the mirror was figured in about five years. In addition to the knife-edge test, Ritchey also used a more quantitative one, the Hartmann test. In this setup, a mask producing an array subapertures (holes drilled at predetermined locations in a metal or wooden plate) is laid on the mirror; a point-like source is placed near the centre of curvature, and the images produced through the subapertures recorded on a photographic plate. The location of each of these images, being directly related to the slope of the mirror at the location of the corresponding subaperture, provides a quantitative way to measure surface misfigure.

With its modern design, unprecedented light collecting power, and in spite of its capricious thermal behaviour, the Hooker telescope would give a long lasting supremacy to the American scientific community and nurture a Californian tradition of technical excellence in astronomy, a tradition still very alive today. Amongst many other major discoveries, it allowed the Andromeda galaxy to be resolved into stars, and its distance to be deduced from light curves of Cepheids [8]. At about the same time, Europeans would focus on positional astronomy and an eventually failed Carte du Ciel, which essentially obliterated European's presence in observational astrophysics until the end of the 20th century. After Ritchey's moving to Paris, in 1924, plans were made for a 7.5 m telescope, which never materialized.

Encouraged by the success of the 1.5 and 2.5 m Mt Wilson's telescope, George Hellery Hale soon started working on his vision of a 5 m class tele-

---

[7]In a Coudé arrangement, the image is relayed through the right ascension axis, thereby providing a fixed observing location where a bulky instrument can be mounted without jeopardizing the balance of the telescope. It is ideally suited for instruments demanding long focal ratios and a relatively small field, e.g. spectrographs.

[8]Henrietta Leavitt, Harvard College Observatory, 1912, established the relation period-luminosity of Cepheid variable stars. Being a woman i.e. confined to the tedious and ridiculously paid work of examining photographic plates for the benefit of an all-male establishment, she would not be awarded credit for her milestone discovery

scope. Initial ideas for a 7.5 m telescope had been put forward, and Hale was aware of Ritchey's initial plans for an even larger aperture, but cautiously -and in view of the technology of the time, wisely- chose to limit extrapolation to a factor two. A six million dollars from the Rockfeller foundation was secured in 1929, a very substantial sum for a science project at that time. Production of the primary mirror blank dragged for almost a decade, until Corning success-fully picked up the challenge after repeated failures by General Electric and produced a 5-m structured, honeycombed one. The casting of the blank was far from a simple affair, and even by today's standard Corning's achievement must be reckoned as a remarkable one [9]. Besides the lightweight structure of the mirror, which implied critical mass savings and lower thermal inertia than that of the Hooker, other novel features were the reliance on low expansion glass (Pyrex), an extremely short focal ratio [10] of f/3.3, and a design of the telescope structure, due to Mark Serrurier, which minimized relative decenters of the primary and secondary mirrors at all orientations of the telescope.

Hale passed away in 1938 and, unfortunately, could not see the superb con-clusion of his work. World War II delayed the completion of the Palomar telescope, which would eventually enter into operations in 1950, after resolu-tion of quite a number of start-up problems, including local re-figuring of the primary mirror to correct for residual edge misfigure. The technical achieve-ment of the Palomar telescope is no better illustrated by the fact that in the decades that followed its commissioning, little attempt was made at exceed-ing it. The only notable exception is the Russian 6 m Bolshoi Teleskop Az-imutalnyi telescope (1974), whose performance is marred by a deceptive site, rapid thermal changes preventing the primary mirror from reaching equilib-rium, and a structure and kinematics optimised for static rather than dynamic loads -hence its poor performance when exposed to wind. Still, this telescope should not be merely dismissed as a failure -as is too usually and unduly done. Its computer-controlled alt-azimuthal mount opened the way to modern tele-scopes and compact enclosures. Hale's telescope dominated the landscape of large telescope-making until the late 1980's, when plans would be drawn for a new generation of even larger ones, dubbed Cathedrals of the 20th Century

---

[9] R. Florence in The Perfect Machine: Building the Palomar Telescope, HarperCollins Publishers, 1994, gives a vivid account of the building of the Palomar telescope. Although Florence at times wrongly attributes major technical innovations to the 5 m telescope designers, his somewhat idealistic but definitely superb account will ring true with many designers of large telescopes.

[10] As noted before, short focal ratios imply severe difficulties in generating the aspherical departure from an ideal sphere. It is, however, directly related to the telescope length, hence to the structure and building sizes, which are major cost positions and performance issues in a telescope project -cost of large structures and buildings, improbable thermal equilibrium hence local turbulence, misalignments and flexures, etc.

[11]. In the meantime, the Bolshoi 6 m telescope, in spite of its originality, did not materialize as an effective challenger, and the period between the Palomar and its successors of the 1990's could be seen as one of consolidation. Several 3- to 4-m class telescopes were built, owing much of their design to the Palomar's one: equatorial mount, Serrurier structure, massive hemispherical dome, and a focal ratio of the primary mirror in the range of 3 to 4. Several reasons contributed to this period of consolidation rather than extrapolation. First, casting large substrates is still a substantial challenge. Even though 8-m class blanks have been made in the early 1990s by Schott (Germany), Corning and the Mirror Lab (Tucson, Az), their production was a daunting task, and few would venture much beyond this size. This particular technological limitation has now been overcome by segmentation, whereby the telescope main mirror is no longer made of a single piece but of segments, whose alignment can be controlled in real time to acceptable accuracy. This technology, pioneered with the 10 m Keck telescopes, underlies every plan for the next generation of extremely large telescopes.

Second, funds have dried up; being it generally perceived that large astronomical telescopes have no other benefit than their long-term and somewhat unpredictable scientific productivity. In this respect, there may be room for cautious hope, as the latest generation of large telescopes in the 8- to 10-m range brought substantial spin-off in terms of industrial return [12], and as future designs, such as that of the ESO's OWL 100-m telescope concept, may accommodate for industrial constraints in a cost-effective manner.

Third, there has been considerable evolution in the area of detectors, with efficiency increasing from a few percents for photographic plates to near-perfect quantum efficiency with modern detectors. On a good site, a CCD on a 1 m telescope would outperform photographic plates taken at the Palomar 5 m. Detectors being now close to the physical limit in terms of sensitivity, the only way up is, again, larger diameter. One should however not hastily conclude that the period between the Palomar 5 m and the 8- to 10-m class telescopes of the 1990s has been void of crucial innovation.

Progress in optical fabrication and metrology allowed the Ritchey-Chrétien design to eventually take over the classical Cassegrain design. In the latter, the primary mirror is parabolic, the secondary hyperbolic, the on-axis image is free of aberrations, but the field of view is limited by field aberrations, mostly coma, astigmatism, and field curvature. In the Ritchey-Chrétien design, both

---

[11] The author wishes to mention that this expression was coined by his own father, an admirer of science but not a scientist himself. Coincidentally, the same expression was used to the same effect by the Director of the European Southern Observatory, R. Giacconi, in the late 1990's.

[12] Although this is rarely advertised, the fabrication and polishing to highest accuracy of the last generation of glass-ceramic mirrors have produced commercial spin-offs in areas as diverse as the manufacturing of cooking plates and consumer electronics.

mirrors are hyperbolic and their combination is free of spherical aberration, as in the classical Cassegrain solution, but also of field coma -and thereby provides a larger usable field of view. Because of its shape, the primary mirror is theoretically more difficult to test while in fabrication. One reason is that a parabola can be tested in auto-collimation against a reference flat [13]. In practice, reference flats are extremely difficult to produce, and the argument against a Ritchey-Chrétien loses its validity beyond sizes on the order of a meter in diameter. Another reason is that the Ritchey-Chrétien design is intrinsically more sensitive to decenter, the correction of the on-axis image being dependent on the proper alignment of both mirrors. Still, the delay in the advent of Ritchey-Chrétien designs -by almost two generations- should be attributed to prejudice and unreasonable conservatism rather than to technical reasons.

Progress in metrology, from qualitative Foucault and cumbersome photographic Hartmann tests to high resolution laser interferometry also allowed deeper aspherization, hence shorter focal ratio of the primary, and more compact structures and buildings. Fig. 16 shows the historical trend towards larger and shorter telescopes, with focal ratios shorter than 3 becoming the standard of the late 20th century. Short focal ratios translate into compact structure and enclosure, the latter no longer being the (absurdly) major cost position of an astronomical telescope. Better metrology and figuring processes also implied faster production and drastic cost reduction factors. A typical polishing run of the Palomar primary mirror took about a week; it took about two days for the 4-m class telescopes of the 1970s and 1980s, and one for the 8-m class telescopes of the 1990s. In terms of square meter per unit of time, the progress is even more impressive: from 20 $m^2$ in 61/2 years (Palomar) to 50 $m^2$ in about 10 months (today's 8-m generation), i.e. a 20-fold increase in productivity.

Evolution in telescope making since the Palomar has not been limited to the area of optical production. The alt-azimuthal mount has become the established solution since the Bolshoi 6 m telescope, for its superior mechanical performance and the compact, cost-efficient enclosure design it allows. Better understanding of the properties of atmospheric turbulence allowed a more accurate characterization of a telescope properties, a more balanced approach towards specifications and error budgeting [14], and a better understanding of the utmost importance of site selection. Any ground-based telescope of appreciable size will be primarily limited by the effect of atmospheric turbulence, not to mention the proportion of photometric nights allowed by weather conditions.

---

[13]In such test, a pinpoint light source placed at the focus of the parabola will provide a collimated beam after reflection on the parabola; this collimated beam can be retro-reflected back onto the parabola by a flat mirror, to eventually provide an aberration-free image of the test source.

[14]The first example is the 4-m class William Herschel telescope, at la Palma, whose optical specifications, drafted by D. Brown, were expressed in terms of allowable wavefront error as a function of spatial frequencies matching those of atmospheric turbulence.

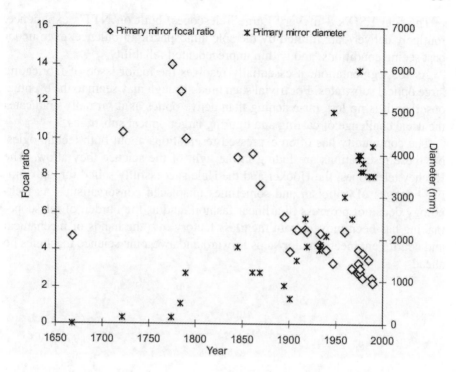

*Figure 16.* Evolution of telescope diameter and primary mirror focal ratio.

Substantial investments in design and manufacturing to tightest specifications must therefore be matched by a rigorous and quantitative site selection process. Recently developed adaptive optics techniques, which allow to compensate for the blurring effect of atmospheric turbulence, do not change this requirement appreciably, as strong or rapid turbulence may impose unacceptably demanding technical requirements.

Last but not least, two epoch-marking technologies have been successfully implemented in 1989 and 1994: active optics, with the 3.5 m ESO New Technology Telescope (NTT), and segmentation, with the 10 m Keck. In the former, real-time adjustments of the primary mirror support forces and of the alignment of the secondary mirror guarantee consistent, optimal performance, and allow relaxation of opto-mechanical fabrication tolerances. These adjustments being derived from wavefront analysis of off-axis stellar sources located outside the scientific field of view, imply minimal operational overheads at the benefit of reliable, substantial performance improvement [15]

---

[15] A fair account of the development of active optics must acknowledge for the fact that in its initial arrangement, the operational concept of the NTT required usually unaware or inexperienced observers to allocate provisions for active corrections in their observing runs. This is no longer the case, as suitable upgrades of control systems and operational schemes have made the system almost completely transparent to the observer.

The four ESO's 8 m Very Large Telescopes, built on NTT's experience, routinely deliver data limited by the sole atmospheric turbulence, even under best seeing conditions, and within unprecedented reliability.

As for segmentation, it essentially resolves the major issue of fabricating large optical substrates. For trivial such breakthrough may seem to the eventual observer, it is no less far-reaching than active optics as it virtually eradicates the usual challenge of casting and figuring larger optical substrates.

The community has often expressed reservations about both technologies, but these reservations are fading in the light of the science they allow. The Ritchey telescopes, the Hooker and the Hale successfully sailed through similar barrages of criticism and sometimes misplaced conservatism. As technology does not progress in a linear fashion, and as the burden of telescope-making has been moved from the glass factory into the hands of mechanical and control engineers, major steps forward and awesome science machines lie ahead.

# MODERN OPTICAL DESIGN

Jerry Nelson and Andy Sheinis

*Center for Adaptive Optics*
*University of California at Santa Cruz*
*1156 High Street*
*Santa Cruz, Ca 95064-1077 USA*
jnelson@ucolick.org

**Abstract**

We present the basics of optical design as it applies to two mirrored telescope systems. We discuss Zernike decomposition of wave-front error and the description of Strehl in terms of small Zernike errors. We also discuss the balancing of aberrations for a two mirror system and present the Ritchey-Chrétien design as an example of a zero coma system.

**Keywords:**    astronomical telescopes, aberrations, telescope designs

## 1.    Ways to study light

### 1.1    Fermat's principle of least time

Fundamental to the understanding of optics is the Principle of Least Time, developed by Fermat ( $\approx$ 1650), which says that light will take the path that gives the shortest time to its destination. One can think of it as though light examines all possible paths to get from one point to another and then takes only and exactly the path that requires the least amount of time. For a more complete description of Fermats Principle of Least Time (Feynman et al., 1963).

### 1.2    Snell's Law

Consider light traveling from one medium to another (Fig. 1). To simplify the problem, we can consider that the media are infinite in extent and that the interface between the media is planar. Note that a more complex geometry will not change the result. Also consider that each medium can be described by an index of refraction, n, which is the apparent speed of the wave through

*R. Foy and F. C. Foy (eds.), Optics in Astrophysics, 37–48.*
© *2006 Springer. Printed in the Netherlands.*

the medium. If we pick an arbitrary point within the medium and another point

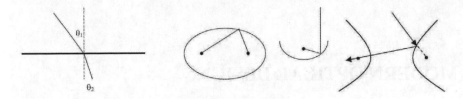

*Figure 1.* Refraction at the interface between two media of different refraction index.

*Figure 2.* Conic conjugates for an elliptical (left), parabolic (center) and hyperbolic (right) mirror.

outside the medium we can ask how does light get from one point to the other point. If we draw every possible path that the light can take in our thought experiment and measure the length of time each path takes we will conclude two things:

1. The shortest time occurs for light travels in straight lines in a constant index media.
2. At the interface between the two media light will "bend" obeying the relation: $n_1 \sin \Theta_1 = n_2 \sin \Theta_2$ for the path taking the shortest time.

The principle of least time can also be used to examine reflection. If we perform a similar experiment to that above except that we make one medium reflective at the boundary we will discover that the only paths for which the flight time of the photon is a minimum are those for which the angle of incidence equals the angle of reflection.

## 1.3 Geometric Optics

We can use the concepts developed above to trace the path of individual rays through an optical system. In doing so we are doing a geometric optical analysis because we are including only the geometry of the light rays to predict their behavior, not their physical, wave-like or electromagnetic properties.

Using the two rules above we can pick an arbitrary point on an object surface and a ray of light coming from that point, and predict it's path through an optical system. A light ray will leave the surface of the object either due to reflection off the object or in the case of warm objects through self-emission. This ray will travel in a straight line until it encounters a change in index of refraction. In general this occurs when it encounters another surface, but the same rules apply if there is a variation in the index of refraction within the media as in the atmosphere.

Once the ray encounters the new surface Snells law of refraction is applied if it is a transparent surface (real index of refraction) or if it is a reflective surface (imaginary index of refraction) the law of reflection is applied. The angles used are the relative to the local normal to the surface.

If the angles are small i.e. close to the optical axis, then the paraxial approximation can be made:

$$\sin \Theta \approx \tan \Theta \approx \Theta \tag{1}$$

For a spherical surface such as a lens or mirror, we are able to determine the angle of refraction, or reflection from the ray height at that surface. The angle the surface normal makes relative to the ray as a function of height h above the optical axis is given by

$$\Theta_f = \Theta_i + \Theta(h), \quad \text{where} \tag{2}$$
$$\Theta(h) = \arctan h/r \tag{3}$$

and $r$ is the radius of the surface.

Thus by a series of calculations of ray height and position we can determine the location of an individual ray as it traverses an optical system. If we trace a series of rays all emanating from a single point on an object and discover that they all intersect at a single point we determine the image location for that point.

By tracing rays from a series of locations on an object we can determine the shape and character of the image, i.e. whether it exists on a planar surface, curved surface or a more complex surface. By tracing a series of different wavelength rays we can determine if the image is the same for all colors or does it focus in different locations for different colors. Furthermore, one can look the locus of all points on rays emitted by a single point on the object and intersecting an image plane. If one measures the number of rays as a function of their radius from the "center of mass" or centroid, one can develop a plot of effect of the imaging system on a single point in object space. This is called the point-spread-function and is a measure of the quality of the image forming ability of an imaging system. Conversely one can measure the number of rays within a circle (or square) relative to the total number of rays hitting the image plane to get encircled (ensquared) energy distribution. This is a prediction of the response one will get in a circular (or square) detector for a single point in object space.

Clearly this process can rapidly become tedious, especially when one thinks about tracing hundreds or thousands of rays from many different points on the object (field points). Fortunately many computer programs have been written that allow us to harness the power of personal computers to do this. Nonetheless, most commercial computer programs use the exact same technique as that which we have described.

(For a complete description of raytracing see Kingslake, 1983; Smith, 2000).

## 1.4     Physical optics calculations

The previous section used a mathematical construct called a ray to predict behavior of light in an optical system. We should emphasize that rays are purely a mathematical construct, not a physical reality. Rays work well to describe the behavior of light in cases where we can ignore its wave-like behavior. These situations are ones in which the angular size of the point-spread-function is much greater than $\lambda/d$, where $\lambda$ is the wavelength of light and $d$ is the diameter of the optical system.

We can also use the wave-like nature of light to describe the behavior of an optical system. If we consider the radiation from a single point in the object, it acts like a wave spreading out in all directions. A portion of this wave interacts with our optical system. The wave can be described mathematically as

$$E = E_0 \exp(-(i\,k\,r - \omega t - \phi)) \tag{4}$$

where $E$ is the observable amplitude of the E or B field, $r$ is the radius vector from the source, $k$ is the propagation vector which has magnitude $2\pi/\lambda$ and the direction of propagation, $\omega$ is the angular frequency, $t$ is the time and $\phi$ is the arbitrary constant phase.

Our coordinate system has $z$ parallel to the optical axis, $y$ is vertical and $x$ comes out of the page.

If we apply Maxwell's equations to this boundary value problem we can derive a complete solution to the amplitude and phase of this field at every point in space. In general, however we can simplify the problem to describing the field at the entrance (or exit) aperture of a system and at the image plane (which is what we are really interested in the end).

At distances that are very large compared to either the source dimensions, or the aperture of our optical system the complex amplitude is just the Fourier transform of the object complex amplitude. Furthermore, it can be shown that the complex amplitude in the image plane of an optical system is just the Fourier transform of the complex amplitude at the aperture plane. (for a complete derivation of this see Goodman, 1996).

We can talk about the wave-front associated with this wave as being a surface of constant phase traveling at the speed of the wave. This surface is related to our previous idea of the rays in that it represents the joined perpendiculars from a series of rays emitted from a single point. A perfectly spherical wave will converge to (or diverge from) a single point in space.

When this wave interacts with an optical system, the system adds a phase shift which is in general a function of x and y.

If we think about what happen to a perfect wave from a point source, it has an arbitrary phase (and amplitude) added to it by our optical system. This phase distribution, $Ae^{\phi'}(x, y)$ can be thought of as producing a wave that is not quite spherical. It is this deviation from a spherical wave that is described by

$$E' = E_0 A \exp\left(-(ikr - \omega t - [\phi(x,y) - \phi'(x,y)])\right) \qquad (5)$$
$$= E_0 A \exp\left(-(ikr - \omega t - \Delta(x,y))\right) \qquad (6)$$

where $\Delta(x,y)$ is this departure from spherical. This departure is called the system aberration and is what produces a fuzzy image.

In the first case, lets consider what happens to a perfect light wave when it impinges on a mirror. In general mirrors can be described as flat (which is less interesting) spherical, conic or aspheric.

## 2. Properties of conics

For conic mirrors there is a unique and interesting situation: There exists a pair of points in space located relative to a conic mirror such that the light travel time for all possible paths from one point to the other are equal and minimum. This means that all light leaving one of these points will intersect the other point if it strikes the conic mirror. These points are called conic conjugates. For any given conic there are only two such points, which are different for different conics.

In terms of wave-fronts the conic mirror will take the spherical wave-front from one conjugate point and modify it to be another perfectly spherical wave-front heading towards the other conjugate point.

For a spherical mirror the conjugate points lie on top of each other at the center of curvature (COC). Light emitted from the COC of the spherical mirror will focus back onto the COC without any additional system aberration.

For an elliptical mirror the two conjugates are real and located at a finite distances on the same side of the mirror. As shown in Fig. 2, light emitted from the one conjugate of the elliptical mirror will focus onto the other conjugate without any additional system aberration.

For a parabolic mirror one conjugate is real and located in front of the mirror, the other is located at infinity. As shown in Fig. 2, light emitted from the one conjugate of the parabolic mirror will focus onto the other conjugate without any additional system aberration. For example, light emitted from an infinitely distant point will come to perfect focus at the finite conjugate.

For a hyperbolic mirror one conjugate is real and located in front of the mirror, the other is imaginary and located behind the mirror. As shown in Fig. 2, light emitted from the real conjugate located in front of the hyperbolic mirror will be reflected to appear as though it comes from the imaginary conjugate located behind the hyperbolic mirror without any additional system aberration.

## 3.    Expansion of optical surfaces

All conics can be expressed via one equation, which describes the position parallel to the optical axis, $z$ as a function of the normalized height above the optical axis $r$.

$$z(r) = \left(k - [k^2 - (K+1)r^2]^{1/2}\right)/(K+1) \qquad \text{or} \qquad (7)$$

$$z(r) = \frac{r^2}{2k} + (K+1)\frac{r^4}{8k^3} + (K+1)^2\frac{r^6}{16k^5} + (K+1)^3\frac{5r^8}{128k^7} + \cdots \quad (8)$$

where $k$ is the radius of curvature. $K$ is the conic constant, either $< -1$ (hyperbola) or $= -1$ (parabola) or $-1 < K < 0$ (prolate ellipsoid) or $= 0$ (sphere) or $> 0$ (oblate ellipsoid).

## 4.    Zernike polynomials

In general a surface or wavefront $z(\rho, \Theta)$ can be expanded in functions that span the space. As an example one can expand in cylindrical polynomials:

$$z(\rho, \Theta) = \sum_{m,n} a_{mn}\rho^m \cos n\Theta + \sum_{m,n} b_{mn}\rho^m \sin n\Theta \qquad (9)$$

where $m \geq n \geq 0$, $m - n$ even (analytic function), $a, b$ are constants, and $\rho = r/R$ (normalized radius).

Conic mirrors produce a perfect spherical wave-front when the light striking the mirror comes from the conic conjugates. In general, most optical systems produce an aberrated wave-front, whose departure from a perfect spherical wave-front, $\Delta(x, y)$ can be described quantitatively.

Since $\Delta(x, y)$ can really be any function bounded by the aperture of the system it is best to use as general description as possible. One such description of this function is to expand $\Delta(x, y) = \Delta(\rho, \Theta)$ about $(\rho, \Theta)$ in an infinite polynomial series. One set of polynomials that are frequently used are Zernike polynomials. Thus one can write $\Delta(\rho, \Theta) = \sum_{m,n} C_{mn}F_{mn}(\rho, \Theta)$.

These polynomials have the very useful property that they are orthogonal over the unit circle such that

$$\int F_{mn}(\rho, \Theta)F_{kl}(\rho, \Theta) = A_{mn}\delta_{mk}\delta_{nl} \qquad (10)$$

The lowest order Zernikes are tabulated in Table 1 along with their common optical name. The next figure shows the first 12 Zernike polynomials with a pseudo-color graphical representation of their value over a unit circle. In the plots, blue values are low and red values are high.

Thus our arbitrary wave-front error can be described as a linear superposition of Zernike polynomials over the unit circle (although for a completely

*Table 1.*   Lowest orders of first Zernike polynomials

| | | |
|---|---|---|
| $Z_{00} = \rho^0$ | $= 1$ | Piston |
| $Z_{1-1}(\rho, \Theta) = \rho \sin \Theta$ | $= y$ | Tip |
| $Z_{11}(\rho, \Theta) = \rho \cos \Theta$ | $= x$ | Tilt |
| $Z_{2-2}(\rho, \Theta) = \rho^2 \sin 2\Theta$ | $= 2xy$ | Astigmatism |
| $Z_{20}(\rho, \Theta) = 2\rho^2 - 1$ | $= 2x^2 + 2y^2 - 1$ | Focus |
| $Z_{22}(\rho, \Theta) = \rho^2 \cos 2\Theta$ | $= x^2 - y^2$ | Astigmatism |
| $Z_{3-3}(\rho, \Theta) = \rho^3 \sin 3\Theta$ | $= 3x^2y - y^3$ | Trefoil |
| $Z_{3-1}(\rho, \Theta) = (3\rho^3 - 2\rho) \sin \Theta$ | $= 3x^2y + 3y^3 - 2y$ | Coma |
| $Z_{31}(\rho, \Theta) = (3\rho^3 - 2\rho) \cos \Theta$ | $= 3x^3 + 3xy^2 - 2x$ | Coma |
| $Z_{33}(\rho, \Theta) = \rho^3 \cos 3\Theta$ | $= x^3 - 3xy^2$ | Trefoil |
| $Z_{40}(\rho, \Theta) = 6\rho^4 - 6\rho^2 + 1$ | $= 6x^4 + 12x^2y^2 + 6y^4 - 6x^2 - 6y^2 + 1$ | Spherical aberr. |

arbitrary wave-front we may require an infinite number of Zernike polynomials to completely describe the error). The Zernike coefficients tell us how much power is in each Zernike term by the following equation normalization equation:

$$\sigma_{mn} = C_{mn} \sqrt{(1 + \delta_{n0}) / (2(m+1))} \qquad (11)$$

which gives the rms values of the $n, m^{th}$ Zernike term.

Thus the quadratic sum of all the Zernike coefficients gives the rms value of the entire wave-front error by:

$$\sigma_{total} = \sqrt{\sum_{mn} \sigma_{mn}^2} \qquad (12)$$

A word of caution: different authors use different conventions for Zernike notation and normalization. We are using the convention of Born and Wolf (1999).

## 5.    One Mirror telescopes

In the conics section we discussed the property of parabolas. Since one of its conjugates is at infinity, it will form a perfect image from the infinite conjugate at focus. This is the basis for a single mirror telescope. This telescope is either used at this prime focus, or sometimes a folding flat mirror (Newtonian telescope) is used to fold the beam to a convenient location for a camera or the

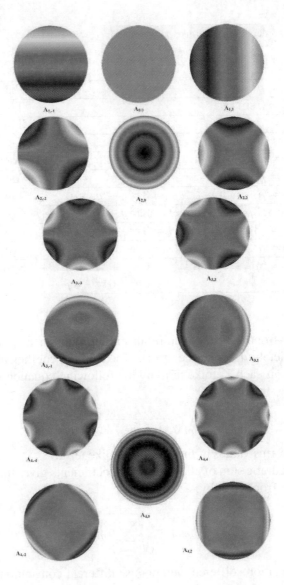

*Figure 3.*   The first 12 Zernike polynomials (see text)

eye. The principle is the same, a Newtonian telescope is really just a parabolic mirror.

## 6.    Two Mirror telescopes

So why aren't all telescopes just a single parabolic mirror? The answer is field coverage. A parabolic mirror develops wave-front aberrations as the object (star) moves away from the infinite conjugate laterally. So if we have two stars in the field we can only get a perfect image of one. For most science objects it is desirable to image a finite field of view (FOV). Any FOV larger than a few arcsecs will require more than a single mirror to correct these off-axis aberrations.

For any two-mirror system we can use geometric optics to derive equations that describe some important parameters. (For derivations see Schroeder, 2000). The effective system focal length can be determined in terms of the individual focal lengths and mirror separation:

$$\frac{1}{f} = \frac{1}{f_1} + \frac{1}{f_2} - \frac{d}{f_1 f_2} \tag{13}$$

Given the focal length of the primary and the required system focal length one can determine the secondary focal length by:

$$f_2 = \frac{-f(f_1 - d)}{f - f_1} = \frac{-f(1 + \Delta)}{m^2 - 1} \tag{14}$$

Conversely the separation can be determined from the primary focal length, the magnification of the secondary and the back focal length.

$$d = f_1(m - \Delta)/(m + 1) \tag{15}$$

Also the diameter of the secondary can be determined from the diameter of the primary, the FOV of the telescope and the spacing between primary and secondary.

$$D_2 = (d + e)D_1/f + 2\,d\,\Theta \tag{16}$$

Conversely the two focal lengths for primary and secondary can be determined given a requirement on the mirror spacing, d and back focal length e by the following two equations.

$$1/f_1 = (e - f)/2\,d\,f \tag{17}$$
$$1/f_2 = (e + d + f)/2\,d\,e \tag{18}$$

Definition of symbols:

$D_1, D_2$     primary, secondary diameter
$F$     final $f$/ratio
$F_1$     primary $f$/ratio
$f$     final focal length
$f_1, f_2$     primary, secondary focal lengths
$d$     primary-secondary spacing
$e$     back focal distance: distance from primary vertex (positive behind mirror)
$\Delta$     $e/f_1$
$m$     magnification by secondary $= f/f_1$
$K_1, K_2$     conic constant of primary, secondary
$\Theta$     angular field radius
$1/\kappa$     focal surface radius of curvature

Furthermore, for any two-mirror telescope in which each mirror is individually corrected for spherical aberration (i.e. composed of two conics) one can solve for the third order aberrations in closed form.

$$ASA = \frac{1}{128F_1^3}\left\{ K_1 + 1 - \frac{1+\Delta}{m+1}\left(\frac{m-1}{m+1}\right)^3\left[K_2 + \left(\frac{m+1}{m-1}\right)^2\right]\right\}$$
(19)

$ASA$: angular diameter of spherical aberration blur circle.

$$ATC = \frac{-3\Theta}{16F^2}\left\{1 + \frac{(m-1)^3(m-\Delta)}{2m(m+1)}\left[K_2 + \left(\frac{m+1}{m-1}\right)^2\right]\right\}$$
(20)

$ATC$: total angular length of angular tangential coma

$$AAST = \frac{-\Theta^2}{2F}\left\{1 + \frac{(m-1)(m-\Delta)}{m(m+1)}\left[1 - \frac{(m-\Delta)(m-1)^2}{4m(1+\Delta)}(K_2+1)\right]\right\}$$
(21)

$AAST$ angular diameter of astigmatic blur circle

Examples of two-mirror designs in which each mirror is individually corrected for spherical aberration are:

    1 - Cassegrain, (parabolic primary and hyperbolic secondary).
    2 - Gregorian, (parabolic primary and ellipsoid secondary)

## 7.     Ritchey-Chrétien design and equations

With these closed form aberration equations it is possible to do something interesting: Rather than use the mirrors at their individual conic conjugates which would correct for spherical aberration in each mirror independently, we can set the overall spherical and comatic aberration to zero and solve for a different set of conics. This is called the aplanatic condition, when $3^{\text{rd}}$ order spherical is balanced against $3^{rd}$ order coma to remove the two. This aplanatic telescope will be composed of a pair of conics that do not have infinity and the focal position as their individual conjugates (i.e. 2 hyperboloids). The telescope will have a much larger FOV because the coma, which grows linearly with $\theta$ will be removed.

The equations for the conic constants required to give zero coma and spherical are:

ASA = 0 (no spherical aberrations)

An aplanatic telescope has ASA=0, ATC=0

$$K_1 = -1 - \frac{2(1+\Delta)}{m^2(m-\Delta)}$$

$$K_2 = \left(\frac{m+1}{m-1}\right)^2 - \frac{2m(m+1)}{(m-\Delta)(m-1)^3} \qquad \text{at RC focus,}$$

$$K_2 = \left(\frac{m+1}{m-1}\right)^2 + (1+K_1)\frac{m+1}{1+\Delta}\left(\frac{m}{m-1}\right)^3 \qquad \text{at other foci.}$$

Another example of an aplanatic telescope is the aplanatic Gregorian(AG), both primary and secondary are elliptical.

Lastly, it is possible to correct for all three major aberrations, Spherical, coma and astigmatism. This gives a system with very large FOV, but with other less desirable features. Examples are:

1. Couder telescope which has 0<m<0.5, a large obscuration, and focus between mirrors
2. The Schwarzshield telescope which has a much larger secondary than primary , a large obscuration and is physically much larger than it's focal length (a big cost increase).

## 8.     Important parameters and cost drivers

Lastly we provide a table that lists some of the real-world parameters one must consider when purchasing or building a telescope.

Optically important

- Size of telescope
- Image quality
- Field of view
- Plate scale
- Field curvature
- Spherical aberration at prime focus

Financially important

- Primary diameter, primary focal lengths (size)
- Primary asphericity (polishing, alignment)

- Secondary diameter (support, polishing, alignment)
- Plate scale, mechanical size of focal surface (size of instruments)
- Available foci: prime, Cassegrain, Nasmyth

## References

Feynman, R., Sands, M., Leighton, R., Addison-Wesley, 1963, *Lectures on Physics*, **1**

Smith, W.J., 2000, *Modern Optical Engineering*, McGraw Hill, 3rd Ed

Kingslake, R., Academic Press, 1983, *Optical System Design*

Goodman, J.W., 1996, *Introduction of Fourier Optics 2nd Ed*, McGraw Hill

Born, M., Wolf, E., 1999, *Principle of Optics 7th Ed*, Cambridge University Press

Schroeder, D.J., 2000, *Astronomical Optics 2nd Ed.*, 2nd edition, Academic Press

Wilson, R.N., 1999, *Reflecting Telescopes Optics (2 vol)*, Springer

# STRUCTURAL MECHANICS

Jerry Nelson

*Center for Adaptive Optics*
*University of California at Santa Cruz*
*1156 High Street*
*Santa Cruz, Ca 95064-1077 USA*
jnelson@ucolick.org

**Abstract**

We present a few basic ideas of structural mechanics that are particularly relevant to the design of telescopes and to the support of related optics. This talk only touches on a very rich and complex field of work. We introduce the ideas of kinematics and kinematic mounts, then review basic elasticity and buckling. Simple and useful rules of thumb relating to structural performance are introduced. Simple conceptual ideas that are the basis of flexures are introduced along with an introduction to the bending of plates. We finish with a few thoughts on thermal issues, and list some interesting material properties.

**Keywords:**     extremely large telescopes, kinematic mount, structural mechanics

## 1.     Kinematics

In this section we discuss the idea of degrees of freedom of motion of simple structures, with emphasis on space frames and trusses. These simple structures are often easily understood and exhibit simple and predictable behavior. They can also be structurally efficient and thus make a valuable group of structures interesting in astronomy.

## 1.1     Structures and degrees of freedom

Space frames are structures made up of nodes and struts. When the struts that connect the nodes together are slender, the axial stiffness of the strut becomes the dominant form of stiffness, and the bending stiffness that might be achieved in the attachment of a strut to a node is usually very much smaller, and can often be ignored.

*R. Foy and F. C. Foy (eds.), Optics in Astrophysics, 49–60.*
© 2006 *Springer. Printed in the Netherlands.*

When the struts constrain all the nodes so there is no relative motion, we talk about the configuration as a structure. When some motion is unconstrained or only very slightly constrained we discuss the configuration as a mechanism or a flexural mechanism.

A *node*, or a structural node is defined by its position in space, hence by its $x$, $y$, $z$ position. It has 3 degrees of freedom in 3-space, and 2 degrees of freedom in 2-space. We often idealize a node as a mathematical point, so its position needs to be defined, but its rotational degrees of freedom are indefinable.

A *strut* is usually viewed as a single degree of freedom constraint. It has a length, and that is its key defining property. A strut or column or beam can connect between two nodes, thus defining the distance between those points. An interesting and important variant of a strut is a cable. This component can only take tension loads, and cannot carry compressive loads.

An *object* is generally a three dimensional construct whose position is defined by its location (3 degrees of freedom- $x$, $y$, $z$) and by its orientation (3 rotations). Thus an object is constrained if six degrees of freedom of the object are constrained. If less than six degrees of freedom are constrained, the object is under constrained and can be viewed as a mechanism. It is also called under-determined. If the object is only considered in two dimensions, then three constraints are needed to define the object ($x$, $y$, rotation). When an object is just constrained it is called determinate or statically-determinate.

In real life nodes are more complex than points, but various approximations to simple nodes can be built. These are often called pin joints, ball joints, or spherical joints- joints that can take axial loads, but cannot carry any torques.

A structure, defined by a set of nodes and a set of struts that interconnect the nodes, may be interconnected in a way that under-constrains the nodes, may determinately constrain the nodes or it may be overdetermined or indeterminate.

## 1.2    Determinate structures

Determinate structures can be thought of as a group of nodes that are inter connected by struts, where there are just enough struts to constrain the nodal positions, but not enough to over constrain any of the nodes. In this idealization, such structures will have no internal stresses, and any changes in length of the struts will cause the structure to change shape in a completely stress free fashion. Yet the structure will resist changing shape in response to any set of externally applied forces.

In this idealization, determinate structures are defined by geometry, and concepts of stress and strain are not needed.

In a structure $N$ denotes the number of nodes and $S$ the number of struts.

## Two-dimensional structures

A two-dimensional structure (one that can be mapped onto a plane) may be determinate only if $S = 2N - 3$. This is a result of 2 constraints needed for each node, and the object as a whole can move in three degrees of freedom.

It is also possible to satisfy this relation and have the structure be underdetermined and locally over determined. Examples of two-dimensional structures are shown in Table 1.

*Table 1.* Two-dimensional structures

| Structure | $N$ | $S$ | $2N - 3$ | Type |
|---|---|---|---|---|
| | 3 | 3 | 3 | determinate |
| | 4 | 5 | 5 | determinate |
| | 4 | 6 | 5 | over determined |
| | 7 | 11 | 11 | both over and under determined |

**Three-dimensional structures** Three-dimensional structures are also quite interesting. A determinate structure must satisfy $S = 3N - 6$. This is a result of 3 constraints needed for each node and the object as a whole can move in six degrees of freedom. As in the case of 2-d, this is a necessary but not sufficient condition.

The simplest determinate 3-d structure has 4 nodes and 6 struts $3N - 6 = 6$ and is a tetrahedron. It is straightforward to make 3-d structures that are over or under determined (Table 2).

In examining 3-d structures, it is clear that each node must be connected by at least three struts. When a node is connected within a structure by only three struts, it can be removed (along with its three struts) without changing the determinate nature of the remaining structure.

*Table 2.*   Three-dimensional structures

| Structure | $N$ | $S$ | $2N - 3$ | Type |
|---|---|---|---|---|
| | 3 | 3 | 3 | determinate, tetrahedron |
| | 4 | 5 | 5 | determinate |
| | 4 | 6 | 5 | determined octahedron |
| | 7 | 11 | 11 | both over-determined |
| | 7 | 11 | 11 | under-determined |

## 1.3    Geodesic structures

It is useful to test our understanding of determinate structures by considering the simple, yet powerful class of 3-d objects called geodesic structures. These are structures that can be mapped onto the surface of a sphere. There is a beautiful structural theorem about geodesic structures that we will now prove.

*Theorem:* A geodesic structure is determinate if and only if all of its faces are triangles.

Recall the elegant Euler's theorem that states that for a 2-d structure the number of nodes minus the number of edges (struts) plus the number of faces = 1, or $N - S + F = 1$. This topology theorem is easy to prove. An examination of Table 3 shows some examples of this theorem and indicates how to prove it.

Next, note that if we take a 2-dimensional faceted structure and fold it into 3-dimensions, we have added a face. Thus we have proven Euler's theorem in 3-dimensions $N - S + F = 2$.

*Table 3.*   Eulers Theorem in 2 dimensions

| Structure | N -S - F |
|---|---|
|  | 3 - 3 + 1 |
|  | 5 - 5 + 1 |
|  | 5 - 6 + 2 |
|  | 6 - 7 + 2 |

If this surface is all triangles $3F = 2S$ (each edge is shared between two triangles).

Combining this with Euler's theorem we obtain $3N - S = 6$, or $S = 3N - 6$, our formula for determinate structures in 3-dimensions.

We have thus established:

- A geodesic structure with all triangular faces is determinate.
- A geodesic structure with any non-triangular faces is underdetermined, since adding the struts needed to make all faces triangles will add constraints.

## 2.   Kinematic Mounts

Kinematic mounts are systems for connecting two "rigid" objects to each other in a determinate fashion without introducing any stresses in either. Sometimes kinematic mounts are very readily detached from each other.

The design and development of kinematic mounts is a rich and complex field, but we only introduce the subject and its basic ideas here. When dealing with optics, stress-free mounts are often essential in order to avoid distorting the optic. Possible disturbances that must be considered include changes in gravity vector and changes in the thermal environment.

The attachment of horses to a wagon forms a simple and interesting example to think through. If we want two horses to pull a wagon, we would like each horse to pull equally, and not have their efforts oppose each other or have one horse carry most of the load. We would also like the horses to be able to walk independently, and not have to require that the horses be identical in size. Of course the two horses need to, on average, walk at the same speed. A solution commonly used is to put a tongue on the wagon, and place a cross bar on the front of the tongue, with a pivot at the connection. Then at the ends of the cross bar we place the horses, so they are pushing against the cross bar ends. Small differences in position are accommodated by rotation of the cross bar, but the forces on the ends of the bar must be equal on average or the bar will continue to rotate. One can continue this logic to handle as many horses as desired. Such connecting arrangements are called whiffletrees.

We list a few examples:

- An object with 3 balls on its surface is positioned on a plate with a small conical hole (3 constraints), a shallow groove (2 constraints) and a flat surface (1 constraint). The balls are placed in the hole, the groove and on the flat.
- An object with 3 balls on its surface is placed on a plate with 3 grooves. Each groove provides 2 constraints. This works unless the three grooves are parallel.
- An object is connected to another with 6 struts (assume pin joints so each strut is a single constraint)
- An object is connected with whiffletrees to another object. Their purpose is to distribute forces according to the geometry (often equal) without constraining the relative positions of the horses

## 3.    Basics of elasticity

We introduce the most basic aspects of elasticity. We begin with Hookes law: the change in length of a strut is proportional to the applied force, or $\delta L = FL/EA$. Note that this is a linear relationship. Restated in a normalized way, $\sigma = E\epsilon$, where $\sigma$ is the stress (Pa or N/m$^2$), E is Young's modulus (Pa) a property of the material, and $\epsilon$ is the strain $(\delta L/L)$ a dimensionless quantity.

A particularly interesting example occurs when we want the elastic response of a system under its self weight. Here, a more massive structure experiences greater forces, but will generally have greater stiffness. Specific stiffness is the resistance to deformations from self-weight and generally scales as $E/\rho$ where $\rho$ is the density of the material. The material property $E/\rho$ is called the *specific stiffness*. Generally, self-weight driven systems have deformations that are independent of the weight of the system.

When a system responds in proportion to the applied forces the system is said to be linear. When the response is linear, one can determine the effect of multiple forces on a system by adding the responses of individual force-response systems. This is the principle of linear superposition.

Non-linear systems exist, so one should be careful to make sure the system of interest is adequately linear before proceeding. Generally non-linear systems are those whose geometry changes during the application of a load. Examples include:

- A pair of spheres pressed against each other
- Belleville washers (shallowly curved washers whose spring constant varies with load because the geometry changes with deformation)
- Large deflections of flat diaphragms (large means deflections larger than the thickness)

For a simple system, such as a rod under compression, one can define the stiffness, or *the spring constant*. If we examine Hookes law, we get $\delta L = L\, \delta F / EA$, where $A$ is the cross sectional area, and $\delta F$ is the applied load. The spring constant $k$ is defined as $dF/dL$, or $k = EA/L$. The basic physics equation $F = kx$ is just a statement of this. For many degree of freedom systems there will be multiple spring constants, each connected to a modal shape.

## 4. Buckling

We have discussed the value of struts or columns in structural mechanics and described their linear elastic properties. They have another characteristic that is not quite so obvious. When columns are subject to a compressive load, they are subject to *buckling*. A column will compress under load until a critical load is reached. Beyond this load the column becomes unstable and lateral deformations can grow without bound. For thin columns, Euler showed that the critical force that causes a column to buckle is given by

$$F_{crit} = n\,\pi^2\,E\,I/L^2 = n\,\pi^2\,E\,A/(L/r)^2 \qquad (1)$$

where
| | | |
|---|---|---|
| $L$ | is the column length | |
| $A$ | is the material cross sectional area | |
| $r$ | is the radius of gyration (defined by the above equation, $r = \sqrt{I/A}$) | |

$r = a/2$ for a cylindrical solid rod of radius $a$
$r = a/\sqrt{2}$ for thin walled tubing of radius $a$
$I$ is the moment of inertia ($= \pi a^4/4$ for a solid cylindrical rod)
$n$ is the end condition
$n = 1$ when both ends can pivot
$n = 2$ when one end is fixed and the other can pivot
$n = 4$ when both ends are fixed
$n = 0.25$ when one end is fixed and the other is free

The shape of buckling is usually that of the column bowing sideways once the critical load has been exceeded. It is worth noting that buckling, although a non-linear condition, may not damage the column, depending on how far the deformations go. Buckling can be used as a design technique, where something will buckle when a load exceeds some design value, and then the result-ing motion may be stopped by another constraint. Of course if buckling is unrestrained, it will ultimately lead to failure of the column.

When the buckling load is smaller than the load that will cause the column to fail under compression, it is called a thin or slender column. If compressive failure occurs first, it is called a short column.

As an example consider a thin walled tube, free to pivot at both ends, where we want it to yield (exceed its elastic limit) just as it buckles. Assume it is a high strength material (yield stress $\approx 0.01E$). We then obtain $L/r = 31$ or $L/a = 22$.

## 5.    Elastic deflections and natural frequencies

In designing and building high performance structures, structural stiffness, gravity driven deflections, and natural frequencies of the structure are often in-teresting and important. Telescopes must adequately resist wind loads, optical alignment must be preserved with varying orientation with respect to gravity, and often active control is needed to move a telescope or its optics, and con-trol bandwidths are often limited by the structural natural frequencies. These characteristics of a structure are interrelated, and it is useful to be able to draw these relationships even if only approximately.

An intuitive beginning point is to examine the self-weight deflections of a structure. These are gravity-driven. Consider the simplest of systems, a mass on a spring. In this system the responses are given by

$$F = mg = k\delta \tag{2}$$

$$\delta = mg/k \tag{3}$$

$$f = \frac{1}{2\pi}\sqrt{\frac{k}{m}} = \frac{1}{2\pi}\sqrt{g/\delta} \tag{4}$$

Thus, if $\delta = 1mm$, $f = 15.8$ Hz. This very simple result is quite useful for approximately evaluating the gravity driven deflections of a structure given its natural frequency, or visa versa. Of course this was derived for a specific and very simple system, so it does not perfectly apply to more complex systems. Still it is a very useful rule of thumb. For a mass on the end of a cantilever beam, the above formula is correct. The lowest natural frequency of a massive cantilever beam is about $1.2\times$ the prediction of the above formula.

## 6.    Flexures

It is a common problem that one wishes to constrain an object from moving in some direction while allowing it to move in other directions. Many mech-anisms have been designed and built that allow quite stiff constraints in one direction while allowing unlimited motion in another. Typically these mechanisms involve parts of the mechanism sliding relative to one another. A variety of designs of bearings allow such motion with only small resistance or friction. However, such mechanisms are subject to wear, and, when lubricated, changes in the lubricant (due to temperature, evaporation, aging, contamination) will cause time varying resistance and ultimately failure to perform as desired.

When the range of travel is limited, it is sometimes practical to design mechanisms that rely on elastic deformation of an element to allow the motion, while providing robust constraint in other directions. Such mechanisms are often called flexures. When lightly loaded these can have infinite lifetimes and no maintenance. There are a large range of such devices that have been designed and used for various applications.

Flexures can allow limited linear travel in one direction while constraining the object in other directions and angles. Flexures can allow rotation in one direction while constraining the object in all directions and two rotations, etc.

We restrict ourselves here to analyzing an extremely simple flexural element, a solid cylindrical rod of radius $a$ and length $L$, well anchored at one end, and unconstrained at the other end. This is shown in Fig. 1.

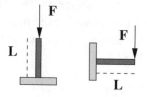

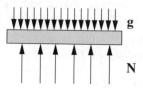

*Figure 1.* Flexural rod, stiff axially and flexible laterally.

*Figure 2.* Thin plate experiencing axial gravity load opposed by $N$ support points.

This element will strongly resist motion along its length and be relatively flexible in opposing motion perpendicular to its length. Axial forces experience a spring constant

$$k_a = \pi E a^2 / L \qquad (5)$$

and lateral forces experience (at the end of the rod) a spring constant

$$k_l = 3EI/L^3 = 3\pi E a^4/4L^3 \qquad (6)$$

The ratio of these spring constants is

$$R = \frac{\delta_{bend}}{\delta_{axial}} = \frac{k_{axial}}{k_{bend}} = \frac{4}{3}\left(\frac{L}{a}\right)^2 \qquad (7)$$

For a slender rod with $L/a = 100$, we get $R = 13,000$. Note that the ratio is independent of $E$, although the load carrying capacity may vary with $E$ and the actual magnitude of deflections will depend on $E$.

Although one can design flexures with quite extreme ratios of stiffness, often the design is also limited by various space constraints, limits to allowed deflections, required load carrying capacity, and the need to avoid buckling.

## 7.    Plate stiffness and deflections

In optics, one frequently wants to know the deflections of an axially loaded plate. Mirrors are often good approximations to thin plates and it is important to understand the deflections of mirrors when loaded by gravity, wind, or other point loads. Fortunately this is a relatively well studied subject, so many solutions exist in the literature. A particularly valuable general reference book is Theory of Plates and Shells, by Timoshenko and Woinowsky-Krieger.

Because there is a large literature, we restrict ourselves to an interesting example, useful for understanding the axial support of thin mirrors. Consider a thin circular plate of radius $a$ and thickness $h$, with elastic constant $E$ and Poissons ratio $\nu$. Let this plate be axially loaded by gravity and assume we will support this plate against this load by $N$ supports. This is shown in Fig. 2.

It is reasonable that the resulting deflections will be a function of the location of the support points and the force applied at these supports. The support point locations and forces can be adjusted in order to minimize the rms gravity induced deflections. When this is done, the rms deflections can be written as (see Telescope Mirror Supports, Nelson et al., 1982)

$$\delta_{rms} = \gamma_N \frac{q}{D} \left(\frac{A}{N}\right)^2 = \beta_N \frac{a^4}{h^2 N^2} \tag{8}$$

where: $\rho$ = density (kg/m3), $h$ = plate thickness (m), $q$ = applied force/area $= \rho\,g\,h$ (Pa), $a$ = plate radius (m), $D = E\,h^3/12(1-\nu^2)(N-m)$ = flexural rigidity, $\gamma_N = 1.2 \times 10^{-3}$ for large $N$ even for $N$ as small as 6, the support efficiency $\gamma$ is only twice the value for $N = \infty$.

## 8.    Thermal effects

Changes in temperature will generally cause dimensional changes in materials. For small changes, most materials expand linearly with temperature change and this expansion will happen in a stress free fashion

$$\delta = \alpha L \delta T \tag{9}$$

$$dL/dT = \alpha L \tag{10}$$

where $\alpha$ is the coefficient of thermal expansion (m/m°C) On the other hand, objects with non uniform $\alpha$ will generally deform with a uniform change of temperature. It is for this reason that when connecting materials with different coefficients of thermal expansion, extreme care must be taken in order to avoid temperature changes deforming the objects.

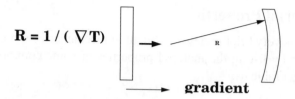

$$R = 1 / (\nabla T)$$

gradient

*Figure 3.* Deformation of a uniform material subject to a temperature gradient.

It is interesting to understand the effects of gradients of temperature as well. Changing thermal conditions generally cause temperature gradients in the materials. A uniform material subject to a temperature gradient will change its shape, and do so in a stress free fashion. The shape changes correspond to isothermal lines becoming curved with a radius of curvature $R$ (Fig. 3).

*Table 4.* Material properties of commonly encountered materials.

| Material | Density $\rho(g/cm^3)$ | Elastic modulus $E(GPa)$ | Thermal expansion $10^6\alpha K^{-1}$ | Specific heat $C(J/kg/K)$ | Thermal conductivity $K(W/mK)$ | $E/\rho$ | $K/\alpha$ |
|---|---|---|---|---|---|---|---|
| Aluminium 6061-T6 | 2.71 | 69 | 23.00 | 960 | 171.00 | 25.5 | 7.43 |
| Beryllium I70 | 1.85 | 304 | 11.20 | 1820 | 220.00 | 164.3 | 19.64 |
| Steel | 7.80 | 193 | 12.00 | 470 | 43.00 | 24.7 | 3.58 |
| Silver | 10.50 | 74 | 19.30 | 230 | 429.00 | 7.0 | 22.23 |
| Copper | 8.94 | 108 | 16.80 | 390 | 401.00 | 12.1 | 23.87 |
| Molybdenum | 10.21 | 324 | 5.00 | 247 | 140.00 | 31.7 | 28.00 |
| Titanium | 4.43 | 114 | 8.80 | 560 | 7.30 | 25.7 | 0.83 |
| Magnesium | 1.85 | 45 | 25.20 | 1000 | 76.00 | 24.3 | 3.02 |
| Lead | 11.34 | 16 | 29.00 | 130 | 35.30 | 1.4 | 1.22 |
| Nickel | 8.90 | 200 | 13.30 | | 90.00 | | 6.77 |
| Invar 36 | 8.05 | 141 | 1.00 | 515 | 10.40 | 17.5 | 10.40 |
| Silicon Carbide | 3.20 | 455 | 2.40 | 650 | 155.00 | 142.2 | 64.58 |
| Glass BK7 | 2.53 | 81 | 7.10 | 879 | 1.12 | 31.9 | 0.16 |
| Glass F2 | 3.61 | 57 | 8.20 | 557 | 0.78 | 15.8 | 0.10 |
| Glass FPL51 | | 73 | 13.30 | | 0.78 | | |
| CaF2 (calcium fluoride) | 3.18 | 110 | 18.90 | 911 | 9.70 | 34.6 | 0.51 |
| Pyrex | 2.23 | 66 | 3.30 | 838 | 1.13 | 29.4 | 0.34 |
| Fused Silica | 2.20 | 73 | 0.56 | 741 | 1.37 | 33.3 | 2.45 |
| ULE | 2.20 | 68 | 0.03 | 766 | 1.31 | 30.8 | 43.67 |
| Zerodur | 2.53 | 91 | 0.02 | 821 | 1.64 | 36.0 | 82.00 |
| Sapphire | 3.97 | 400 | 5.60 | 753 | 30.00 | 100.8 | 5.36 |
| MgF | 3.18 | 169 | 14.00 | 1004 | 21.00 | 53.1 | 1.50 |
| Diamond | 3.51 | 1050 | 0.80 | 108 | 2600 | 299.1 | 3250.00 |

## 9.      Material Properties

We close this very brief review of structural mechanics as it relates to astronomical optics by listing the material properties of some commonly encountered materials (Table 4).

## References

*Theory of Plates and Shells*, Timoshenko, S. and Woinowksy-Krieger,S., 1959, Ed. McGraw-Hill

*Telescope Mirror Supports: Plate Deflections on Point Supports.* Nelson, J., Lubliner, J. and Mast, T., 1982, SPIE. **332**, 212

# SEGMENTED MIRROR TELESCOPES

Jerry Nelson

*Center for Adaptive Optics*
*University of California at Santa Cruz*
*1156 High Street*
*Santa Cruz, Ca 95064-1077 USA*

jnelson@ucolick.org

**Abstract**

We present the basic structural and optical issues that cause us to be interested in segmented mirror telescopes as a path for building large telescopes. The history of segmented mirror telescopes is briefly reviewed. Segmentation geometries that might be suitable are reviewed with their advantages and disadvantages. A key issue, the asphericity of individual segments is reviewed and its impact outlined. Diffraction effects of segments are summarized as are the infrared impact of segmentation.

**Keywords:**    segmented mirrors, extremely large telescopes

## 1.    Motivation for segmented mirror telescopes

The size of telescopes is critical to the scientific capability of the telescope. Many, if not most, observations done by telescopes are limited by the amount of light collected from the astronomical target. Angular resolution is also often a critical limitation to understanding the nature of the astronomical targets.

For seeing-limited observations, the benefit of larger telescopes (measured in needed observing time to reach a given signal-to-noise ratio) varies as $D^2$ where $D$ is the diameter of the telescope. This is just the obvious advantage of more collecting area.

When one is making diffraction-limited observations of a point source in the presence of large backgrounds (the norm for most interesting, faint targets), the benefit grows as $D^4$. This comes from the collecting area advantage combined with the diffraction limited image shrinking (size $\approx \lambda/D$), thus reducing the amount of background "behind" the image.

*R. Foy and F. C. Foy (eds.), Optics in Astrophysics, 61–72.*
© 2006 *Springer. Printed in the Netherlands.*

The unique advantage of giant telescopes is the improvements in angular resolution that allow one to better understand the structure and morphology of non-point sources. For ground based telescopes this will generally require adaptive optics (AO) to remove the blurring caused by the atmosphere. AO use is rapidly increasing on the largest telescopes (where the benefits are greatest) and AO use is now commonly applied for wavelengths as short at $1\mu$ m (see Ch. 13).

Figure 1 shows scientifically interesting regions vs wavelength and angular resolution, to provide some general motivation for the scientific value of higher angular resolution. Near infrared is generally $1 - 3\mu$m and mid IR is generally $5 - 28\mu$m.

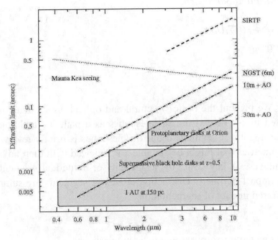

*Figure 1.*    Science targets as a function of wavelength and angular resolution

For the largest telescopes being developed ($\approx$ 30m) the diffraction limit at a wavelength of $1\mu$m is about 50 parsecs at $z = 2 - 8$, or about 1AU at 150pc (the nearest star forming regions).

## 1.1    Challenges for monoliths

Building a giant telescope from a monolithic mirror has many difficulties. These difficulties typically grow rapidly with the increasing size, and quickly make monolithic mirrors impractical. We simply list the key issues, and leave a more detailed discussion of each to engineering discussions.

- Reduced availability of mirror blank material
- Passive support of mirror will result in large optical deflections
- Very expensive mirror produces high risk of breakage from mishandling
- Larger mirrors are subject to larger deformations from thermal changes
- Vacuum chamber for mirror coatings becomes very large and expensive

- Tool costs for all parts (fabrication and handling) are large
- Shipping is difficult

An obvious solution to these and other problems is to compose the primary mirror from smaller segments, rather than a single large mirror. Although many important problems are greatly reduced by building the primary from small segments, there are a number of unique issues, concerns, and problems that arise with segments, and must be understood and dealt with before one can confidently proceed with building a giant telescope with a segmented primary.

## 1.2 Challenges for segments

Although one might imagine many issues that arise with segments, they can generally be grouped into four categories.

- Segments are difficult to polish
  - There are many segments
  - Segment surfaces are off-axis pieces of parent (not locally axisymmetric)
- Segments need active position control
- Segment edges add to diffraction effects and thermal background effects
- Telescope will have more parts leading to an increase in complexity

The major issues associated with polishing segments will be dealt with in this presentation. The methods for providing adequate active control are described in the literature (Mast and Nelson, 1982a; Mast and Nelson, 1982b; Nelson et al., 1982; Chanan et a., 2004). The impact of segment edges is discussed in this presentation.

The increase in complexity is evident. How important this is, or alternately, how to successfully deal with it, is a more complex and often subjective topic. We experience many complex systems that we take for granted, ranging from automobiles to power grids, to space vehicles. In general the increase in system complexity with segmented mirror telescopes is restricted to large multiplicity of identical components. It is thus possible (and important) to deal with this issue by making components that are highly reliable, and at the same time providing sufficient spares. Since these components must generally all be working, it is also important that replacing defective components with spares can be done quickly and easily. Designing the observatory to support this becomes a key issue in the observatory design.

Another issue only indirectly raised is the passive support of the optics. Here, segments have a great advantage since gravity deflections increase as $D^4$. Thus the support systems for segments are vastly simpler than those for a corresponding monolithic mirror. This great advantage for segments must be combined with the necessity for active control in comparing support systems.

*Figure 2.* Archimedes fired Roman navy

*Figure 3.* Horn d'Arturo mirror

*Figure 4.* Pierre Connes 4.6m Telescope

## 2.    History of segmented mirror telescopes

People have divided regions into segments for ages, ranging from bathroom tiles to modern segmented mirror telescopes. Even the application of segmentation to optics is old. The first recorded use of segmented optics was by Archimedes, who in 212BC had an array of mirrors focused on the attacking Roman navy in order to defend Syracuse (Fig. 2).

More recently, Horn d'Arturo in Italy made a 1.5 m segmented mirror in 1932 (Fig. 3). It was only used vertically, and was not active.

In the 1970's Pierre Connes in France made a 4.2 m segmented mirror telescope for infrared astronomy (Fig. 4). It was fully steerable, and active. Unfortunately, the optical quality was too low to be useful for astronomy.

Another type of segmented mirror telescope (actually multiple telescopes on a common mount) was developed in the 1970's and completed early in the 1980's. This was called the Multiple Mirror Telescope (MMT), and was built in southern Arizona. The telescope was made of 6 1.8 m primary mirrors, each axisymmetric. Although this telescope worked, it suffered from a number of problems, and was not viewed as entirely successful. An opportunity arose to replace it with a 6.5 m telescope with a monolithic primary and this was done in $\approx$ 2002.

In the late 1970's a very ambitious project to build a 10-m diameter segmented mirror telescope was begun, called the Keck Observatory. This project was formally begun in 1984 and completed and began science observations in 1993.

This segmented mirror telescope was quite successful and due to its success, funds were acquired to make a second Keck telescope, and it was positioned to allow the two Keck telescopes to be used for interferometry as well as for individual telescope observing.

The second telescope was completed in 1996 and began science observations in that year. Each of these telescopes has a suite of science instruments

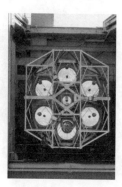

*Figure 5.* The Multiple Mirror Telescope (credit : MMTO)

*Figure 6.* Keck 10-m primary made of 36 segments

and each has an adaptive optics system. Interferometric science observations began in 2003.

The success of the Keck telescopes led to a Spanish project Gran Telescopio Canarias (GTC) to build a 10-m telescope on the Canary Islands in La Palma. This telescope is similar to the Keck, and should be operational in 2006.

The Keck telescopes have a hyperbolic primary mirror that leads to the requirement of polishing off-axis mirror segments. However, this is not the only possible approach. Telescopes with spherical primary mirrors can also be designed, and in 1998 the Hobby-Eberly Telescope was completed in Texas. The HET has 91 spherically polished primary mirror segments and is effectively an 8-m telescope. Along with suitable correcting optics it makes good images. It had some problems with the active control of its primary mirror segments, that now seem resolved, and it is actively engaged in astronomical research. With a very similar design, the South African Large Telescope (SALT) is nearing completion in 2004.

Future projects for segmented mirror telescopes include the 30-m CELT, GSMT and TMT projects (the merger of CELT and GSMT) and the 100-m developments by ESO for OWL.

## 3. Segmentation geometries

There are many ways one can imagine dividing up a primary mirror into smaller optical elements. A few include:

- Independent telescope arrays
- Independent telescopes on a common mount
- Random subapertures as part of a common primary
- Annular segmentation of a common primary
- Hexagonal segmentation

*Figure 8.* The Keck Observatory with two 10-m telescopes

*Figure 7.* The Hobby-Eberly Telescope

Independent arrays of telescopes have been discussed for decades but have generally not been successful, except for radio telescopes, where interferometry is a key virtue, aided by the fact that the individual telescope signals can be amplified and combined while preserving phase information. This is not practical in the optical, thus there are significant inefficiencies in sensitivity by coherently combining the light from an array of optical telescopes. Instrumentation for an array of telescopes has also been a cause of difficulty. Perhaps the best known successful array has been the VLT with four 8-m telescopes, each with its own suite of science instruments, and the capacity to combine all telescopes together for Interferometric measurements.

As mentioned before, the MMT is an example of independent telescopes on a common mount, with special added optics to combine all the light to a common focus.

In special applications people have built telescopes with very sparse arrays of mirrors in order to sample the resolution space with minimum number of mirrors. Systems like this may give greater angular resolution (longer baseline) but have less sensitivity due to the smaller collecting area.

Dense segmentation is generally considered when one is attempting to functionally replicate the optical properties of a single monolithic mirror. These kinds of mirrors have the best central concentration of light (leading to the smallest instruments and best signal-to-noise ratio measurements, and for a given collecting area, are generally the most compact and thus least expensive.

Dense segmentation can be done with annular patterns (Fig. 9), hexagonal patterns (Fig. 10), and many others. A regular pattern such as a hexagonal pattern can only be done with regular hexagons if the mirror is flat. Tessellating a curved surface can only be done with non-regular polygons.

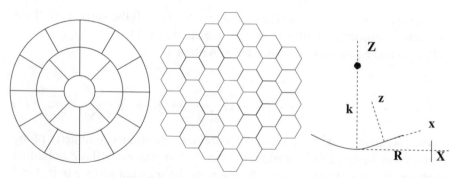

*Figure 9.* Dense mirror segmentation with an annular pattern

*Figure 10.* Dense mirror segmentation with a pattern hexagonal

*Figure 11.* Coordinate system for conic equations

Each of these patterns has its advantages and disadvantages, specific to the fabrication techniques planned. Annular rings provide maximum identical segments that may encourage replication techniques, and in any case reduce the number of spare segments needed. Hexagons make best use of the material typically produced in round boules of glass, and are easier to polish because the corners are less severe. They are also likely to be easier to support against gravity and attach to motion actuators (3 per segment) because of their symmetry.

## 4. Segment surface asphericity

Two-mirror telescopes are the most common optical design for ground based telescopes. These systems require a parabolic or hyperbolic primary mirror. As mentioned before, more complex optical systems can accommodate a spherical primary with its attendant simplifications, but several additional mirrors are needed to correct the spherical aberration, and the light loss and alignment complexity makes this configuration less commonly used. Here we will assume that a non spherical primary is needed and we will discuss the resulting surface shapes that segments will have.

The primary mirror is typically a figure of revolution, but since it is not spherical, pieces of the primary will not look identical, and will not be figures of revolution about their local centers. This basic fact introduces significant complexity for segmented mirrors. The polishing of off-axis segments, that are not figures of revolution, is generally much more difficult than the polishing of spheres. Second, off-axis optics must be carefully aligned in all six rigid body degrees of freedom, and the larger the segment asphericity (deviation from a sphere), the tighter the alignment tolerances become.

In this section we will describe in mathematical detail the surfaces of these segments. Consider a coordinate system shown in Fig. 11.

The general equation of a conic can be written as

$$z(r) = \frac{r^2}{2k} + (K+1)\frac{r^4}{8k^3} + (K+1)^2\frac{r^6}{16k^5} + (K+1)^3\frac{5r^8}{128k^7} + \dots \quad (1)$$

When we look in the local segment coordinate system, the symmetry of the equation seen in the global coordinate system is lost, and we will see azimuthal variations. We wish to express the equation for the segment surface in its local coordinate system

$$z(\rho, \Theta) = \sum_{mn} \alpha_{mn}\rho^m \cos n\Theta + \sum_{mn} \beta_{mn}\rho^m \sin n\Theta \quad (2)$$

where $\rho, \Theta$ are the local coordinate system polar coordinates, $k$ the parent radius of curvature, $K$ the conic constant, $a$ the segment radius, $R$ the off-axis distance of the segment center, $\epsilon = R/k$

Symmetry will cause $beta_{mn} = 0$, and analyticity requires m,n=0, m=n, m-n divisible by 2. This expansion has been made (Nelson and Temple-Raston, 1982) and yields

$$\alpha_{20} = \frac{a^2}{k}\frac{2 - K\epsilon^2}{4(1 - K\epsilon^2)^{3/2}} \simeq \frac{a^2}{2k} + \frac{Ka^2\epsilon^2}{2k} + \frac{9K^2a^2\epsilon^4}{16k} + \dots \quad (3)$$

$$\alpha_{22} = \frac{a^2}{k}\frac{K\epsilon^2}{4(1 - K\epsilon^2)^{3/2}} \simeq \frac{Ka^2\epsilon^2}{4k} + \frac{3K^2a^2\epsilon^4}{8k} + \dots \quad (4)$$

$$\alpha_{31} = \frac{a^3}{k^2}\frac{K\epsilon[1 - (K+1)\epsilon^2](4 - K\epsilon^2)}{8(1 - K\epsilon^2)^3} \simeq \frac{Ka^3\epsilon}{2k^2} + \frac{(9K - 2)Ka^3\epsilon^3}{8k^2} + \dots \quad (5)$$

We work out an example to show the asymmetry and the value of these equations. Consider Keck Observatory: $a = 0.9$ m, $k = 35$ m, $K = -1.003683$, $D = 10.95$ m.

$$\alpha_{33} = \frac{a^3}{k^2}\frac{K^2\epsilon^3[1 - (K+1)\epsilon^2]^{1/2}}{8(1 - K\epsilon^2)^3} \simeq \frac{K^2a^3\epsilon^3}{8k^2} + \dots \quad (6)$$

$$\alpha_{40} = \frac{a^4}{k^3}\frac{8(1 + K) - 24K\epsilon^2 + 3K^2(1 - 3K)\epsilon^4 - K^3(2 - K)\epsilon^6}{64(1 - K\epsilon^2)^{9/2}} \simeq \quad (7)$$

$$\frac{(1 + K)a^4}{8k^3} + \dots$$

For the outermost segment $R = 4.68$ m, and

- $\alpha_{20} = 11376\mu m$ (spherical shape, varies slowly from segment to segment)
- $\alpha_{22} = -101.1\mu m$ (astigmatism)
- $\alpha_{31} = -38.1\mu m$ (coma)
- $\alpha_{33} = 0.17\mu m$
- $\alpha_{40} = 0.09\mu m$ (spherical aberration)

As the relative size of segments shrink, the higher order terms become smaller quickly, and for "small" segments, the astigmatic term is the only term that matters.

In this example, the astigmatism dominates with amplitude of $101\mu m$. For a good optical surface, one wants the surface errors to be small compared with the wavelength at which the optic is to be used. For a wavelength of $0.5\mu m$ (visible light) we see that the astigmatism is $200\lambda$, a VERY significant aberration. Polishing out this shape is a unique challenge for segments, and requires special polishing techniques.

## 5. Segment Polishing

We discuss very briefly the key issues for polishing segments. There are two major areas of concern. The first is the challenge of polishing aspheres (see also Ch. 8). The second is the issue of polishing the surface properly right out to the edge of the segment.

Polishing aspheres is difficult because polishing only works well when the polishing tool fits the glass surface to within typical distances of $\leq 1\mu m$. Since polishing tools move in a random motion to produce the desired smoothness, this means the tool must be spherical in shape, and thus the contact area is limited by the asphericity. There are three main approaches to polishing. The first, the one used for the Keck segments, is stressed mirror polishing (Lubliner and Nelson, 1980; Nelson et al., 1980) where the mirror is deformed into a sphere so large tools can be used to polish the mirror. The second approach is to dynamically deform the tool so it always fits the mirror as it moves around the surface (stressed lap polishing), and the third method is to used suitably small tools so the tool fit is adequate, and make raster scans of the mirror with this small tool.

In addition to these approaches, two methods are currently in use that do not require a good fit between the tool and the part. These are ion figuring, and magnetic rheological figuring (MRF). Both methods are relatively slow in their material removal, but both are quite accurate, and are being used commercially. Ion figuring was used as the final production step in making the Keck segments.

Edge effects must be considered carefully since there are so many interior edges with a segmented mirror telescope. Again several approaches have been considered. The approach used for the Keck segments was to polish the mirrors as rounds, and when the main polishing was complete, cut the rounds into the desired hexagonal segments. This introduces no local effects, but introduces some global deformations. For the Keck segments this deformation was removed by ion figuring. A second approach is to add small "shelves" of material to the edges in order to support the polishing tool when it is near the edge. It is the polishing near the edge that typically rolls the edge in an optically objectionable fashion. The third approach is to polish near the edges with smaller and smaller tools to control the size of the rolled edge. A variant on the first method is to polish spheres with a planetary polisher. In this case the polishing tool is much larger than the part and the part is placed face down on the polishing tool. This produces roll free edges.

## 6.     Diffraction

Segment edges will introduce diffractive effects in the image that are in addition to the diffractive effects caused by the total aperture itself.

For the simplest of mirrors, circular apertures, the effects of diffraction cause the diffraction-limited image to be an Airy pattern, and for large distances this pattern falls as $1/\Theta^3$ and will be azimuthally symmetric.

Mirror segments (like Keck) will concentrate the diffracted energy into lines perpendicular to the edges, thus producing a diffraction pattern that is brighter or darker in some places than that of a circular aperture.

The amount of energy in the diffraction pattern, and the angular scale of this pattern is set by the scale of the segment size and the size and scale of the segment gaps. In general, diffracted energy is spread out over an angular scale set by $\lambda/w$ where $w$ is the linear scale of the pattern of interest. This could be the segment gap, the segment diameter, or the primary aperture for example. The amount of energy in the diffraction pattern is equal to the blocked area of the segment gaps. Hence, if 1% of the primary is in fact segment gaps or segment edge bevels, then 1% of the light will go into a diffraction pattern characteristic of the segment gaps.

Again using the Keck telescope as an example, the segments have 2mm bevels on the edges and a 3mm air gap between segments, for a 7mm total non-optical strip between segments. This is about 0.7% of the area of the primary mirror, thus there is an added diffraction pattern that has about 0.7% of the flux that the central image has. At $1\mu m$ this energy will be diffracted into angular scales of about 30 arcsec. Thus in this example, the diffracted energy is small and is spread over a very large region, making its local effects even less significant. For comparison, the structural spiders that typically support the secondary mirror often block about 1% of the light going to the primary, and with widths of 1-4 cm, this diffracted energy is slightly larger than that of segment edges, and somewhat more centrally concentrated.

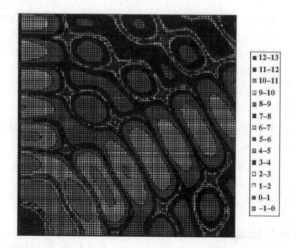

*Figure 12.* Diffraction pattern from a single hexagon. Each contour is a factor of 10 lower than the last.

The net diffraction pattern from an array of hexagons very much resembles the diffraction pattern from a single hexagon. We show the diffraction pattern for a single hexagon in Fig. 12.

## 7. Infra-red properties

In the infrared region longer than $\approx 2.2\mu$m the thermal emission from the environment (including the telescope and optics) becomes an important source of background. In the K band ($1.9 - 2.5\mu$m) at spectral resolutions high enough to resolve the strong atmospheric OH emission, thermal backgrounds can become important. Since the thermal backgrounds are the short wavelength tail of the blackbody spectrum, this background rises rapidly with increasing wavelength.

All telescopes suffer from this thermal background, depending on the temperature of the telescope and its optics. In practice, telescopes with clean and freshly applied mirror coatings (such as silver) have emissivities $\geq 1\%$ per surface at wavelengths beyond $1\mu$m. Of course as the optics degrade with time, dirt, etc. theemissivity will grow.

The gaps between segments will generally have much higher emissivities, and they may be close to 1. Thus, for a telescope like Keck, the segment edges will add an additional 0.7% thermal background flux beyond the $\geq 3\%$ that the 3-mirror Nasmyth configuration provides. When one includes the typical effects of dirt and aging, and the backgrounds from the atmosphere, and from additional warm optics that might exist in the beam train (such as windows or warm AO mirrors) the added background from segment edges is a real but typically small effect. Of course it is worthwhile to make efforts to minimize the size of segment gaps. Bevels are typically added to mirrors to avoid chipping of edges, and the air gaps between segments are there to avoid mirrors touching each other during installation and removal.

## References

*Stressed Mirror Polishing: A Technique for Producing Non-axisymmetric Mirrors.* Lubliner, J., Nelson, J., 1980,Applied Optics **19**, 2332

*Figure Control For a Fully Segmented Primary Mirror.* Mast, T., Nelson, J., 1982, App. Optics **21**, 2631

*Figure Control for a Segmented Mirror.* Mast, T., Nelson, J., 1982, Keck Report **80**

*Stressed Mirror Polishing: Fabrication of an Off-Axis Section of a Paraboloid.* Nelson, J., Gabor, G., Hunt, L., Lubliner, J., Mast, T., 1980, Appl. Optics **19**, 2341

*Telescope Mirror Supports: Plate Deflections on Point Supports.* Nelson, J., Lubliner, J., Mast, T., 1982, SPIE **332**, 212

*Off-axis Expansions of Conic Surfaces.* Nelson, J., Temple-Raston, M., 1982, Keck Report **91**

*Control and alignment of segmented-mirror telescopes: matrices, modes, and error propagation* Chanan, G., MacMartin, D., Nelson, J., Mast, T., 2004, Applied Optics **43**, 1223

# HECTOMETRIC OPTICAL TELESCOPES
# ESO'S 100-M OWL TELESCOPE CONCEPT

Philippe Dierickx

*European Southern Observatory*
*Karl-Schwarzschild str. 2*
*D-85748 Garching bei München2*
pdierick@eso.org

**Abstract**    The quest for larger telescopes is a never-ending story. Astronomers having very few other ways to perform experimentation, this quest has closely followed the path of technology development, and even promoted it more than once. Traditional design approaches, and above all technology limitations, have limited gains in size to a mere factor 2 between successive generations. We argue that such limitations have recently shifted from the realm of poorly reliable glass-making and opto-mechanical alignment to that of predictable control systems and large-scale mechanical structures. Novel design approaches taking full benefit of such evolution will, therefore, enable the design, construction and operation of facilities hugely more powerful than currently available. A crucial assumption is that giant telescopes could take full benefit of adaptive optics, a technique currently under rapid development. We elaborate on the design of ESO's concept for a 100-m class, diffraction-limited telescope.

**Keywords:**    optical telescopes

## 1.    Introduction

After preliminary work had hinted that giant telescopes might be within the reach of existing technology (Gilmozzi et al., 1998), in 2000 ESO committed a phase A study for a 100-m class optical telescope, dubbed OWL for its keen night vision and for OverWhelmingly Large. Although proposals for telescopes with diameters much larger than deemed possible had been laid down as early as 1978 (Meinel, 1978; Barr, 1979; Ardeberg et al., 1992; Mountain, 1996), this was, perhaps, the first occasion where not only possible design so-

*R. Foy and F. C. Foy (eds.), Optics in Astrophysics*, 73–86.

lutions but also plausible fabrication technologies were introduced. Since then, concepts of 20- to 100-m telescopes have been promoted worldwide[1]

Several factors support an unprecedented increase in telescope diameter: optical segmentation, active wavefront control, adaptive optics. Optical segmentation, implemented in the Keck and Hobby-Eberly telescopes, bypasses the critical difficulty of producing large, homogeneous optical substrates (see Ch. 6). Active wavefront control, which allows monitoring and controlling telescope errors in real time, guarantees optimal performance while permitting substantial relaxation of fabrication and stiffness tolerances. Finally, adaptive optics allows the geometrical extent to remain compatible with realistic instrumentation and, last but not least, dramatically increases optical efficiency by freeing the telescope from the adverse effect of atmospheric turbulence.

A substantial relaxation of figuring constraints is also allowed by spherical primary mirror designs, albeit at the cost of more complex optical solutions. Classical Ritchey-Chrétien-designs, however, are less than optimal for extremely large telescopes, the difficulty of manufacturing and aligning a convex, aspherical secondary mirror becoming prohibitive, not to mention the awkward manufacturing of off-axis aspherical segments of the primary mirror.

Bar adaptive optics, the feasibility of the crucial components of a giant telescope is no longer questioned. All industrial studies commissioned by ESO within the framework of its OWL design study confirm that the technological risk underlying the production of thousands of segments and structural elements is intrinsically lower than the risks taken less than two decades ago with 8- to 10-m class telescopes. Feasibility per se is evidently not sufficient; novel design approaches are required to guarantee cost-effective, streamlined production, integration and maintenance of the subsystems, and to ensure that the telescope works as a coherent and reliable system. The challenge of making 8- to 10-m class telescopes fell primarily on the shoulders of industry, optical manufacturers being on the front line, then on system designers, who had to ensure reliable performance. This can be reversed with giant telescopes, with low-risk in components manufacturing balanced by a substantial increase of system complexity, in particular in the area of control systems.

The design of the Owl 100-m telescope relies extensively on proven fabrication technologies, in particular on mass- or serial-production schemes, and incorporates several distinct wavefront control loops. The overall characteristics of the current design are listed in Table 1.

---

[1] see www.eso.org/projects/owl/Links for a partial list.

*Table 1.* Owl overall characteristics.

| Optics | | 6-mirror, f/7.5, 6,900 mš collecting area, near-circular outer rim |
|---|---|---|
| | M1 | Spherical dia. 100m, f/1.2 |
| | M2 | Flat, dia. 25.6 m |
| | Corrector | 4 elements, dia. 8, 8, 3.5, 2m |
| | FOV | 10 arc min. seeing-limited; 6 focal stations (rotation of M6) |
| | | > 2 arc min.diffraction-limited (vis.) |
| Structure | | 2.1 Hz eigenfrequency, 14,200 tons moving mass; |
| | | Standard steel beams, serially produced joints; |
| Kinematics | | Distributed low-cost friction drives, self-adjusting; |
| Control | | Multi-stage, distributed wavefront control: phasing, pre-setting, field stabilization, focusing, fine centering, dual conjugates active optics, adaptive optics |

## 2. Optical Design

The question of optical design is the subject of lively discussions. Die-hard proponents of classical design solutions, namely two-mirror ones based on the Ritchey-Chrétien formula, will go at length justifying the performance merits of simple, two-mirror solutions. In theory, the Ritchey-Chrétien design can hardly be surpassed, as it provides a relatively large field of view and a minimum number of reflections, hence minimal emissivity in the infrared. Figuring and testing of the off-axis aspherical segments of the primary, however, is a complex task that is bound to increase cost and risk. Spherical primary mirror solutions, on the contrary, are ideal from the point of view of manufacturing, all segments being identical, and easy to figure and test according to proven technologies. The performance cost, however, is far from negligible, as the spherical aberration of the primary implies substantially more complex designs, with subsequent loss of throughput and increase of emissivity. Unfortunately, so far discussions focused on the segments fabrication issue, which can only be part of a more complex trade-off at system level. A large telescope is a controlled opto-mechanical assembly; the consequences of optical solutions cannot be narrowed to the fabrication of the segments. At an earlier stage of the OWL study, the ESO has evaluated a fairly broad range of opto-mechanical designs (Dierickx et al., 2000) in relation to cost, risk, feasibility, performance, maintainability and operations. These evaluations consistently pointed towards spherical primary solutions, in view of the sheer size of the telescope and its exposure to inevitable excitations (gravity, wind, thermal change), its versatility, functional and performance requirements, and its cost ceiling (1 billion Euros in capital investment).

Still, the discussion is not closed, and will continue producing fruitful results, with proponents of different designs working hard on developing cost-

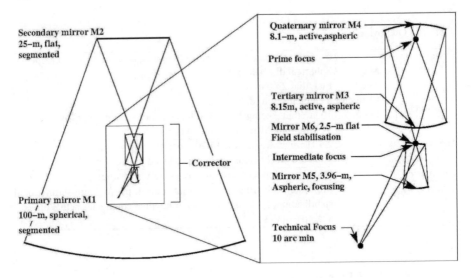

*Figure 1.*   Owl optical design layout.

effective technologies supporting their designs. In brief, ESO's decision to opt for a baseline spherical primary mirror solution is, ironically but very usefully, leading to substantial technology development in rival aspherical segments production technologies, and subsequent cost reduction. Hard numbers are still missing; if the author is allowed to rely on informal discussions with industrial suppliers, the cost increase from spherical to aspherical segments may have gone, in the past five years, from a factor 4-5 to less than two.

The OWL optical design is shown in Fig. 1. It is based on a spherical and flat folding secondary mirrors, with a four-element corrector providing for the compensation of spherical and field aberrations as well as advanced active optics and dual-conjugate adaptive optics. A complete discussion would exceed the scope of this report; we shall however mention a few key arguments supporting this solution:

- Low sensitivity to lateral decenters (hence to gravity, wind and thermal excitations), the alignment of the primary and secondary mirrors being quite inconsequential, and stiffness at the location of the critical subsystem (the corrector) being fairly high[2];
- Low number of surfaces (6) for the complete range of wavefront control functions: field stabilization, active focusing and centering, actively deformable surfaces, dual conjugates adaptive optics;

---

[2]Tolerances at the level of the corrector are comparable to those applying to the VLT 8-m telescopes, and easier to achieve in view of the increased design space. It shall be noted that in view of the size, and inevitable exposure to wind, sensitivity to wind excitation is of crucial importance.

- Low cost and fabrication risk for long-lead items (primary mirror segments);
- Ease of integration and maintenance; interchangeability of all segments);
- Large, well-corrected field of view (10 arc minutes seeing-limited, 2 arc minutes diffraction-limited in the visible [3]);
- Baffling, reduction of stray light implied by the availability of intermediate foci and pupil images;
- Availability of an intermediate focus (after the quaternary mirror) for the calibration of adaptive optics interaction matrices.

The main disadvantages of the design are the number of reflections, the implied emissivity[4], and the difficulty to manufacture the quaternary mirror[5].

The properties of the design, including its ability to rely on Laser Guide Stars (LGS), have been extensively described in the literature and will not be recalled herein (Dierickx et al., 2000; Dierickx et al., 2000; Dierickx, 2001).

## 3. Optical Fabrication

The very first question that comes to mind when dealing with giant telescopes is the cost-effective feasibility of its optics. Assuming classical materials for the segments blanks, however, there is no need for a very substantial increase in production capacity from existing suppliers [6], provided that the segment size [7] remains below 2-m. Moderately lightweight Silicon Carbide is also considered as a serious and potentially cost-effective candidate, for its superior thermal performance and specific stiffness.

With spherical segments, optical figuring and testing is a proven and reliable process, well suited for mass-production. Serial production of diffraction-limited, large optics is already under way for laser fusion projects, with European suppliers increasing their capacity to approximately 1,000 m2 per year [8]. Aspherical segments would certainly be feasible as well, but the inherent risk and potentially lower quality need to be properly evaluated. In figuring optical

---

[3]For comparison, a classical Ritchey-Chrétien solution of similar dimensions would offer about 20 arc seconds diffraction-limited field of view.

[4]For IR applications demanding a very small field of view only, this disadvantage could be mitigated by exchanging the corrector for a simpler one, thereby limiting the number of hot surfaces to 4, perhaps 3.

[5]In the design presented here, the compensation of the enormous spherical aberration of the primary mirror inevitably falls on the quaternary one. With a f/1.42 spherical primary mirror, the aspheric departure to be figured into the quaternary is as large as 14 mm.

[6]This statement is supported by contracts placed by ESO with CORNING, SCHOTT, and LZOS for the serial production of segments blanks in ULE (Ultra-Low Expansion glass), Zerodur and Astro-Sitall, respectively.

[7]Another compelling reason to limit segment size is the cost of transportation; beyond 2.3-m flat-to-flat segment size, transport in standard 40 ft containers is excluded and transport costs increase beyond reasonable limits.

[8]A driving factor is the production of 600 mm amplifier plates for the Megajoule project.

surfaces, polishing and testing play equally important roles. The final shape is basically determined by two references: the figuring tool and the metrology. Spherical segments can be polished with large-size, stiff tools, which provide a stable, passive reference, and tested against a unique matrix in a Fizeau interferometric arrangement. Aspherical segments, on the contrary, require that either the tool be flexible or small to accommodate for the aspheric departure, or that the part be bent with a dedicated warping harness and polished spherical, the aspheric shape being then attained when the bending moments are removed. In both cases the passive reference is lost, either because of the flexibility or small size of the tools, or because of residual internal stresses in the segment [9]. Metrology is an issue as well, the aspheric departure of individual segment, hence the characteristics of the metrology setup, being dependent of the segment's radial position. In brief, stringent quality control requirements and potentially lower yield make aspherical segments ill-adapted to cost-effective serial, reliable production.

As for the Owl 100-m telescope, the segments size (primary and secondary mirrors) is in the range 1.3-2.3-m flat-to-flat with a likely cost optimum at 1.6-m, corresponding to about 3000 segments for the primary and 250 for the secondary mirrors. The upper limit is set by the compatibility with cost-effective transport in standard containers. The lower limit, somewhat less clearly defined, is set by the implied cost and complexity of the control system. Three possible technologies have been considered for the optical figuring: replication, planetary polishing, and individual computer-controlled robots equipped with large, stiff tools.

With optical replication, the segments need to be shaped to a few microns accuracy; the final surface is obtained by injecting epoxy between the segment and a polished, negative master, which is removed after hardening of the epoxy. The technique has been proven on 1-m size optics (Assus et al., 1994) but unpredictable surface stresses in the epoxy and an expected deterioration of the master beyond a few dozen replication cycles make it less attractive than it seems. Planetary polishers, possibly complemented by local corrections with ion-beam or computer controlled, small tool polishers for final touches, seem ideal in view of the relative stability and high yield of the process. In this scheme, three to four 8- m class machines used serially would be required to achieve the required yield of about 1.5 segments per day (Fig. 2) Planetary polishers are however less attractive with large segments as the cost of large machines may become higher than that of a sufficiently large number of classical robots working parallel one piece at a time.

---

[9]This problem can be alleviated by stringent requirements on the residual internal stresses of the segment blanks, however at a higher cost of the substrates.

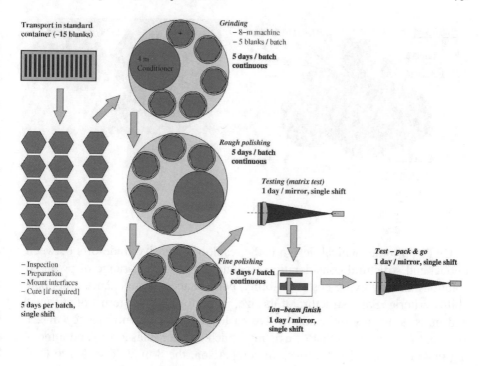

*Figure 2.* Polishing on planetary polishers.

Besides the increase of reflections implied by Owl optical design, a price to pay for the spherical primary mirror solution is the difficulty to compensate for its spherical aberration, and in particular the horrendous aspherization of the quaternary mirror (which is conjugated to the primary). A possible test setup has been identified and the state of current technology allows for cautious hope; industrial studies are however still required to confirm feasibility and evaluate implied cost and schedule.

## 4. Opto-Mechanical Design

The telescope has been designed as a coherent opto-mechanical system to ensure a reasonable balance between constraints. In particular, the optical solution has been tailored to avoid critical centering tolerances at locations where the structural design could not realistically provide high stiffness. Hence the use of a flat secondary mirror, a rather inefficient feature in terms of optical design, but an ideal one in relation to decenters. Critical tolerances inevitably show up, but they are now applying to the surfaces within the corrector, and to a lesser extent to the corrector itself. High structural stiffness is rather easily achieved at these locations, and the tolerances turn out to be comparable to those applying to ESO's Very Large Telescope (VLT), while allocations for design space and mass are far more generous with Owl.

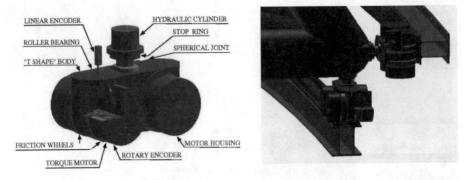

*Figure 3.*   Friction drives.

Owl's opto-mechanical design is the result of extensive trade-offs between optical and structural considerations spanning over a wide range of potential solutions and involving cost, feasibility, performance, maintenance and oper-ability criteria. For a structure of its size, its performance in terms of stiffness and mass is already quite impressive: a locked rotor eigenfrequency of 2.1 Hz, i.e. comparable to that of existing, much smaller telescopes, and a mov-ing mass of "only" 14,000 tons. In comparison, the 8-m VLT scaled to Owl dimensions would be nearing 1,000,000 tons. In addition, Owl's structure is made of standard steel beams and serially produced nodes, thereby allowing low cost, fast supply and integration schemes. The alt-az design has been aptly named "fractal", with a six-fold symmetry and substructures cloned from main structural modules at scales that are integer multiples of the segment size. As a result, the design ensures ideal, nearly all-identical interfaces to the main optics, optimal load transfers, ideal maintainability and low cost, in addition to a near-circular pupil and a Point Spread Function exclusively dominated by 6-fold symmetry features.

The design is supported by stress, fatigue, buckling and safety analysis, and relies on standard steel. Only in localized areas would higher quality (albeit still standard) steel be needed. Carbon fiber instead of steel pre-tensioning cables may provide higher performance at a marginal cost increase, but are not strictly required.

The kinematics is provided by friction drives (bogies); hydraulic pads have been briefly considered but the cost and complexity implied by kilometers of high-accuracy tracks were deemed unacceptable. ESO's experience with the VLT Coudé rotating tables clearly showed that micron-accuracy could be read-ily achieved with properly designed friction drives. Figure 3 shows a notional design of Owl's friction drives and their interfacing to tracks. Several hundreds of individual units would be required, but the cost is modest in comparison to that of hydraulic pads, with a far better load distribution hence relaxed founda-

*Figure 4.* Layout of the Owl observatory.

tion and alignment requirements [10]. Hydraulic connections between individual bogies are foreseen as a way to periodically re-adjust load distributions to cope e.g. with seasonal changes.

A possible layout of the facility is shown in Fig. 4. The telescope would be operated in open air, a co-rotating enclosure would be substantially more expensive and, with a 100-m slit, would provide little wind shielding anyway. Computer Fluid Dynamics (CFD) simulations, measurements on radiotelescopes and wind tunnel testing are planned to verify that the range and bandwidth of the kinematics and wavefront control systems are sufficient to cope with wind buffeting in the structures and on the mirrors. The design being insensitive to lateral primary-secondary mirror decenters, and the corrector being located in a stiff area, the crucial issue is less tracking performance than rapidly varying deformations of the segmented mirrors. A preliminary CFD analysis has shown that quasi-static (<1 Hz) errors would be well within the range and bandwidth of the active optics and field stabilization control sys-

---

[10]Relaxed foundation requirements may allow for poorer soil properties, hence allowing a broader site selection.

tems. At higher frequencies, the results are suspiciously optimistic, and can no longer be relied on - the complexity of a sufficiently high-resolution full-scale CFD model becoming unreasonably high.

The sliding enclosure concept is based on the design of an existing hangar built by SIAT South of Berlin to house giant Zeppelins (SIAT's hangar could accommodate two Owls). It was erected in about 1.5 years at a reported cost of 70 M Euros. In spite of being smaller, Owl's enclosure may not be significantly cheaper as more demanding site-dependent constraints may apply. Owl enclosure would provide passive thermal cooling only; local air conditioning is foreseen in primary and secondary mirror covers as well as in the corrector and the 6 instruments racks. The primary mirror covers, retracted from the telescope and stacked on top of each other, can be seen on fig. 4, in front of the enclosure.

## 5.    Control Systems

The OWL concept is built on the VLT experience, with extensive wavefront control capability. The number of degrees of freedom is however substantially larger, as shown in Tab. 2.

The first stage, pre-setting, is aimed at ensuring that optical surfaces are centered to within an accuracy allowing subsequent loops to be closed rapidly; in practice this means that the corrector shall be centered to a few mm accuracy, and surfaces within the corrector to typically 0.1 mm and about one arc second. Neither the metrology nor the actuation mechanisms required to achieve such tolerances do represent substantial challenges; a system currently being implemented in the South African Large Telescope (SALT) would most likely meet requirements, and an alternative solution based on fiber extensiometers will be evaluated in-situ in existing facilities.

Phasing of the segmented mirrors is a more complex issue, with several thousands of segments to be kept aligned within a few nm. The baseline solution is to mount position sensors delivering real-time measurement of the differential piston between adjacent segments. Noise propagation is well characterized, thanks to the Keck experience. Periodic on-sky calibration of the sensors is required; Gary Chanan showed that the Keck solution could be readily extrapolated to several thousands of segments (Chanan et al., 2000). Alternative methods have been proposed (Schumacher et al., 2001) and will be tested on-sky in the forthcoming years. In the case of Owl, independent calibration of the primary and secondary mirrors is an added complexity, which favors in-pupil phasing sensors and subsequent Fourier filtering to disentangle the individual segmentation patterns.

Field stabilization and active optics are essentially based on VLT experience, with the added complexity of controlling not one but two actively de-

*Table 2.*   Outline of Owl wavefront control functions.

| Wavefront control function | Objective, metrology and active elements |
|---|---|
| Pre-setting ........ $\Longrightarrow$ | *bring optical system into linear regime* |
| | Metrology: internal, tolerances 1-2 mm, 5 arc secs |
| | Correction: re-position Corrector, M3 / M4 / M5 |
| Phasing .......... $\Longrightarrow$ | *keep M1 and M2 phased within tolerances* |
| | Metrology: Edge sensors, Phasing WFS |
| | Correction: Segments actuators |
| Field Stabilization .. $\Longrightarrow$ | *cancel "fast" image motion* |
| | Metrology: Guide probe |
| | Correction: M6 tip-tilt (flat, exit pupil, 2.35-m) |
| Active optics ....... $\Longrightarrow$ | *finish off alignment / collimation* |
| .... $\Longrightarrow$ | *relax tolerances, control performance and prescription* |
| | Metrology: Wavefront sensor(s) |
| | Correction: Rotation & piston M5; M3 & M4 active deformations |
| Adaptive optics .... $\Longrightarrow$ | *atmospheric turbulence, residuals* |
| | Metrology: Wavefront sensor(s) |
| | Correction: M5, M6, ... |

formable surfaces (the tertiary and quaternary mirrors). This requires several guide stars, in theory three and in practice five to seven, and a corresponding number of wavefront sensors. Sky coverage is not an issue in view of the low spatial sampling required, hence limiting source magnitude (v 18-20) and of the available field of view (10 arc minutes). It should be noted that the availability of two deformable surfaces provides for extended correction capability, not longer limited to the field position of a unique wavefront sensor.

Adaptive optics (AO) undoubtedly constitutes the most daring challenge. It is also absolutely essential; seeing-limited observations with a 100-m class telescope would imply impossibly short focal ratio of the instrumentation, and immediate saturation by sky background. A smaller on-sky resolution element is the only way out, which implies at least some degree of adaptive compensation for atmospheric turbulence.

The most advanced astronomical AO systems today pack around 700 degrees of freedom and their efficiency in normal operation is still to be assessed; although there are promising avenues for development, the implementation plan for an Owl-class AO facility must follow a careful, gradual path. The first stage will be implemented in the sixth mirror (M6, diameter 2.5-m) and provide for ground-layer correction [11] or classical AO with natural guide stars in the infrared. Today's technology would allow for 1.8 mm thick, 2-m class

---

[11]i.e. relatively wide-field seeing reduction to 0.2 arc seconds, depending on turbulence intensity and structure.

mirror shells and an actuator interspacing of 30 mm. For Owl first AO stage, the likely requirements become 2.5-m diameter, 1mm thick (possibly less), and 10-15 mm actuator interspacing. Although far from trivial, the implied technological development over a decade is incremental and does not seem unreasonably challenging. Moderate segmentation of the in-pupil AO mirror is a backup solution. Wavefront sensing will require large format (512x512), fast, low read-out noise detectors which also represent a challenge, albeit moderate: experts consider a 256x256 detector as probably feasible today. On the positive side, the large aperture will probably allow for a better sky coverage with Natural Guide Stars (NGS). The reason is purely geometrical: with a small aperture, quite a number of guide stars are needed to probe the volume of turbulence across the field of view. With a sufficiently large aperture, this number should decrease as the volume probed with individual guide stars in relation to the total volume to be probed increases. Simulations indicate that in the K band, a sky coverage of about 30% down to the galactic pole may be possible.

For extreme AO (high level correction), Laser Guide Stars (LGS) will most likely be required. Even though the underlying science cases will probably imply rather small field of view, several LGS are needed to cope with the cone effect (Ch. 15). In addition, the LGS being at finite distance, their imaging through the telescope does not provide for sufficient image quality, even after refocusing them. In brief, imaging a source at 90 km with Owl is doing macro-photography, something it can hardly be designed for. It has been shown that NGS will still be required to sense the lowest wavefront modes (Dierickx, 2001) but the implied limitation in terms of field coverage are comparable to those applying to 8- to 10-m class telescope for the sensing of wavefront tilt. Alternative ways to do wavefront sensing on LGS are being explored by Ragazzoni et al, but at this day no proof-of-concept is yet available [12].

Although a clear objective, extreme AO at visible wavelength will certainly require more technology development, delaying its implementation to later stages. MOEMS (Micro-Opto-Electronical Mechanical devices) constitute a promising avenue, however not the only one (see Ch. 10). Piezo-stacks, combined with thin shells and moderate segmentation in a pupil, may offer a more conservative baseline.

In conclusion, adaptive optics is the only control system that still requires extensive but not unreasonable technology development. The project is cautiously allocating more than 10% of its total budget to adaptive optics. Last but not least, an aptly named MAD (Multi-conjugate Adaptive optics Demonstrator) instrument is currently under construction at ESO to demonstrate the

---

[12]The idea is to select light beams coming from the elongated LGS in such a way that taken individually, they have the same properties as if they were coming from an infinite distance, as seen by the wavefront sensor. The concept is called PIGS for Pseudo-Infinite Guide Stars and should be tested on-sky in 2003.

viability of relatively wide-field AO correction (Marchetti et al., 2002). On-sky results are expected early 2005 and will plausibly constitute a go-no-go milestone for Owl.

## 6. Cost and Schedule Estimates

Cost and schedule estimates for OWL are periodically updated with the progress of the design, analysis, and industrial studies. So far these studies have unequivocally proven the enormous benefits of a design favorable to in-dustrialized, serial- or mass-production processes. The current estimate is 940 million Euros in capital investment for final design and construction. Roughly half of that figure is already supported by industrial studies, and a few more are planned to consolidate the final estimate. It assumes the eventual site to be at moderate altitude, to have low seismicity, low average wind speed, favorable soil properties, and to be reasonably close to adequate, pre-existing infrastruc-tures. Probably none of these factors are required to ensure feasibility, but each could have a significant cost impact.

Elaborate schedule estimates have been drawn on the basis of industrial studies and past experience. As the aperture will be gradually filled, science operation may start ahead of the completion date. According to current plans, commissioning could be completed with provisional, non-adaptive M5 and M6 units (see Fig. 1 for the numbering of the surfaces). Once the aperture would be filled over a diameter of 60m, the provisional M6 unit could be replaced by the final one, thereby allowing an early start of science with near-IR, classical adaptive optics capability and wide-field seeing reduction. The second stage of adaptive optics (M5) would be integrated with the aperture reaching 80-m. Following a possible final design and construction schedule science operations could start in 2017. It is based on the assumption that an extensive phase B would be required over the period 2006-2010 before obtaining full funding for construction. If funding were readily available at the time of writing of this article, first light could occur in 2012. To some extent the mentioned schedule has been optimized to avoid unreasonably sharp peaks in cash-flow require-ments, and to maximize R&D time for the AO modules. It also includes more than a year engineering time before the telescope is handed over to the sci-entific community. Again, site properties may have a significant impact; the assumptions underlying the cost estimates apply to the schedule as well.

## References

Ardeberg, A., Andersen, T., Lindberg, B., Owner-Petersen, M., Korhonen, T., Søndergård, P., 1992, *Breaking the 8m Barrier - One Approach for a 25m Class Optical Telescope*, ESO Conf. **42**, 75

Assus, P. et al, 1994, *Performance and potential applications of replica tech-nology up to the 1-m range*, SPIE **2199**, 870

Barr, L.D., 1979, *Factors Influencing Selection of a Next Generation Telescope Concept*, SPIE **172**, 8

Chanan, G., Troy, M., Ohara, C., 2000, *"Phasing the primary mirror segments of the Keck telescopes: a comparison of different techniques"*, SPIE **4003**, 188

Dierickx, P.,2001, *Optical design and adaptive optics properties of the OWL 100-m telescope*; ESO Conf. *Beyond Conventional Adaptive Optics, Venice*, (http://lenin.pd.astro.it/venice2001/proceedings/)

Dierickx, P., Beletic, J., Delabre, B., Ferrari, M., Gilmozzi, R., Hubin, N., 2000, *The Optics of the OWL 100-m Adaptive Telescope*, Proceedings Bäckaskog Workshop on Extremely Large Telescopes, p97

Dierickx, P., Delabre, B., Noethe, L., 2000, *OWL optical design, active optics and error budget*, SPIE **4003**, 2000

Gilmozzi, R., Delabre, B., Dierickx, P., Hubin, N., Koch, F., Monnet, G., Quattri, M., Rigaut, F., Wilson, R.N., 1998, *The Future of Filled Aperture Telescopes: is a 100m Feasible?*, SPIE **3352**, 778

Marchetti, E., Hubin, N., Fedrigo, E., Brynnel, J., Delabre, B., Donaldson, R., Franza, F., Conan, R., Le Louarn, M., Cavadore, C., Balestra, A., Baade, D., Lizon, J.L., Gilmozzi, R., Monnet, G., Ragazzoni, R., Arcidiacono, C., Baruffolo, A., Diolaiti, E., Farinato, J., Viard, E., Butler, D., Hippler, S., Amorim, A., 2002, *MAD the ESO multi-conjugate adaptive optics demonstrator*, SPIE *Adaptive Optical System Technologies II*

Meinel, A.B., 1978, *An overview of the Technological Possibilities of Future Telescopes*, ESO **23**, 13

Mountain, M., 1996, *What is beyond the current generation of ground-based 8-m to 10-m class telescopes and the VLT-I?*, SPIE **2871**, 597

Schumacher, A., Montoya, L., Devaney, N., Dohlen, K., Dierickx, P., 2001, *Phasing ELTs for Adaptive Optics: preliminary results of a comparison of techniques*, ESO Conf. *Beyond Conventional Adaptive Optics, Venice* (http://lenin.pd.astro.it/venice2001/proceedings/).

# OPTICAL FABRICATION FOR THE NEXT GENERATION OPTICAL TELESCOPES, TERRESTRIAL AND SPACE

Robert E. Parks

*Optical Perspectives Group, LLC*
*Tucson, AZ 85718*
*USA*

reparks@optiper.com

**Abstract**      After a recall of the constraints in the construction of new large telescopes and of the shape of mirrors and associated aberrations, one addresses methods of polishing as dwell time, and bend and polish ones.

**Keywords:**      optical telescopes, off axisaspheres, polishing methods

## 1.      Introduction

Optical fabrication covers a huge variety of disciplines and crafts. This discussion will be limited to those areas that are specifically related to the questions likely to arise in the planning and building of the next generation of large optical telescopes. This means limiting the discussion to aspheric mirrors, and, generally, off axis aspheres. In addition, the discussion will be very practical. Because space is limited, there will not always be room to explain why a particular approach is suggested but the reader can be assured that it is a practical method.

The first part of the presentation will deal with the fabrication of aspheric mirrors of varying degrees of difficulty and explaining how the difficulty is related to the optical parameters of the surfaces. In general, the same difficulties arise in testing of aspheres in the same order and for the same reasons, so this chapter will serve as somewhat of an introduction to the optical testing chapter as well.

After covering the difficulties of manufacture, a series of practical approaches to manufacturing will be discussed. These will start with traditional techniques

*R. Foy and F. C. Foy (eds.), Optics in Astrophysics, 87–96.*

and proceed to some that are in the development phase. There are no hard and fast solutions to manufacture. Some optical manufacturers are more familiar and comfortable with certain techniques than others, and will tend to want to stay with these. Other manufacturers are more adventuresome and will take more risks in trying newer methods, but then there are risks for all involved. The ultimate choice of method of manufacture will depend on the intended uses of the telescope and the degree of flexibility wanted in its uses, the budget, facilities, talents and charisma of the project team.

## 2.    The design of Next Generation Telescopes

Monoliths have reached their practical limits of size in the 8 m diameter range. Larger telescope primaries will necessarily be segmented while secondaries will remain monolithic. The segments of the primary will be segments of a hyperboloid of revolution, one that is very close to a paraboloid of the same vertex radius. The segments will be thin, solid meniscus sheets of a low thermal expansion glassy material such as Zerodur or ULE. While it would be desirable to make the segments lighter weight to reduce the mass of the primary, the lightweighting process adds so much to the cost of the mirror that this approach is practical only for space optics, although there is research going on into less expensive, very low expansion lightweights.

The segments will be made of a very low expansion glassy material because there are no other materials that can duplicate the physical properties of the glasses that are necessary for a high resolution optical telescope. Glassy materials are temporally stable, very homogeneous, have a low coefficient of thermal expansion, are not affected by humidity, are perfectly elastic up to the breaking point and are transparent so they can be easily inspected for impurities and strain. Further, glass can be polished to a smoothness of less than 1 nm rms. About the only property that would improve glass as a mirror substrate material would be if it had a higher thermal conductivity.

## 3.    Primary mirror construction

The primary mirror is really a structured component. The actual mirror is the thin (roughly 100 nm thick) reflective coating that is deposited on the glass substrate. The highly smooth and highly accurate shape polished into the upper surface of the glass is itself supported on a multiplicity of loosely coupled bearing points of the steel support structure. The glass serves as the mechanical coupling between the continuous smooth, accurate upper surface and the rear that is supported by many localized bearing points or actuators. In performing this function of carrying its self-load to these individual support points, glass has mechanical properties of nearly the density and Young's modulus of aluminum.

The two natural choices for the segment shapes of the primary are concentric rings of trapezoids or an array of virtually identical hexagons. Each has pros and cons. The polished surface of every trapezoid in any ring is identical to any other in that ring while the surface shape of hexagons depends both on the radial and azimuthal position in the aperture although symmetry means that many hexes with the same radial position will have the same shape. Hexes are more easily polished than trapezoids because their outline is more nearly circular and all the vertices are obtuse angles. Since the physical outlines are nearly identical for all hexes, the tooling required for handling, polishing, testing and coating are the same for all hexes. For arrays with large numbers of segments, hexes are more appealing the larger the number of segments.

## 3.1    The shape of conic primaries

Before describing off-axis aspheric segments, a moment should be taken to say that spherical surfaces can be polished to very high figure accuracy with very little effort. In fact, it is hard to polish anything using a random stroke and not have it come out spherical. This is why the Hobby-Ebberly telescope at the University of Texas has a spherical primary made of segments (Hobby-Ebberly). It is an inexpensive way to build a telescope, but the telescope has limited performance and flexibility. Most astronomers do not want the limitations inherent in spherical primary designs in spite of their ease of fabrication.

The secret is to build a primary that is aspheric to get the desirable optical properties of these designs while keeping the segments as close as possible to a shape that is spherical or can be polished almost as easily as a sphere such as a toroid, a surface with two constant radii. To study this idea further, consider a mirror that is a parabola of revolution. We use a parabola because the more realistic hyperboloid is only a few percent different from the parabola but the equations are simpler and thus give more insight into the real issues of fabrication. The sagitta, or sag, of a parabola is its depth measured along a diameter with respect to its vertex, or

$$sag_{parabola} = \left(\rho^2/2R_{vertex}\right) \tag{1}$$

where $\rho$ is the radial distance and $R_{vertex}$ is the vertex radius of the paraboloid. The sag of a sphere of the same vertex radius is

$$sag_{sphere} = R_{vertex} - \sqrt{R_{vertex}^2 - \rho^2} \tag{2}$$

Because the difficulty of polishing the parabola is proportional to its departure from spherical, the difference in sag between the two, to first order, is

$$\Delta_{sag} = \left(\rho^4/8R_{vertex}^3\right) \tag{3}$$

Several things are immediately obvious; the problem is much worse at the edge of the parabola than near the center and gets worse as $(\rho/R_{vertex})^3$. Since $(\rho/R_{vertex})$ is proportional to the $f/\#$ of the primary, the difficulty as the primary gets faster is clear. Of course, this must be traded off with the cost of the dome and other lesser issues that depend on the $f/\#$ of the primary.

Something that is not obvious from the simple sag equation is how the local radii of curvature change with radial distance, $\rho$. Using analytical geometry one can show that the local radius of curvature in the radial direction goes as

$$R_{radial} = R_{vertex} \left[ 1 + (\rho/R_{vertex})^2 \right]^{3/2} \tag{4}$$

while the radius at the same location but perpendicular to this is

$$R_{tangential} = R_{vertex} \left[ 1 + (\rho/R_{vertex})^2 \right]^{1/2} \tag{5}$$

In other words, the two radii are different with the radial radius being increasingly longer than the tangential radius as the radial distance increases. The two differing radii means that the surface of the parabola is locally astigmatic (or saddle shaped) to a greater or lesser degree depending on the radial position. This is one reason a parabola is difficult to polish; if a sub-diameter lap is a perfect fit to the surface at some location, rotating the lap 90 degrees now makes it touch the surface in just two places at the high places of the saddle.

## 3.2    The shape of off-axis conic segments

Following on the above, we consider the departure of off-axis paraboloidal segments from a sphere of the same vertex radius. Figure 1 shows the geometry of an arbitrary segment, taken to be circular for ease of argument.

$\Delta_{sag}$ for the symmetric paraboloid is known from Eq. 2. To find $\Delta_{sag}$ over the off-axis aperture a coordinate transformation is performed to shift the center of the coordinate system to the center of the off-axis segment and to scale the off-axis aperture radius to a. Then we have

$$\Delta_{sag} = [\kappa/8R_{vertex}] \left[ (ax' + h)^2 + (ay')^2 \right]^2 \tag{6}$$

If this expression for the aspheric departure from spherical is expanded and recast in terms of commonly accepted optical aberrations, we find the departure is made up of a linear combination of aberrations and alignment terms shown in Tab. 1. The $\kappa$ is the conic constant if the asphere is not a parabola and the $c_v$ below is $1/R_{vertex}$.

The $b_0^0$ term is piston and is included only for completeness since it has no effect on the measurement. The $b_1^1$ term is the tilt of the off-axis paraboloidal

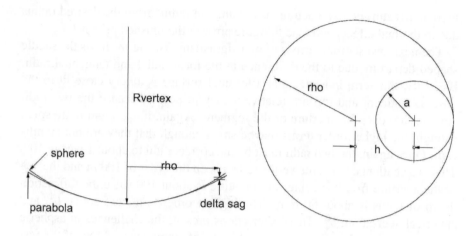

*Figure 1.* Geometry of an off-axis segment of a paraboloid.

*Table 1.* Departures of a parabola from a sphere in terms of common optical aberrations.

| | |
|---|---|
| $b_4^0 = \kappa c_v^3 a^4 / 48$ | $b_3^1 = \kappa c_v^3 a^3 h / 6$ |
| $b_2^2 = \kappa c_v^3 a^2 h^2 / 4$ | $b_2^0 = \kappa c_v^3 (4a^2 h^2 + a^4) / 16$ |
| $b_1^1 = \kappa c_v^3 (3ah^3 + 2a^3 h) / 3$ | $b_0^0 = \kappa c_v^3 (3h^4 + 6a^2 h^2 + a^4) / 24$ |

surface relative to the sphere at a radial distance of $h$. The $b_2^0$ term shows that the local radius of curvature of the paraboloid increases as the center of the sub-aperture moves outward in the full aperture. The $b_2^0$ term is the average change in focus, or radius, over the aperture of radius, $a$, displaced a distance, $h$, from the center of the symmetric paraboloid.

To this point, there are no non-spherical aberration terms. Provided $c_v$, $a$ and $h$ are small enough, the higher order terms might be small enough that the off-axis segment could be a pure sphere with a slightly longer radius than the vertex radius of the paraboloid. What we are really saying is that if the $f/\#$ of the off-axis segment is large enough, there is no need to make the segments aspheric. However, this brings out another point. The radius of the segment must be correct for the radial position of the segment in the aperture because in the case of a segmented primary, an error in radius is a figure, or surface, error. This is also the reason very low expansion glasses are needed for the segments.

Particularly during manufacture, any change in radius from the desired radius due to thermal effects will lead to figure error in the finished primary.

The next lowest order term is $b_2^2$ or astigmatism. This term shows the saddle shaped departure due to the difference in the local radial and tangential radii. Basically, this term indicates that a toroidal surface is a very close fit to the off-axis segment and tori are relatively easy to polish because the two radii are constant over the aperture of the segment. Again, if $c_v$, $a$ and $h$ are small enough, the higher order terms maybe small enough that they are not significant. Once again, the two radii must be the correct radii to about 1 part in $10^5$. This is a challenge when the vertex radius is on the order of 100 m and the $f/\#$ of the aperture from the center of curvature is about 100 since the diffraction depth of focus is about 50 mm. The highest order terms are coma, $b_3^1$, and spherical aberration, $b_4^0$. These aberrations present the challenges in aspheric fabrication when they are large enough to be of significance. Most of the rest of this article will discuss methods of producing surfaces with these shapes as well as the astigmatism.

## 4. Methods of polishing aspheres

There are two classes of polishing aspheres, one a traditional method of using a sub-aperture tool and rubbing longer where more material must be removed to get the desired surface shape. This category will be called a dwell time technique. The other method mechanically deforms either the asphere or the lap until there is a near perfect match of the asphere to the lap so that polishing proceeds as though one were polishing a sphere. The method was first applied by Schmidt in making corrector plates for Schmidt telescopes. This category will be called bend and polish methods. Within the two categories, there are various implementations, some well tested and some that are still under development. We will go through the methods in chronological order of development.

### 4.1 Dwell time methods

In dwell time methods, the desired surface is converted into a computer topographic map that is adjusted in tilt so that it touches the three highest points on the surface. Then a raster pattern is created so that the tool remains longest in the places where the most material must be removed to achieve the desired asphere. When we speak of tool, the traditional tool would be a sub-diameter pitch, polishing tool, quite possibly with a flexible backing (Kodak polish). Other "tools" include an ion beam gun for ion polishing (Kodak-ion), a magneto-rheological fluid for MRP (QEDMRF), a RF plasma torch for reactive chemical assisted polishing or RAP (Carr, 2001), or possibly hydrofluoric acid and a mask.

Each "tool" has a unique footprint or wear pattern that is a function of its diameter. This wear pattern is convolved with the topographic map of material to be removed to create the raster pattern with a dwell time at each raster position. In some implementations, these dwell time methods work very well. A factor of 10 improvement in surface figure is common for the ion beam and MRP methods but there are some definite issues as well.

One issue with the contact methods of traditional polishing and MRP concerns the figure, or shape, at the edge of a segment. As the tool begins to cross over the edge of the workpiece, the footprint or wear pattern changes because the tool is not supported in the same way it was when it was entirely on the work. The effect is to "roll" the edge of the work down. The partial solution is to use a smaller tool at the edge of the work so the tool center can get closer to the edge. The non-contact methods such as ion milling and RAP do not have edge problems.

Another problem with all the dwell time methods is that since a specific diameter tool is often used to do the polishing, the finished surface has a "roughness" of a spatial frequency associated with the tool diameter. This rather coherent roughness can produce diffraction artifacts in the image produced by the telescope. A partial solution to this problem is to use several tool sizes and do the figuring in stages rather than all at once.

At least two of the methods are quite slow; MRP and ion milling. MRP is too slow to be practical for terrestrial primaries but may be practical for space optics. These slower methods are particularly well suited to aspheres that are traditionally polished to within a micrometer or so of the correct figure prior to using the dwell time method. This is a particularly cost effective way of using these two methods. Traditional dwell time polishing is also slow so the preferred method is to grind or lap in the desired aspheric surface to within a few micrometers prior to polishing. Then use the dwell time polishing to not only remove the pitted surface due to abrasive grinding but also complete the final figuring. This method works well and is fast (for aspheric polishing but not fast relative to polishing a sphere) but has some issues relative to testing of both ground and polished surfaces because different test methods are needed for ground and polished (or specular) surfaces.

The RAP and etching methods are under development. RAP is a chemically assisted method that works under ambient laboratory conditions and has shown a wide dynamic range in terms of glass removal rates. The method also appears to achieve a polish from a ground surface. The footprint is readily defined and unaffected by workpiece edges. The method shows great promise but to date has not been used on work larger than a few centimeters.

The etching method is one where the work is masked with a resist such as wax and successively etched as the masking is extending from the regions that are to remain the highest to the lowest. The contour steps initially might be on the order of 0.5 micrometer and would be polished out with a flexible lap. After a first cycle, the segment would be tested and a new set of masks designed to correct residual errors using a finer contour step. This method is in early stages of development and probably will not work on glass ceramics where the hydrofluoric acid derivative works more aggressively on the glassy phase than the crystalline phase. With development, the method might hold promise.

## 4.2    Bend and polish methods

The original bend and polish idea is due to Bernard Schmidt who built what has become known as a Schmidt telescope with a spherical primary and an aspheric corrector plate that introduces exactly the right amount of spherical aberration to cancel the spherical aberration that would be introduced by the spherical primary (Schmidt). The corrector has to be thicker at the edge in proportion to the radial distance in the aperture to the fourth power to provide the needed correction.

It turns out that the bending equation for a thin parallel plate that is simply supported around the edge and subjected to a uniform load bends with a fourth order radial component. By supporting the glass window around the edge and using the correct vacuum for uniform loading, the corrector plate can be bent concave and then polished spherical. When the vacuum is released, the plate comes back to its original flat shape but the upper surface has the residual fourth order correction needed to cancel the spherical aberration (Everhart, 1966a; Everhart, 1966b). Numerous Schmidt telescopes have been built using this technique.

When ideas for what were to become the Keck Telescopes were forming in Jerry Nelson's head, he was trying to think of a relatively inexpensive method of making the segments. When he realized that the shape of the segments could be well approximated by the linear combination of low order aberration basis functions listed in Tab. 1, it occurred to him that the plate bending equation, $\nabla^4 w = p$, could be solved entirely in terms of boundary conditions of forces and moments on the edge of a circular segment. This led to the method adopted for the production of the Keck segments (Nelson et al., 1982).

Each circular segment was uniformly supported on its back. Twenty-four equally spaced metal blocks were cemented to the edge of the segment and loaded with precise forces or moments to bend the segment to the reverse of the final shape. The segment was then ground and polished spherical. When the edge loads were removed, the blank assumed the desired figure to about half a micrometer out of a total correction of nearly 400 micrometers for the edge segments. The edges of the segments were then sawn off to make them hexagonal. This process relieved some internal stresses that further warped the surface. This residual error was removed by one cycle of ion polishing in most cases.

The advantages of the bend and polish idea were recognized by Roger Angel at the University of Arizona. He was trying to make monolithic 3.5 m diameter primaries that were about $f/1.25$. This is far faster than traditional techniques produce cost effectively. He made a 1.5 m diameter polishing lap with 24 edge actuators to bend the lap into a shape that fit the mirror whatever the radial position of the lap. The actuators were also synchronized with the rotation of the lap so the astigmatic part of the bending stayed aligned radially with the mirror. Thus, the lap was dynamically deformed using the same bend and polish idea as Nelson used as the actively stressed lap was moved about the mirror surface (Martin et al., 1990). The same lap was used on the $f/1.25$, 6.5 m mirror used in the MMT upgrade. The polishing method is very fast by traditional standards, nearly as fast as polishing a sphere, but some hand figuring was required to reach the final figure accuracy. The MMT now produces sub arc second images (MMT).

A final application of bend and polish may be applied to new telescopes with hundreds of segments. Plane mirrors are economically made using a continuous polisher, or CP machine. The machine consists of an annulus of pitch about 2.5 times the diameter of the maximum size workpiece to be polished. The work is placed face down on the pitch and constrained from rotating with the pitch annulus by rollers touching the edge of the work. A large flat mirror can be fully polished out in one eight hour shift and the machine holds three pieces at a time (Kodak-CP).

Obviously, telescope mirrors need power but this type CP machine has also been used to polish concave spheres by contouring the annulus of pitch into a convex spherical surface. The idea now is to take segments that have already been ground to the correct spherical radius and attach a bending frame to load the blank into the reverse of the desired figure. Then segment and frame would be set face down on the CP machine to polish the segment spherical. The details of the process have not been worked out and it might be necessary to grind in the aspheric curve using the same strategy on a CP grinding machine, but the whole idea seems quite practical when faced with the task of producing hundreds of segments in a short time and for a reasonable price.

For all its advantages, there is a drawback to applying bend and polish to the mirror segments. The mirror blank must be sized so it is thin enough to allow bending to the desired figure without breaking yet thick enough so that it does not distort too much under its own weight when used in the telescope. This is a delicate trade-off during the telescope design. The actively stressed lap does not share this drawback.

## 5.     Conclusion

Something implicit in the above discussion but not stated explicitly is how much our present computing power has to do with the design and operation of large telescopes. The same can be said for advances in the fabrication of mirror segments but the manufacturers of segments are not moving as fast taking advantage of this computing power as the scientists using the telescopes. There is room for a further push to bring a more detailed and sophisticated set of software tools to the aid of the mirror makers.

Techniques for understanding the shapes of off-axis segments have been explained. When these shapes are close to spheres or torroids, the polishing of the surfaces is relatively easy using traditional methods. As the components of coma and spherical aberration become larger as the segments become larger or the $f$ number of the primary becomes faster, it is difficult to make the segments without resorting to dwell time or bend and polish techniques. Examples of both methods have been given with the more useful technique being the bend and polish approach to either the segment itself or the polishing lap.

## References

*www.as.utexas.edu/mcdonald/het/het$_g$en$_0$1.html*

*www.kodak.com/US/en/government/ias/optics/polish.shtml*

*www.kodak.com/US/en/government/ias/optics/ion.shtml*

*www.qedmrf.com/technology/how.shtml*

Carr, J. W., 2001, *Atmospheric Pressure Plasma Processing for Damage-Free Optics and Surfaces*, Engineering Research Development and Technology - FY-99. Lawrence Livermore National Laboratory, p. 3:1

*www.fvastro.org/articles/schmidtp2.htm*

Everhart, E., 1996, *Making Corrector Plates by Schmidt's Vacuum Method*, Appl. Optics **5**,713

Everhart, E., 1996, *Errata*, Appl. Optics **5**, 1360

Nelson, J., Lubliner, J. and Mast, T., 1982, *Telescope mirror supports; plate deflections on point supports*,SPIE **332**,12

Martin, H.M., Anderson, D.S., Angel, J.R.P., Nagel, R.H., West, S.C., Young, R.S., 1990, *W. Progress in the stressed-lap polishing of a 1.8-m f/1 mirror*, SPIE **1236**,682

*www.mmto.org/mmtconv_pix/firstlight.jpg*

*www.kodak.com/US/en/government/ias/optics/plano.shtml*

# OPTICAL TESTING FOR THE
# NEXT GENERATION OPTICAL TELESCOPES

Robert E. Parks

*Optical Perspectives Group, LLC*
*Tucson, AZ 85718, USA*

reparks@optiper.com

**Abstract**     This lecture addresses the optical testing of concave aspheric mirrors. Examples of measurements of low order aberrations are shown. There are noises and bisases due to environmental effects, such as air turbulence, mirror temperature. Methods of interferometric testing are discussed.

**Keywords:**     optical testing, mirror segments, aspheric mirrors

## 1.     Introduction

The discussion of optical testing will be limited to concave aspheric mirrors such as are likely to be segments of a large segmented primary mirror. Because these mirror segments will have long radii of curvature and subtend a relatively slow $f/\#$ cone, they present some unique, practical optical testing problems. Since so much of the general topic of optical testing is the same for whatever type of optic one is testing, this discussion will concentrate on the aspects of testing that bear on primary mirror segments.

There are two fundamental yet intimately related problems in testing these segments for figure error; first, the phase or wavefront error must be measured normal to a reference surface and, second, the lateral position of the segment relative to the reference must be accurately known because the radius of the surface changes with lateral position. The first part of the paper will discuss measuring the phase and the environmental issues that go along with this aspect of the measurement. In the second part of the paper the lateral alignment of the reference to the test piece will be treated.

97

*R. Foy and F. C. Foy (eds.), Optics in Astrophysics, 97–105.*
© 2006 Springer. Printed in the Netherlands.

## 2.      Measurement of surface or wavefront error

Because of the commercial availability of interferometers that have a re-peatability around 1 nm peak-to-valley (P-V) at any pixel location in the de-tector, the discussion will be limited to the use of interferometric tests but the principles apply to any type of optical test device. Using this 1 nm repeatability as a benchmark, it will be easy to demonstrate where some of the other testing problems occur long before we hit the repeatability benchmark.

We will use a specific example to demonstrate the problems likely to be en-countered when testing. Assume we have an extremely large telescope with around 1000 - 1 m hex segments whose prime focal length is 50 m, or ver-tex radius is 100 m. The inner ring of segments will have their centers at an aperture radius $h$ of 3.5 m and the outermost ring at 14.5 m.

The departure of these segments from spherical is given by the terms set out in the Tab. 1 in Ch. 8. Table 1 simply gives a similar Table of the peak-to-valley departure for the inner and outer rings of segments for the aberrations using those equations (and multiplying by 2 to get P-V except for spherical aberration where the multiplication factor is 1.5).

*Table 1.* Peak-to-valley departure from a sphere for the inner and outer rings of segments for the aberrations using equations of Tab. 1 in Ch. 8

| Error term | Inner ring ($h = 3.5$ m) | Outer ring ($h = 14.5$m) |
|---|---|---|
| Spherical aberration | .003 $\mu$m | .003 $\mu$m |
| Coma | .146 | .604 |
| Astigmatism | 1.531 | 26.281 |
| Focus, or power | 1.531 | 26.281 |
| Tilt | 21.73 | 1525.52 |

Starting with tilt (since piston cannot easily be measured, nor does it mat-ter) on the inner ring, this says that if an interferometer were focused at the center of curvature of a 100 m radius sphere looking at a 1 m diameter seg-ment 3.5 m from center of the sphere and a paraboloid were substituted for the sphere and their vertices were coincident, the interferometer would have to be moved laterally 2173 $\mu$m to be aligned with the chief ray of the segment of the paraboloid. At this point there would be no tilt fringes visible in the interfer-ometer. For the outer ring, the interferometer would have to be moved 152.552 mm to eliminate the tilt fringes. Figure 1 shows the effect of removing the tilt by shifting the interferometer for the inner ring. The Figure 1 also shows the residual error after shifting the interferometer for the outer ring; there are too many fringes to show the case before shifting.

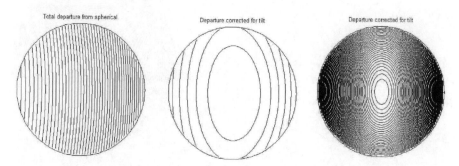

*Figure 1.* Departure from spherical on the inner ring of segments before correcting for tilt (left), after correcting for tilt (middle), and after correcting for tilt on the outer ring (right). Contour lines are approximately 1 $mum$ surface error.

Because the radius of curvature of the paraboloid also increases as the segment is moved from the center to the edge, the interferometer would have to be moved away from the vertex by 122.48 mm to remove the defocus in the case of the inner ring segment and 2108.05 mm for the outer segment. Once this had been done, the interferometer would sense a 1.531 $\mu$m surface error due to astigmatism, a 0.146 $\mu$m surface error due to coma and an insignificant error due to spherical aberration. In all likelihood, a commercial interferometer could measure this error without any auxiliary optics, and if there were no environmentally induced errors, make the measurement accurately to nearly the repeatability level of the interferometer. Figure 2 shows the departure after correcting for focus on the inner and outer rings of segments. The large difference in departure between inner and outer rings of segments is obvious.

The 28 $\mu$m of astigmatism in the surface of the outer segment is sufficient that most of the light returning from the segment would not make it through the interferometer to the detector. The rays would be blocked by apertures in the interferometer. It would require some sort of auxiliary optic or a different test set up to test this surface. The departure after removing the astigmatism is less than 1 $\mu$m and is not even shown here except in the asymmetry of the largely astigmatic contour pattern. The errors in Fig. 2 are the real concern in both fabrication and testing in this example. In any reasonable telescope design with a segmented primary, a similar situation will hold true.

## 2.1    Environmental effects

The first environmental problem encountered would be air turbulence. The light travels a 200 m path from center of curvature to the mirror and back. To give an example of the potential problem, assume the temperature of the air

Departure corrected for tilt and power

Departure corrected for tilt and power

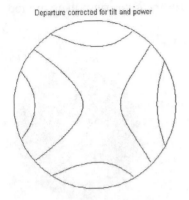
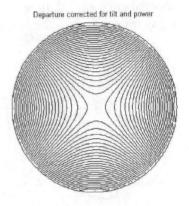

*Figure 2.*   Departure after correction for tilt and focus on the inner ring of segments (left) and the outer ring (right). Contours are approximately 1 $\mu$m surface error.

along the optical path to one side of the mirror segment is 0.01 C different than on the other side. Since the index of air changes as the ratio of the absolute temperature, the difference in index due to the temperature difference is $\Delta n = 3.4 \times 10^{-5}$. Over the $2 \times 10^5$ mm optical path, the difference in optical path length is 6.8 mm, orders of magnitude higher than the level of accuracy one would seek. This is an unrealistically high estimate since it assumes the temperature difference is constant along the entire optical path, but it also indicates how serious a problem long air paths are.

Another problem with the long distances comes about when trying to measure the radius of curvature. We previously showed that the radius of curvature of the segments would have to be correct to about $1 : 10^5$. Since the coefficient of expansion of most common materials is larger than this, holding the distance from the interferometer to the mirror over these distances is difficult.

There are two other related sources of environmental disturbances, seismic and acoustic vibration. Optical testing facilities should be located far from highways and railroad lines. The test hardware should be isolated from the building structure to avoid seismic disturbances. Acoustic disturbances are often overlooked and can be a bigger problem than seismic. Loud noises of any sort are a problem. The sub normal hearing frequency noise made by many air conditioning and handling units is often an unforeseen problem.

The conclusion of this discussion is that the reference test surface should be as close as possible to the segment under test. Further, to aid in keeping a constant radius of curvature reference, the reference should be made of a very low expansion material such as Zerodur or ULE.

# 3.    Possible methods of practical interferometric testing

Providing the departures from spherical are not too great and the segments not too large, the situation in this example, the segments can be tested against a test plate of a slightly smaller convex radius than the vertex radius of the primary. The test would look much like Fig. 3. A point source of coherent light would illuminate the steep side of the test plate made of ULE or Zerodur. The steep side would be polished as a hyperboloid with a conic constant of around -2.25 to make the rays strike the opposite side with a 100 m convex radius at normal incidence.

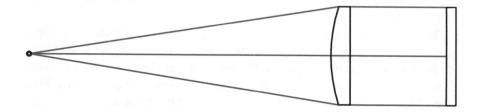

*Figure 3.*    Test plate method used to check the figure of a segment of the primary mirror.

A nearly perfect diverging wavefront would exit the test plate appearing as though it came from a source 100 m away. A segment would be positioned so that its mean center of curvature was coincident with that virtual source 100 m away. In the worst case in our example, the un-equal air path would be about 4 m rather than 204 m. Interference would take place between the wavefront reflected off the 100 m radius side of the test plate and the segment. The roughly 3 m back to the source and beamsplitter is common path and will not affect the interference pattern.

The test plate method also eliminates, up to a point, the problem of all the returning rays getting to the detector. As long as the beamsplitter and detector are large enough to collect all the return rays with the detector placed beyond the caustic produced by the aspheric segment, all the light will reach the detector. Also, it is not impractical to think of test plates as large as 2 m in diameter if larger segments are desired. Homogeneity is important but not critical since the glass in the test plate is in the common path. In the same way, the asphere on the test plate has to be smooth but does not have to be accurate to better than a micrometer or two.

Eventually, as the primary mirror $f/\#$ gets faster the segments will have so much astigmatism that looking at the segments straight on will not work. With the addition of a 1 m convex mirror, the astigmatic part of the correction can be cancelled. Think of the region at the end of the minor semi-axis of an ellipsoid of revolution as shown in Fig. 4. The radius of curvature of that surface in the plane perpendicular to the figure is the length of the minor semi-axis. The radius of curvature in the plane of the Figure is the length of the semi-minor axis squared divided by the semi-major axis length, or $b^2/a$.

Since we know the segment of the ellipsoid will image perfectly a source at one focus to the other focus, we have a null test to cancel the astigmatism. In order to keep the un-equal optical path of the interferometer short, a second convex spherical mirror is used to reflect the light back to the segment and test plate. This test has a second advantage over the straight on test because the light reflects twice off the segment making the test is twice as sensitive to errors in the segment. The second reflection also means that substantially less light is reflected back to the detector but the segment can be spray silvered to improve its reflectivity (Armstrong, 1977).

There is another method of using the test plate straight on. A hologram that cancels the astigmatic error can be written on the long radius side Burge, 1994. Although the references cite the use of this method for testing secondary mirrors, it would work just as well for primary mirror segments. In addition, one of the inventors, Jim Burge, suggested that the hologram does not have to be on the test plate itself but could be in the return beam near the detector. In this case, the hologram would be no larger than the bundle of return rays where they come to focus. Because the hologram would be small, a different hologram for each segment shape could be made that would entirely cancel the errors for a particular segment. Then a whole series of holograms could be written on a photomask and the photomask repositioned for each new segment being tested.

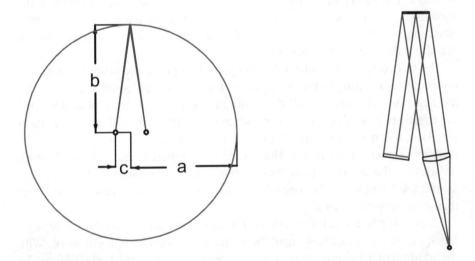

*Figure 4.*   Segment at the end of the semi-minor axis, b, of an ellipsoid (left), and use of a test plate and auxiliary convex mirror to shorten the optical path of the test (right).

Using the test plate method to test the primary mirror segments substantially helps the index of refraction of air problem cited above. It also makes it possible to control the radius of curvature of the segments because the test plate is made of very low expansion material. This coupled with the test plate being near the segment make the measurement of the distance between the test plate and segment easy. By positioning the segment axially relative to the test plate so the interferometer shows best focus and then measuring the distance from test plate to segment gives the radius of the segment to the required accuracy. Since all segments are tested against the same test plate, there is a high degree of control of the radius of all segments. The absolute vertex radius of the primary is not critical but once this has been established, all segments should match via this single reference to avoid a figure error due to variations in the radius.

## 3.1 Lateral position of the segment

Thus far we have only talked about the axial position of the segment during testing. There are five other degrees of freedom to locate the segment relative to the test plate. The two tilt degrees of freedom about the axes parallel to the face of the segment are constrained by the interferometer. When the interferometer shows no tilt in the fringes, these two degrees of freedom are properly constrained.

Tilt or rotation about the $z$ axis, or the axis between the test plate and segment, is controlled by analysis of the residual astigmatism. If the segment is properly rotated about the $z$ axis, the astigmatism will be aligned in the radial direction. Any component of astigmatism not aligned in the radial direction indicates a rotational misalignment of the segment and can be corrected.

The final two degrees of freedom, translation in the $x$ and $y$ directions, are the real concerns. With no loss a generality, assume the center of the segment lies on the $x$-axis of the parent parabola. Then it is clear from the formulae for coma and astigmatism in Ch. 8 that the departure from spherical depends on the distance the segment is from the center of the parent parabola, $h$. The best fit radius for the segment depends on $h$ as well to virtually the same degree as astigmatism.

For the example we are working with these displacements from the theoretical position do not make too much difference in the final test results but they can be important in other designs. For the record, the change in the coma component of the departure is

$$\Delta b_3^1 = \kappa a^3 / 3R_{vertex}^3 \Delta h = 4.2 \times 10^{-8} \Delta h \tag{1}$$

For $\Delta h = 1$ mm the error in $b_3^1$ would be only 40 pm, completely in the noise. Notice this result is independent of where the segment is located in the aperture of the primary. Clearly, this becomes a bigger problem when the diameter of the primary is taken into account. Holding a segment at the edge of the aperture to the correct lateral position to 1 mm over the temperature swings the steel structure is likely to see changes the problem from inconsequential for testing to a possibly significant one for the finished telescope mirror segment support system.

Following similar reasoning, the error in the astigmatism component of the departure is

$$\Delta b_2^2 = a^2 h / R_{vertex}^3 \Delta h = 8.75 \times 10^{-7} \Delta h \qquad (2)$$

for the inner ring of segments with $h = 3.5$ m and $3.63 \times 10^{-6} \Delta h$ for the outer ring. A 1 mm error in $h$ will produce a roughly 1 nm error in the inner ring or a 3.6 nm error in the outer ring. Again small errors but close to the limit of the peak-to-valley systematic errors to be allowed in the testing of the outer segments.

The test plate method of performing the testing provides a reasonably convenient method of assuring lateral alignment since the test plate and segment are never more than about 2 m apart independent of the ring in which the segment is. Unfortunately, the outer segments are the farthest away but it is not hard to align to 1 mm over a 2 m distance. Care is required in the design of the test set up but no unusual steps need be taken.

As before, if the segments are larger or the primary is faster, all the alignment tolerances will get tighter. However, for this example, which is typical of some designs under discussion, the testing does not appear to be a major problem. Like any test where a few nm rms precision is required it will not be easy and close attention will have to be paid to the environment, but nothing looks unreasonable.

For the case where the departure is more severe than this example, the holographic approach would probably be applied to the test plate method. Particularly if small holograms placed near the focus of the returning rays were used, and this would certainly be the less costly method, alignment features could be built into the hologram that would help keep the segment in the correct lateral position and rotational orientation about the $z$ axis relative to the test plat and hologram.

## 4.    Conclusion

We have shown that there is a straightforward method for testing off-axis segments of a very large telescope. That is not to say there may be other, perhaps better, methods, but often knowing a solution exists leads to further insights into the entire project design. The major reason the testing is straight-forward is that the individual segments subtend such a slow $f/\#$ cone of light even though the primary is very fast. The test procedure takes advantage of this narrow light cone.

We have emphasized unequal path interferometry because it is the most fa-miliar method of optical testing these days. However, we have also shown how sensitive this type of interferometry is to environmental disturbances. There are a number of other types of optical testing including common path methods such as shearing interferometry and Shack-Hartmann testing Malacara, 1992. As we pointed out in the fabrication section, optical testing has no more kept pace with the benefits of increased computing capability than fabrication has.

There are many potential paths for revitalizing old methods of test using new array detectors and powerful processors. The only reason most testing is done by unequal path interferometry is due to the invention of the HeNe laser. If its development had been delayed a year or two and computing capabilities had been a little more advanced, most optical testing would be done with slope measuring, common path techniques that are less sensitive to the environment.

## References

Armstrong, B. ,1977, *Chemical spray silvering*, Appl. Optics **16**, 2785

Burge, J. H., Anderson, D. S., 1994, *Full-aperture interferometric test of con-vex secondary mirrors using holographic test plates*, SPIE **2199**, 181

U.S. Patent 5,737,079.

Malacara, D., 1992, *Optical Shop Testing* 2nd Ed., J. Wiley

# MOEMS, MICRO-OPTICS FOR ASTRONOMICAL INSTRUMENTATION

Frédéric Zamkotsian

*Laboratoire d'Astrophysique de Marseille, FRANCE*

Frederic.Zamkotsian@oamp.fr

**Abstract**    Future astronomical instrumentation requires multiplexing capabilities and high spatial resolution. Emerging key technology are able to achieve these aims. Based on the micro-electronics fabrication process, Micro-Opto-Electro-Mechanical Systems (MOEMS) are compact, scalable, replicable and may be customized for specific tasks. They will be widely integrated in next-generation astronomical instruments, such as Adaptive Optics systems and Multi-Object Spectrographs. Applications are foreseen for NGST, VLT-2, NG-CFHT and OWL.

**Keywords:**    Micro-optics, MOEMS, multi-object spectroscopy, adaptive optics, astronomical instrumentation

## Introduction

Scientific breakthroughs often follow technological breakthroughs permitting to make significant steps forward. Astronomical research of this decade is related to the quest for our Origins: How did Galaxies form? How did Stars and Planetary Systems form? Can we detect Life in other Planets? The science requirements provided by those topics are very constraining for future astronomical instrumentation, calling for multiplexing capabilities and high spatial and spectral resolutions.

Based on the micro-electronics fabrication process, Micro-Opto-Electro-Mechanical Systems (MOEMS) have not yet been used in astronomical instrumentation, but this technology will provide the key to small, low-cost, light, and scientifically efficient instruments, and allow impressive breakthroughs in tomorrow's observational astronomy. Two major applications of MOEMS are foreseen:

- **the programmable multi-slit masks for Multi-Object Spectroscopy.**
Thanks to its multiplexing capabilities, Multi-Object Spectroscopy (MOS)

107

*R. Foy and F. C. Foy (eds.), Optics in Astrophysics*, 107–122.

is becoming the central method to study large numbers of objects (see Ch. 12). However, it is impossible to use traditional ground-based MOS in space. New methods need to be defined and technologies developed. A possible option would be to use Integral-Field Spectrographs, which observe spectroscopically all the pixels within a given field of view. Indeed, the concept is very interesting for specific observations: galaxy clusters, stellar clusters where the density of targets is high. However, for one of the most central astronomical program, deep spectroscopic survey of galaxies, the density of objects is low and it is necessary to probe wide fields of view. MOEMS provides a unique and powerful way of selecting the objects of interest (whatever the criteria, distance, color, magnitude, etc.) within deep spectroscopic surveys. This saves time and therefore increases the scientific efficiency of observations.

- **the Micro-Deformable Mirrors for Adaptive Optics.** To reach the faintest objects, we must get the best Point Spread Function (PSF) with the minimum of energy scattered within the outer areas of the PSF. Also sharp PSFs will allow to reach the best spatial resolution (limit of diffraction) and therefore to potentially resolve objects such as remote interacting building blocks in their way to become giant galaxies, star-forming regions within nearby galaxies or disks around forming planetary systems. The wave front perturbations are mainly due to the atmosphere and could be corrected by the use of Adaptive Optics systems. MOEMS devices should enable the correction of wave front perturbation in next generation telescopes and instrumentation.

In Laboratoire d'Astrophysique de Marseille, we are engaged since several years in the development of these types of MOEMS components.

## 1.    MOEMS

MOEMS are designed for a wide range of applications like sensors, switches, micro-shutters, beam deflectors, and micro-deformable mirrors. The main advantages of micro-optical components are their compactness, scalability, and specific task customization using elementary building blocks. As these systems are easily replicable, the price of the components is decreasing dramatically when their number is increasing. They will be widely integrated in next-generation astronomical instruments, for both ground-based and space telescopes, as they allow remote control. The two major applications of MOEMS are programmable slit masks for Multi-Object Spectroscopy and deformable mirrors for Adaptive Optics systems. MOEMS technology is closely linked to the micro-electronics fabrication process. Various materials are deposited on the surface of a substrate, and, using masks, their localization on the substrate is precisely defined in order to ensure their specific tasks. In microsystems, there are two kinds of layers: structural layers and sacrificial layers. The structural layer materials are polysilicon or metal, and the sacrificial layers are silicon oxides or organic materials. The sacrificial layers are chemically dissolved at the end of the fabrication process in order to create air gaps between the remaining structural layers. A great level of sophistication in the micro-electronics technology ensures excellent tolerances on layer thickness and patterning precision.

Micro-mechanical actuation is obtained using electrostatic, magnetic or thermal effects, but the most advanced actuation is achieved by the electrostatic effect. The electrostatic actuator consists of two electrodes of metal or polysilicon heavily doped with phosphorous, isolated from each other by a gap of thickness g filled with dielectric medium, usually air. When a voltage V is applied between the electrodes, an attractive force F is generated, and if one of the electrodes is mobile, it moves towards the other. By neglecting edge effects and bending of the electrodes, i. e. assuming good stiffness of the electrodes, the instantaneous, nonlinear electrostatic force F is defined by the following relationship of key parameters:

$$F = \frac{A\epsilon V^2}{2g^2} \qquad (1)$$

where A is the overlapping electrode area and e is the dielectric constant in the inter-electrode spacing.

The accessible motion is generally limited to one third of the inter-electrode initial gap. At this limit, the nonlinear electrostatic force increases more rapidly than a linear restoring force, applied for example by springs attached to the upper electrode. The electrostatic effect then becomes unstable and the mobile electrode drops toward the fixed electrode and sticks to it.

In the design of MOEMS components, various parameters have to be tuned. These parameters differ according to the functionality of the component. We will consider two different family of devices, programmable slits for Multi-Object Spectroscopy, including Micro-Mirror Arrays (MMA) and Micro-Shutters Arrays (MSA), and Micro-Deformable Mirrors (MDM) for Adaptive Optics systems.

MMA and MSA are an arrays of electrostatically driven bistable mirrors or shutters, with a size of a few tens of micrometers, which can occupy two discrete positions, ON and OFF, with switching times of a few microseconds. For MMA, the two positions are obtained when the mirror hits physically the substrate, and for MSA, shutters are open or closed. Specific parameters for these devices are the tilt angle (MMA) and the actuation voltage. The tilt angle determines the separation between the input beam and the output beam and therefore the possible numerical aperture of the instrument. A large tilt angle reduces also the scattered light of the array entering the output pupil. However, the fabrication process is not compatible with the use of very thick sacrificial layers on top of the substrate, limiting the tilt angle. The two discrete positions are typically rotated $\pm 10°$ with regard to the substrate plane, i.e. a $40°$ beam separation. The measurement and the analysis of this scattered light are under way (Zamkotsian et al., 2003). The second parameter is the actuation voltage: its value has to be low enough for matching the performances of integrated driving electronics.

MDM are constituted by a reflective surface with an underlying set of actuators. For MDM, key aspects are inter-actuator spacing, inter-actuator coupling, actuator bandwidth and low driving voltage. High order wave front correction needs a large number of actuators. The size of conventional Si wafers limits the maximal size of single deformable mirrors, increasing the required density of actuators. Typical inter-actuator spacing could be in the range 200-1000 $\mu$m. Inter-actuator coupling factor can be defined by the ratio of the motion of an OFF-actuator to the motion of a neighboring ON-actuator. If this factor is close to 1, there is too much redundancy, the effective actuator number is drastically reduced and high order deformations cannot be corrected properly. If this factor is zero, sharp slopes are present on the surface, usually on top of the actuator location, resulting in larger residual wave front errors. If this factor is 20-30 %, the surface has a smooth overall shape with low residual errors.

For both devices, MMA and MDM, three additional parameters are of interest: the optical surface quality, the driving electronics and the actuator bandwidth. As these micro-optical components include mirrors, their surface quality must be excellent. An original method based on Foucault's knife-edge test for the surface characterization of individual MMA micro-mirrors has been developed (Zamkotsian and Dohlen, 1999). The driving electronics is also a challenge as the high number of actuators integrated on a semiconductor substrate leaves individual actuator driving impossible. The driving circuit has to be integrated on the wafer or directly bonded to the optically active elements. The driving voltage must then be as low as possible, a few tens of volts will be preferable to the several hundreds of volts often used in present-technology devices. The design of driving circuits requires high attention in order to match the needed bandwidth, and to be compatible with the technology employed to realize the mobile mirrors on top of the electronic driving circuit. Finally, the bandwidth of these components must match the specific requirements.

Cryogenic-operation is not yet demonstrated for any MOEMS. Specific problems have to be studied in cryogenic conditions: the residual stress in the component layers and the sticking effects. The bending of the mirrors is due to residual stresses, especially when an additional reflective coating is required with different thermal-expansion characteristics. This effect has to be minimized because the coating stress is a major issue during cooling, with a high risk of breaking or pealing. Sticking effects in cryogenic conditions have also to be studied with care. In MMA, there is a contact between the two electrodes of micro-mirrors, and sticking is avoided by using a thin insulating layer and a reset electrical pulse. At low temperature, sticking effects are dramatically enhanced and efficient solutions must be found.

## 2. Programmable multi-slit mask for Multi-Object Spectroscopy

In order to obtain spectra of hundreds of objects simultaneously, the 1-5 $\mu$m Near-Infra-Red Multi-Object Spectrograph (NIRSPEC) for NGST requires a programmable multi-slit mask (PMSM). Conventional masks or complex fiber-optics-based mechanisms are impracticable in space. A promising solution is the use of MOEMS devices such as micro-mirror arrays (MMA) or micro-shutter arrays (MSA), allowing remote control of the multi-slit configuration in real time (Burg et al., 1998; Burgarella et al., 1998).

Our group is involved since several years in ESA's studies of the NIRSPEC instrument. We have focused our work on three main topics: MMA and MSA modeling, characterization of the MMA and MSA, and optical design for the MOS (Zamkotsian et al., 1999; Zamkotsian et al., 2000a).

### 2.1 Principle

By placing the PMSM in the focal plane of the telescope, the light from selected objects is directed toward the spectrograph, while the light from others objects and from the sky background is blocked in a light trap (Fig. 1). PMSM allows remote control of the multi-slit configuration in real time.

Actually, the only device available is the MMA designed by Texas Instrument for video projection. This is an array of 1K×1K, 17 $\mu$m pitch bistable mirrors (Hornbeck, 1995). A picture of the micro-mirrors is shown in Fig. 1. These electrostatically driven mirrors can occupy two discrete positions (ON and OFF), rotated ś10° with regard to the substrate plane. The switching time is a few microseconds. Using this MMA, any required slit configuration might be obtained with the capability to match point sources or extended objects (Fig. 1). In the park position, i.e., without driving voltage applied, the micro-mirrors are undeflected, parallel with the substrate. In action, the micro-mirrors in the ON position direct the light toward the spectrograph and appear bright, while the micro-mirrors in the OFF position are dark.

MSA are under development at the NASA's Goddard Space Flight Center, and has been selected to be the multi-slit device for NIRSpec. They use a combination of magnetic effect for shutter opening, and electrostatic effect for shutter latching in the open position (Moseley et al., 2002).

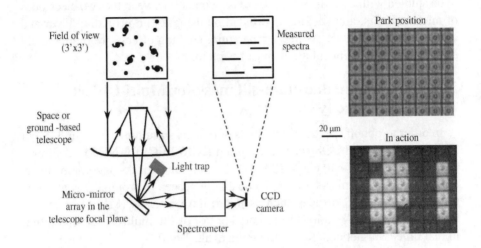

*Figure 1.* Principle of Multi-Object Spectroscopy using a Micro-Mirror Array. MMA array in the park position and in action

## 2.2    Modeling

Two key parameters have been found to determine the ultimate capabilities of the MOEMS-based MOS: contrast and spectral photometric variation (SPV). Contrast is defined as the total amount of non-selected flux of light passing through the multi-slit device; a contrast requirement of 2000 has been established in order to avoid pollution of the spectra by strong sources (spoilers) and sky background. The SPV is the unpredictable photometric variation due to the random position of the sources on the PMSM grid; SPV requirement has been fixed to <10%.

Using Fourier theory, we have modeled the SPV for MSA (Zamkotsian et al., 2002a; Zamkotsian et al., 2003). We consider two slit configurations, $2 \times 2$ shutters for the 1-1.75 $\mu$m and 1.75-3 $\mu$m bands and $3 \times 3$ shutters for the 3-5 $\mu$m band. The SPV is strongly affected by the object position in the slit and wavelength. This generates geometrical and diffraction effects. We obtain the surfaces shown in Fig. 2, where upper surfaces are the SPV due only to the geometrical effect, and lower surfaces are the SPV when the diffraction effect is added. Normalization has been done at the maximum of the geometrical effect at each wavelength. For each wavelength hills and valleys are observed. Valleys are located on the edges of the surfaces when the object is at the limits of the slits, reducing drastically the amount of light passing through the slit. However, for short wavelengths, a valley is also present in the middle of the slit, along the grid between two adjacent micro-shutters. We notice two major effects: the location of the maxima is strongly dependent on the wavelength, moving from the center of the micro-shutters at 1 and 2 $\mu$m to the center at the slit at 3 $\mu$m. The second effect is the decrease of transmitted flux due to diffraction. At short wavelengths, the diffraction effect is negligible, but at longer wavelengths, diffraction reduces the flux by up to about 20%. By moving point sources along the slit, we note a final SPV of from 22% to 65%, for the 1-1.75 $\mu$m and 1.75-3 $\mu$m bands, and a SPV of from 28% to 36% for the 3-5 $\mu$m band. These values are exceeding by far the 10% required variation. We have therefore proposed a dithering strategy able to solve this problem, referred as **blind dithering** (Zamkotsian et al., 2002a).

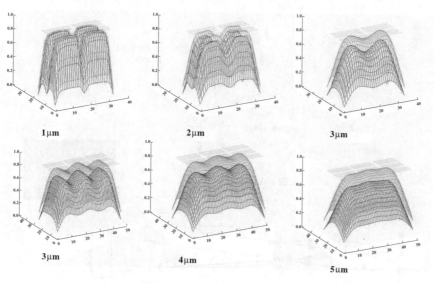

*Figure 2.* Spectral photometric variation versus source position (1-3$\mu$m and 3-5$\mu$m ranges). Upper surfaces are only affected by the geometrical effect of the slit geometry, and lower surfaces include the diffraction effects.

## 2.3    Surface characterization

Due to diffraction effects of micron-sized mirrors in a regular array, commonly used techniques for surface characterization based on interferometry are inefficient. To overcome the diffraction effects we have developed a novel surface characterization method with an incoherent light source, based on the Foucault's knife-edge test (Zamkotsian and Dohlen, 1999). Since Léon Foucault introduced the knife-edge test in the last century (Foucault, 1859), it has been widely used for testing optical surfaces (see Ch. 3). The test offers a simple way of obtaining easily understandable, qualitative information of the surface shape.

Our experimental set-up is shown in Fig. 3. Incoherent white light emanates from a slit source. After re-imaging of the slit in the focal plane F of the lens L2, the light is collimated and illuminates the array surface. The reflected beam is refocused back onto the plane F and an enlarged image of the micro-mirrors is formed on a CCD camera. By moving a knife-edge through the image slit while observing the micro-mirror illumination, we are able to determine for each pixel of the field of view the position of the knife-edge for which the illumination of this pixel is blocked. The position of the knife-edge determines the local slope on the micro-mirror surface. A picture of the studied field of view is shown in Fig. 1.

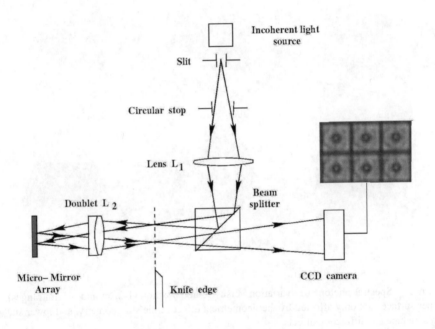

*Figure 3.*    Experimental bench for micro-mirror surface characterization using the knife-edge method

A slope map for 8×6 micro-mirrors (Fig. 4) reveals classical knife-edge test figures on each micro-mirror. The slopes are gray-scale coded with values ranging from -0.8 mrad (black) to +0.7 mrad (white). These small slope values and their uniform distribution over the field of view indicate a satisfying flatness of the surfaces and a good uniformity of the array. This good surface quality is a result of the use of the mature micro-electronics realization process. Slight mirror-to-mirror shape variations indicate variable amounts of tilt and astigmatism (Zamkotsian and Dohlen, 1999).

By integration of the local slopes, we have reconstructed the micro-mirror surface. An example is shown in Fig.4, along the line indicated by an arrow on the slope map. The surface deformations do not exceed 1 nm along the studied profile. Although surface shapes vary from mirror to mirror, deformations in the nanometer range demonstrate the remarkable quality of this device.

Assuming axi-symmetrical deformation, simulation of the complete micro-mirror surface has been found to have a "palm-tree" shape, with typical maximum deformation less than 2 nm. This shape can be explained by strain relaxation in the thin aluminum layer constituting the mirror surface (Zamkotsian and Dohlen, 1999).

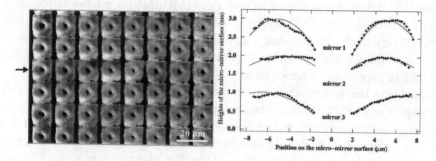

*Figure 4.* Slope map for a 8x6 micro-mirrors field, and surface reconstruction

We have developed another bench for the measurement of the contrast value. Contrast measurement have been carried out on the MMA fabricated by Texas Instrument, in order to establish the test procedure (Zamkotsian et al., 2002a; Zamkotsian et al., 2003). We can address several parameters in our experiment, as the size of the source, its location with respect to the micro-elements, the wavelength, and the input and output pupil size. In order to measure the contrast, the micro-mirrors are tilted between the ON position (towards the spectrograph) or the OFF position (towards a light trap). Contrast exceeding 400 has been measured for a 10° ON/OFF angle. Effects of object position on the micro-mirrors and contrast reduction when the exit pupil size is increasing have also been revealed.

## 3. Micro-Deformable Mirror for Adaptive Optics

Different research groups around the world are currently involved in the design of highly performing adaptive optical (AO) systems as well as for next generation instrumentation of 10m-class telescopes than for future extremely large optical telescopes. Four different types of AO systems are foreseen: Classical Adaptive Optics, Multi-Conjugate Adaptive Optics, Low-order Adaptive Optics, and High-order Adaptive Optics (see Ch. 13). Numerous science cases will use these AO systems, classical AO will provide accurate narrow field imagery and spectroscopy, MCAO, wide field imagery and spectroscopy, low-order AO, distributed partial correction AO, and high dynamic range AO, detection and study of circumstellar disks and extra-solar planets. Corrected fields will vary from few arcsec to several arcmin.

These systems require a large variety of deformable mirrors with very challenging parameters. For a 8m telescope, from a few 10 up to 5000 actuators are needed; these numbers increase impressively for a 100m telescope, ranging from a few 1000 to over 250 000, the inter-actuator spacing from less than 200 $mu$m to 1 mm, and the deformable mirror size from 10 mm to a few 100 mm. Conventional technology cannot provide this wide range of deformable mirrors. The development of new technologies (MOEMS) is promising for future deformable mirrors. The major advantages of the micro-deformable mirrors (MDM) are their compactness, scalability, and specific task customization using elementary building blocks. This technology permits the development of a complete generation of new mirrors.

However this technology has also some limitation. For example, pupil diameter is an overall parameter and for a 100 m primary telescope, the internal pupil diameter cannot be reduced below 1 m. According to the maximal size of the wafers (8 inches), a deformable mirror based on MOEMS technology cannot be build into one piece. New AO architectures are under investigation to avoid this limitation (Zamkotsian and Dohlen, 2001).

The European Southern Observatory (ESO) has an important role in the conception of extremely large telescopes. Within the framework of a Research Training Network supported by the European Community, ESO is leading a group of seven laboratories engaged in the Adaptive Optics for Extremely Large Telescopes project (AO-ELT). These laboratories are ESO, ON-ERA (France), Observatories of Marseille (France) and Arcetri and Padova (Italy), the Max Planck Institute for Astronomy in Heidelberg (Germany), and GRANTECAN (Spain).

This technology is also foreseen for the development of the AO systems of other extremely large telescope, as the Next Generation Canada-France-Hawaii-Telescope. Our group propose the replacement of the 3.6m telescope by a 20-30m class telescope (Burgarella et al., 2000). However, future instrumentation for the VLT will also benefit from this technology providing replicable and compact mirrors for various AO systems. We are involved in the development of the second generation of VLT instrumentation (Zamkotsian et al., 2001), in the FALCON project (Hammer et al., 2001) and the VLT-Planet-Finder project.

## 3.1 DM parameters

The main DM parameters for Extremely Large Telescopes are summarized in Table 1. Whatever the size of the internal pupils, the number of actuators equals at least the number of $r_0$ patches in the telescope pupil (primary mirror). For ELT's ranging from 20m to 100m, the number of actuators is between 5 000 and over 250 000 in the near infra-red; for visible, even larger number of actuators are required. In order to correct the highest atmospheric perturbations, the number of actuators is constant for a given size of the telescope. The inter-actuator spacing is therefore proportional to the internal pupil size. Actuator stroke is not yet well determined, in particular due to the uncertainty of the outer scale effect, but it is generally admitted that 5 - 10 $\mu$m is appropriate. For high-order corrections, the bandwidth must be of the order of 1 kHz. For a smooth overall shape of the surface with low residual errors, the inter-actuator coupling should be 20-30%.

*Table 1.* Deformable Mirror parameters for Extremely Large Telescopes

| | |
|---|---|
| Number of actuators | 5000 to > 250 000 |
| Inter-actuator spacing | 200 $\mu$m - 1 mm |
| Actuator stroke | 5 - 10 $\mu$m |
| Inter-actuator coupling | 20 - 30 % |
| Bandwidth | 1 kHz |

## 3.2 MDM architectures

In Fig. 5 are shown the three main MDM architectures under study in different laboratories (Zamkotsian et al., 2000a). First, the **bulk micro-machined continuous-membrane deformable mirror**, studied by Vdovin from Delft University, is a combination of bulk silicon micromachining with standard electronics technology (Vdovin et al., 1997). This mirror is formed by a thin flexible conducting membrane, coated with a reflective material, and stretched over an electrostatic electrode structure. Second, **the segmented, micro-electro-mechanical deformable mirror** realized by Cowan at Ohio Air Force Research Laboratory consists of a set of segmented piston-only moving surfaces, fabricated in dense array (Roggeman et al., 1997; Cowan et al., 1998). Third, **the surface micro-machined continuous-membrane deformable mirror** made by Bifano at Boston University is based on a single compliant optical membrane supported by multiple attachments to an underlying array of surface-normal electrostatic actuators (Bifano et al., 1997). The efficiency of this device has been demonstrated recently in an AO system (Perreault et al., 1999). The third concept is certainly the most promising architecture, but low actuation efficiency and mirror surface quality need further improvement.

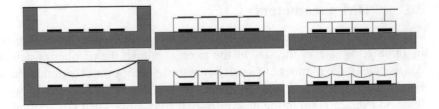

*Figure 5.* Different MDM architectures

Since September 2000, we have engaged an active collaboration with a French laboratory expert in micro-technologies, the Laboratoire d'Analyse et d'Architecture des Systèmes (LAAS) in Toulouse, France, for the conception and the realization of micro-deformable mirror (MDM) prototypes. Our design is based on three elementary buildings blocks (Fig. 6). From top to bottom:

- **the mirror surface.** It has to be continuous and with the highest optical quality in order to minimize straylight and diffraction effects. Mirror material without stress and perfect planarization are the main goals.
- **the actuation mechanism.** We have chosen an electrostatic actuation and there are two key parameters, the driving voltage which has to be as low as possible and the stroke, which has to be as large as possible. These actuators must have the potential to reach acceptable voltages for a conventional micro-electronics circuit.
- **the driving electronics.** It has to underlie the deformable mirror structure and be integrated in the substrate. The success of the Micro-Mirror Array developed by Texas Instruments indicates that integration of the optical architecture and the driving electronics is the most reliable approach. The electronics and the optical architecture process have to be compatible.

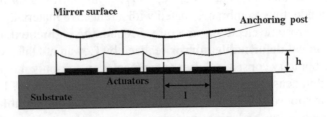

*Figure 6.* Schematic view of our Micro-Deformable Mirror architecture.

As a high optical quality mirror is the most challenging building block for this device, we have first focused our work on the realization and the characterization of the mirror surface.

## 3.3 MDM realization

After conception of the test structures by LAM and LAAS researchers, several samples have been realized at LAAS. Our first goal is the realization of a high quality and flat surface on top of a "mountainous" structure, in order to make the mirror layer. This layer is of first importance as it sees the photons of the instrument and it has to be completely planarized.

The originality of our approach lies in the elaboration of a sacrificial layer and of a structural layer made with low-temperature processes, suitable for integration of the optical architecture on top of the driving circuit realized in the substrate (Zamkotsian et al., 2002b).

A piston actuator structure etched with a structural layer on top of the sacrificial layer (500 $mum$ square piston area) is presented in Fig. 7. Before etching of the sacrificial layer, the top surface is completely flat.

*Figure 7.* Piston actuator structure etched with a structural layer on top of the sacrificial layer (500 $\mu$m square piston area); general view and close-up.

The critical operation is the etching of the sacrificial layer. With a proper wet etching process, the remaining structural layer shows a high quality surface (Fig. 8). The 10 $\mu$m thick sacrificial layer has been etched and the layer stays with a perfect plane shape. Stiffness of the structural layer is also visible in Fig. 8b where the substrate have been cleaved near a structure. Attachment points are 500 $\mu$m away and there is no bending of the structure.

## 3.4 MDM characterization

Characterization of the mirror surface are conducted along three scales: large, medium, and small scales. Large scale defects describe the **flatness** of the structure on distances longer than 1 mm, typically longer than the inter-actuator spacing. If the range is below 1 $\mu$m, these defects can be compensated by setting an offset driving voltage on the adjacent actuators, in order to generate a nominal flat surface. Medium scale defects describe the **print-trough effect**. The different layers of the structure are masked and etched with specific

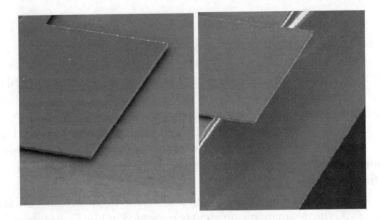

*Figure 8.*   Structural layer after sacrificial layer etching: (a) on the substrate; (b) above the edge of a cleaved sample. Structure is 500 $\mu$m wide and attachment points 500 $\mu$m away.

patterns. As they are deposited one above the previous one, the resulting top surface is usually non-planar. This effect has to be minimized as much as possible. Print-trough effect are in the vicinity of the underlying structure and the typical size is therefore in the order of 100 $\mu$m or below. Small scale defects are also called **roughness**. They are measured in small areas (1-10 $\mu$m) and values are in the nm range.

The sample is measured with a confocal microscope, by using a chromatic coding of the height, and with a contact-less optical "needle" focused on the surface. We determine by this way the surface topography and its roughness.

Measurements have been performed on these samples. With this process, very efficient planarization has been obtained: the long-distance flatness is below 0.2 $\mu$m, the print-through of localized 9 $\mu$m steps is reduced to below 0.5 $\mu$m and a rms roughness of 15 nm has been measured over the surface ( Zamkotsian et al., 2002b). Major efforts are going into reducing print-through effect and roughness. The integration of this mirror surface on top of an actuator array is under investigation.

## 4.    Conclusion

Micro-Opto-Electro-Mechanical Systems (MOEMS) will be widely integrated in new astronomical instruments for future Extremely Large Telescopes, as well as for existing 10m-class telescopes. The two major applications are programmable slit masks for Multi-Object Spectroscopy (see Ch. 12) and deformable mirrors for Adaptive Optics systems. First prototypes have shown their capabilities. However, big efforts have still to be done in order to reach the requirements and to realize reliable devices.

The Laboratoire d'Astrophysique de Marseille is developing since several years an expertise in the design and characterization of micro-optical components, as well as in their integration in astronomical instruments.

A strong collaboration between micro-optics and astronomy will certainly lead to reach the best scientific return for the lowest cost in next generation astronomical instrumentation for ground-based and space telescopes.

## Acknowledgments

The author would like to thank K. Dohlen from LAM for fruitful discussions, H. Camon, N. Fabre and V. Conédéra from Laboratoire d'Analyse et d'Architecture des Systèmes (Toulouse, France) for sample realization, and P. Lanzoni, G. Moreaux from LAM for device characterization.

## References

Bifano, T.G., Mali, R.K., Dorton, J.K., Perreault, J., Vandelli, N.,Horenstein, M.N., Castanon, D.A., 1997, Opt. Eng. **36**, 1354

Burg, R., Bely, P.Y., Woodruf,f B., MacKenty, J., Stiavelli, M., Casertano, S., McCreight, C. Hoffman, A., in Space Telescopes and Instruments V, Eds P.Y. Bely and J.B. Breckinridge, 1998, SPIE **3356**, 98

Burgarella, D., Buat, V., Dohlen, K., Zamkotsian, F. and Mottier P., 1998, in Proceedings of the 34th Liege International Astrophysics Colloquium (International Workshop on the Next Generation Space Telescope "Science drivers and technological challenges")

Burgarella, D., Buat, V., Dohlen, K., Ferrari, M., Zamkotsian, F., Hammer, F., Sayède, F., 2000, 21st CFHT Anniversary Workshop, Hawaii

Cowan, W.D., Lee, M.K., Welsh, B.M., Bright, V.M., and Roggeman, M.C., 1998, *Opt. Eng.* **37**, 3237

Foucault L. M., 1859, *Mémoire sur la construction des télescopes en verre argenté*, Ann. Obs. Imp. Paris **5**, 197

Hammer, F., Sayède, F., Gendron, E., Fusco, T., Burgarella, D., Buat, V., Cayatte,, V., Conan, J.-M., Courbin, F., Flores, H., Guinouard, I., Jocou, L., Lançon, A., Monnet, G., Mouhcine, M., Rigaud, F., Rouan, D., Rousset, G., Zamkotsian, F., 2001, Proc. Scientific drivers for ESO future VLT/VLTI instrumentation

Hornbeck L. J., 1995, SPIE **2639**, 2, Color reprint available from Texas Instruments Digital Imaging Group

Moseley, H., Aslam, S., Baylor, M., Blumenstock, K., Boucarut, R., Erwin, A., Fettig, R., Franz, D., Hadjimichael, T., Hein, J., Kutyrev, A., Li, M., Mott, D., Monroy, C., Schwinger, D., 2002, SPIE **4850**

Perreault, J.A., Bifano, T.G. Levine, B.M., 1999, SPIE **3760**, 12

Roggeman, M.C., Bright, V.M., Welsh, B.M., Hick, S.R., Roberts, P.C., Cowan, W.D., Comtois,J.H., 1997, Opt. Eng. **36**, 1326

Vdovin, G., Middelhoek, S., Sarro, P.M., 1997, Opt. Eng. **36**, 1382

Zamkotsian F. Dohlen K., 1999, Appl. Optics **38**, 6532

Zamkotsian F., Dohlen K., Burgarella D., Buat V., 1999, in proceedings of NASA colloquium on NGST Science and Technology Exposition, ASP Conf. **207**, 218

Zamkotsian, F., Dohlen, K., Buat, V., Burgarella, D., 2000, SPIE **4013**, 580

Zamkotsian, F., Dohlen, K., Ferrari, M., 2000b, SPIE **4007**, 547

Zamkotsian, F. Dohlen, K., 2001, Conf. *Beyond conventional Adaptive Optics*

Zamkotsian, F., Burgarella, D., Dohlen, K., Buat, V., Ferrari, M., 2001, Proc. Scientific drivers for ESO future VLT/VLTI instrumentation

Zamkotsian, F., Gautier, J., Lanzoni, P., Dohlen, K., 2002a, SPIE **4850**

Zamkotsian, F., Camon, H., Fabre, N., Conédéra, V., Moreaux G., 2002b, SPIE **4839**

Zamkotsian, F., Gautier, J., Lanzoni, P., 2003, SPIE **4980**

# OPTICAL AND INFRARED DETECTORS FOR ASTRONOMY

## Basic principles to state-of-the-art

James W. Beletic

*Rockwell Scientific Company*
*5212 Verdugo Way*
*Camarillo, Ca 93012 U.S.A.*
jbeletic@rwsc.com

**Abstract**     Detectors play a key role in an astronomical observatory. In astronomy, the role of the telescope and instrument is to bring light to a focus - in effect, the telescope-instrument act as "spectacles". The detectors, meanwhile, have the critical role of sensing the light - the detectors are the "eyes" of an observatory. The performance of an astronomical observatory is directly dependent upon the performance of its detector systems.

In many ways, today's optical and infrared detectors are nearly perfect, with high quantum efficiency, low readout noise, high dynamic range and large arrays of pixels. However, as good as the detectors are, there are limitations that must be understood and respected in order to produce the best astronomical instruments and thereby, the best science.

This chapter explains how optical and infrared detectors work, from basic principles to the state-of-the-art. The role of optical and infrared detectors in an observatory is presented, and the state-of-the-art is related to an ideal detector. An overview of the detector physics is presented, showing that the detection of light is a 5 step process. Each step in this process is explained in detail in the subsequent sections. The chapter concludes with references for further information.

**Keywords:**     detectors, optical, infrared, quantum efficiency, noise

## 1.     The role of detectors in astronomy

Astronomy is a vibrant science, with significant new discoveries about the universe being made every year. Fueling these scientific breakthroughs are technological advances in many areas, from the new generation of 8-10 meter

*R. Foy and F. C. Foy (eds.), Optics in Astrophysics, 123–153.*
© 2006 *Springer. Printed in the Netherlands.*

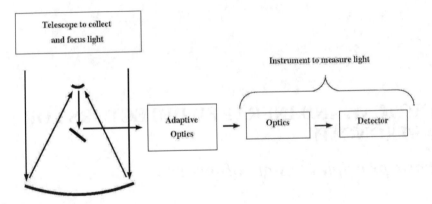

*Figure 1.* A ground-based astronomical observatory can be depicted as having four major part: telescope, adaptive optics (optional), instrument optics, and the detector.

telescopes to innovative instrumentation concepts. Detector technology is key to the entire endeavor, since most astronomical information is derived from the detection of light. The majority of astronomical science is performed using optical and infrared light, a region of the electromagnetic spectrum that can be used to investigate most of the universe. This chapter will concentrate on the detector technology that is used at ground-based telescopes to sense the portion of the spectrum that propagates through the atmospherically transparent window of 0.32 to 20 $\mu$m. Wavelengths shorter than 320 nm are absorbed by the atmosphere (which is good for us, else we would all get skin cancer), and wavelengths much longer than 20 $\mu$m require significantly different technology.

To understand the role of detectors in astronomy, we can divide an astronomical observatory into four major parts as shown in Fig.1.

The telescope collects and focuses the light. For a ground-based telescope, the image is always distorted by the Earth's atmosphere. For a large telescope, the atmospheric blurring (or "seeing", see 1) can spread the light over an area that is up to 1000 times larger than the diffraction-limited spot. Adaptive optics (see 13) is now used at many ground-based telescopes to negate the seeing, although the technology is presently limited to providing good results in the infrared for relatively bright objects, and the correction is only effective over relatively small fields of view ($\simeq$ 1 arc min diameter).

The instrument, which is placed at the telescope focal plane, consists of optics and a detector to measure the light. As depicted in Fig. 2, the instrument attempts to measure a three-dimensional data cube - intensity as a function of wavelength ($\lambda$) and two spatial dimensions on the sky (right ascension and declination).

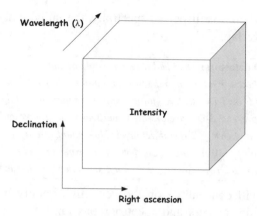

*Figure 2.* Three-dimensional data cube that is probed by an astronomical instrument: the intensity is a function of two spatial directions on the sky (right ascension and declination - analogous to longitude and latitude) and the wavelength dimension.

A major challenge for instrumentation comes from detectors being "colorblind". All optical and infrared detector arrays, from the human eye and photographic film to the state-of-the-art used on astronomical telescopes, can only measure intensity and not wavelength. Every photon detected by an optical or infrared detector produces an electron, and all electrons look the same, whether produced by a short or long wavelength photon. In a similar way, the rods and cones in the human eye only sense intensity. Color perception is due to subsets of rods and cones having sensitivity to specific colors. Thus, a focal plane array can not simultaneously measure the full three-dimensional data cube shown in Fig. 2. A significant challenge for an instrument builder is designing optics that efficiently sample a 2-D portion of the 3-D data cube. There are several ways to select a 2-D portion of the data cube:

**Imaging:** ▪ uses a <u>filter</u> to take an image in a limited wavelength band.

**Spectroscopy:** 3 types are used in astronomy:

- ▪ a <u>slit spectrograph</u> which places a slit (*e.g.*, 1 arc sec wide by 30 arc sec long) at the focal plane to filter out a 1-D slice of the sky and disperses the spectrum in the orthogonal direction,
- ▪ a <u>multi-object spectrograph</u> that spatially filters the image with several small slits in the image plane and disperses the spectra in the orthogonal direction (in many cases, none of the slits can overlap in the spatial direction),
- ▪ an <u>integral field spectrograph</u> samples a contiguous 2-D array of spatial pixels and uses lenses, fibers or mirrors to align the light so

that when it is dispersed spectrally, the individual pixel spectra do not overlap.

---

**The promise of 3-D detectors:** *The ideal detector would be able to measure the wavelength dimension of each photon in addition to the two spatial dimensions - a "3-D detector". There is progress being made on 3-D detectors - for example, the superconducting tunneling junction (STJ). However, at present, 3-D detectors are experimental and have only achieved limited wavelength resolution ($\lambda/\delta\lambda \leq 25$) with relatively slow count rates (few kHz per pixel) on a small number of pixels ($\simeq 100$). Since these devices are experimental and are not being used in a facility instrument at any telescope, we do not discuss them further in this chapter.*

---

This chapter will concentrate on the very high quality detectors that are needed in scientific imagers and spectrographs, and other applications that require high sensitivity, such as acquisition and guiding, adaptive optics and interferometry. We limit our discussion to focal plane arrays – large two-dimensional arrays of pixels – as opposed to single pixel detectors (*e.g.*, avalanche photodiodes).

Note that there are many other applications of detectors in an astronomical observatory, including detectors that are used for active optics, site monitoring (seeing, cloud cover), surveillance and safety monitoring.

It is easy to conclude that detectors are one of the most critical technologies in astronomy. Thus, for an observatory to take a leading role in astronomical research, it must ensure that it has expertise and adequate resources allocated to this area.

## 2.     The ideal detector and the state-of-the-art

Ideally, an observatory would install perfect detectors in the focal plane of its instruments. What makes a perfect detector? The attributes of an ideal detector and the performance achieved by today's technology are given in Table 1. Optical and infrared detectors are nearly ideal in several ways:

- Quantum efficiency (QE) - The sensitivity of a detector can be nearly perfect, with up to 99% of the photons detected at the wavelength for which the detector is optimized.
- Delta function response - Over most of the wavelengths of interest, optical and infrared detectors produce one photoelectron for every detected photon, which provides a one-to-one correspondence between detected photons and photoelectrons. This means that the detector response is exactly linear to the intensity incident on the detector – an attribute that allows astronomers to precisely remove sky background and electronic bias to accurately measure the intensity of the astronomical object.
- Large number of pixels - A very large number of pixels can now be integrated on a single focal plane. The largest array presently in operation in

astronomy is the mosaic of 40 CCDs with 377 million pixels in the wide field of view imager (Megacam) at the Canada-France-Hawaii Telescope (CFHT) on Mauna Kea, Hawaii.

*Table 1.* Attributes of an ideal detector and performance achieved by the state-of-the-art detectors in astronomy.

| Attributes of ideal detector | Performance of state-of-the-art |
|---|---|
| Detect 100% of photons | √ Up to 99% detected |
| Photon detected as a delta function | √ One electron for each photon |
| Large number of pixels | √ Over 377 million pixels |
| Time tag for each photon | ⊘ No – framing detectors |
| Measure photon wavelength | ⊘ No – provided by optics |
| Measure photon polarization | ⊘ No – provided by optics |
| No detector noise | ⊘ Readout noise and dark current |

However, there are several ways that optical and infrared detectors fall short of being ideal:

- Limited temporal resolution - Focal plane arrays are all inherently framing detectors and knowledge of the arrival time of photons is limited to the frame time of the detector. While the frame time can be quite short for adaptive optics detectors ($\sim$1 ms), in most astronomical instruments the frame time is on the order of seconds or minutes, adequate for most astronomical science, but not all.
- Colorblind - As stated above, the focal plane arrays in use today can only detect intensity – wavelength must be provided by the optics of the instrument.
- Polarization blind - Polarization cannot be detected, but must be provided by the optics.
- Detector noise - The two most significant noise sources of a detector are readout noise and dark current.

  - Readout noise is the noise that comes from the amplification of the small amount of electrical charge that is produced by the light.
  - Dark current comes from the thermal excitation of electrons in the detector material - thermally generated electrons can not be distinguished from photoelectrons.

- Other detector imperfections - Detectors also exhibit a wide range of other "features" that make these devices less than perfect. A major attribute that describes performance is cosmetic quality. Due to defects in the material or fabrication errors, some pixels of an array can exhibit

# Detector zoology

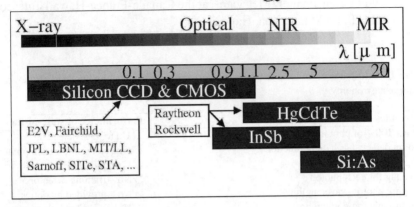

*Figure 3.* Optical and infrared detector "zoology". The wavelength region is stated on the first row, with corresponding wavelength (in $\mu$m) shown on the second row. The type of detector material and associated manufacturers are shown in the boxes below, which also depict the wavelength coverage possible with each kind of detector material.

lower QE, and in the worst case, be completely dead. Also, some pixels will exhibit excessively high dark current. In a CCD, an extremely bad pixel can make an entire column useless, and thus, the primary measure of cosmetic quality for a CCD is the number of bad columns.

## 3.     Detector zoology

Optical and infrared detectors can be organized into the "zoology" presented in Fig. 3. The wavelength regions included in the graph are x-ray, optical (0.3 - 1.1 $\mu$m), near infrared (NIR, 1-5 $\mu$m) and mid-infrared (MIR, 5-20 $\mu$m). For detection of optical light, silicon (Si) is the best material, and there are several designers and manufacturers of silicon-based detectors. In the near infrared, two materials are used for focal plane arrays: Mercury-Cadmium-Telluride (HgCdTe), also known as "Mer-Cad-Tel", and Indium-Antimonide (InSb). In the mid-infrared, arsenic doped silicone is used. In the infrared, there is limited choice of manufacturer; only Raytheon Vision Systems and Rockwell Scientific have a significant presence.

## 4.     Detector architecture and operation: 5 steps of light detection

In this section, an overview of detector architecture and operation is given. Although optical and infrared detectors are often thought of as very different beasts, in practice they are more alike than different. Thus, optical and infrared

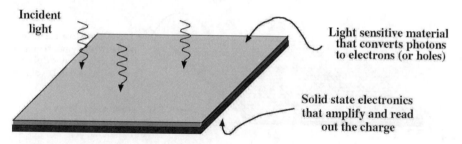

Incident light

Light sensitive material that converts photons to electrons (or holes)

Solid state electronics that amplify and read out the charge

*Figure 4.* Simplified schematic of an optical/infrared focal plane array. The detector is a thin wafer of light sensitive material that is connected to a thin layer of solid state electronics - the connection is made either by direct deposition (CCD) or bump bonding (IR detector). The solid state electronics amplify and read out the charge produced by the incident light.

detectors will be explained using the same five steps of light detection, with the differences between optical and infrared detectors noted when appropriate.

First, let us begin with basic detector architecture, as shown in Fig. 4. An optical or infrared focal plane is simply a thin wafer of light sensitive material that is connected to a thin layer of solid state electronics – these electronics amplify and read out the charge produced by the light.

In order to provide resolution in two-dimensions, the light sensitive material is subdivided into an array of pixels, as shown in Fig. 5. These pixels are defined by electric fields that are created within the light sensitive material. The electric fields can be generated one of two ways: (1) permanently defined by implanting (*doping*) a very small amount of another material, or (2) programmably defined by electric fields produced by wires in the solid state electronics.

Ideally, any photoelectric charge that is generated within a pixel is captured by that pixel. Unfortunately, due to Brownian motion, there are several microns of lateral charge diffusion, which causes blurring of the image that can degrade the optical resolution of the instrument. With proper design, the amount of charge diffusion blurring can be made negligible when compared to the optical blurring.

The wafer of light sensitive material is very thin compared to the total size of the detector. The light sensitive material is typically 10 to 15 $\mu$m thick (about one-tenth the diameter of a human hair), but can be up to 300 $\mu$m thick. The detectors are made as large as possible in the lateral dimensions to provide a large focal plane. However, the yield of good quality is inversely proportional to area, so there is a limit in size beyond which it is not economically prudent to produce detectors. The standard size for scientific grade optical detectors is a 2Kx4K pixel array, with each pixel being 15 $\mu$m square. (*The unit "K", when used for specifying number of pixels, denotes the binary value 1024.*) Thus, the standard optical detector is 2048×4096 (15 $\mu$m) pixels, with total

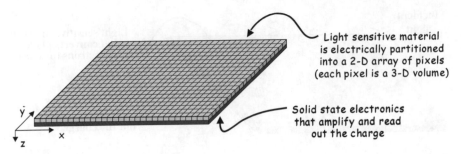

*Figure 5.* Simplified schematic of the 2-D array of pixels in a focal plane array. The thin wafer of light sensitive material is partitioned into a two-dimensional array of pixels that collect the electric charge produced by the light. Each pixel is a three-dimensional volume that is defined by electric fields within the light sensitive material.

dimension of $3.1 \times 6.1$ cm$^2$, and a thickness of 15 $\mu$m. The largest infrared detectors manufactured today are 2K$\times$2K, 18 $\mu$m pixels ($3.7 \times 3.7$ cm$^2$) and 10 to 15 $\mu$m thick.

Since the photosensitive material and the electronics layer are very thin, the detector is mounted on a mechanical package for structural integrity. This package is thermally matched to the detector so that the detector will not be stretched or compressed during the large transition from room temperature to operating temperature.

With this basic understanding of the geometry of a detector, we can present the five basic steps in optical/IR photon detection. The steps listed below follow the framework taught by James Janesick (2001) in his classic "CCD course".

1  Get the light into the detector. This may sound trivial, but unless a good anti-reflection coating is used, a significant fraction of the light will be reflected at the surface of the photosensitive material.

2  Charge generation. Once the light is within the volume of the photosensitive material, the photon energy must be absorbed and converted to charge. The photon energy creates electron-hole pairs.

3  Charge collection. Electric fields within the photosensitive material collect charge into pixels. The detector can be designed to collect either electrons or holes.

4  Charge transfer. In the infrared, no charge transfer is required. For an optical CCD, the charge is moved to the edge of the detector where the amplifiers are located.

5  Charge amplification. Very small amounts of charge must be amplified before it can be digitized and transfered to a computer. Amplification is a noisy process. At present, CCD amplifiers exhibit lower noise than infrared amplifiers.

The first two steps, getting light into the detector and charge generation, affect quantum efficiency. The point spread function (PSF) of a detector is affected by the $3^{rd}$ and $4^{th}$ steps (charge collection and charge transfer). Unless proper attention is paid to the PSF of a detector, poor PSF can negate the heroic efforts of optical designers who strive to maintain a tight PSF through the instrument optics. All five steps affect the sensitivity of a detector, with different steps dominating the sensitivity at different light levels. At high light level, the sensitivity is primarily a function of quantum efficiency and PSF. At low light level, the noise due to amplification dominates the sensitivity of a detector.

The next sections will expand on the physics of the five steps listed above. As much as possible, the explanation of optical and infrared detectors will be combined.

## 5. Getting light into the detector: anti-reflection coatings

The velocity of light is a function of the material through which it travels. The speed of light in a vacuum, denoted by $c$, is $2.998 \cdot 10^8$ m·s$^{-1}$. The velocity of light $v$ in a medium other than vacuum is given by,

$$v = c/n \tag{1}$$

where $n$ is the index of refraction for the medium ($n = 1$ for a vacuum). When light encounters a change in the index of refraction, some of the light is transmitted and some is reflected, as shown in Fig.6 (left). For propagation perpendicular to the interface, the fraction of incident energy that is reflected is given by,

$$Reflected\ energy = ((n_t - n_i)/(n_t + n_i))^2 \tag{2}$$

The index of refraction and amount of light reflected at an interface with air ($n_{air} \sim 1.00$) is given in Table 2. Note that the index of refraction of silicon varies significantly over the 0.32-1.1 $\mu$m range, as shown in Fig.10; $n_{Si} \sim 4$ applies for 0.5-0.9 $\mu$m.

As shown in Table 2 and Eq. 2, a significant amount of light is lost due to reflection at an interface when there is a large change in the index of refraction. Whenever light undergoes a sudden change in velocity, energy is

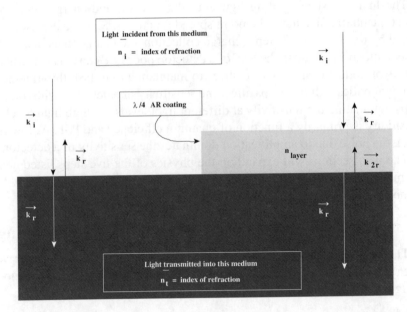

*Figure 6.*   Reflection of light at an interface and the physics of an anti-reflection (AR) coating. The propagation vector (**k**) denotes the direction and amplitude of the incident ($\mathbf{k}_i$), transmitted ($\mathbf{k}_t$), and reflected ($\mathbf{k}_r$) waves for light propagating from the incident medium with index of refraction $n_i$ to a medium with index of refraction $n_t$. There is no AR coating on the left and a single layer, $\lambda/4$ thick AR coating on the right. In this figure, the AR coating is chosen so that the two reflected waves $\mathbf{k}_r$ and $\mathbf{k}_{2r}$, are equal amplitude and 180° out of phase, and thus they exactly cancel. Thus, with the AR coating on the right, there is no reflected light – all energy will be transmitted into the second medium.

*Table 2.*   Reflection of light at an interface: Index of refraction and percentage of energy reflected at an air-medium interface for glass, silicon and HgCdTe.

| Material | Index of refraction | % Energy reflected at air interface |
|----------|---------------------|-------------------------------------|
| Glass | 1.5 | 4 |
| Silicon | ~4 | 36 |
| HgCdTe | 3.7 | 33 |

reflected. The amount of energy that is reflected increases as the change in velocity increases. The amount of light that is transmitted and reflected is rigorously computed using the laws of electromagnetic propagation and continuity of fields – on a more intuitive level, the amount of energy reflected at an interface can be understood by analogy. Nearly all physical systems, as well as human interactions, do not adapt well to a sudden change in conditions – the greater the change in direction or speed, the greater the amount of energy that is "reflected".

The amount of energy that is reflected can be significantly reduced by use of an *anti-reflection coating* (AR coating). An AR coating provides a smoother transition from one medium to another, and can be used to provide zero reflection at a selected wavelength (see 19). The simplest example of an AR coating is the single layer coating shown on the right side of Fig. 6. If the AR coating is $\lambda/4$ thick, then the amount of energy reflected is given by,

$$Reflected\ energy = \left( \frac{n_i n_t - n_{layer}^2}{n_i n_t + n_{layer}^2} \right)^2 \tag{3}$$

where $n_{layer}$ is the index of refraction of the AR coating. For an AR coating where $n_{layer}^2 = n_i n_t$, there is no energy reflected - all of the light is transmitted into the medium! An intuitive way to understand the physics is as follows. Light is indeed reflected at both interfaces, but the amount of energy reflected is the same at each interface. Since the AR coating is $\lambda/4$ thick, the two reflected waves are $180°$ out of phase and they exactly cancel each other, resulting in zero energy being reflected back into the incident medium.

Unfortunately, 100% transmission only occurs at a single wavelength, since the AR coating can only be $\lambda/4$ thick for a single wavelength. Some manufacturers use multi-layer coatings to obtain better results over a broader bandpass but, for CCDs, the maximum is typically a two-layer AR coating. When selecting an AR coating material, the optical designer tries to achieve the following goals: the coating is 100% transmissive for the wavelengths of interest, the index of refraction of the coating is intermediate between air and the detector material, and the AR coating can be applied with the coating technology available. The most popular AR coating for silicon ($n_{Si} \sim 4$) is hafnium oxide ($HfO_2$, $n_{HfO_2} \sim 1.9$). Figure 13 presents the results of detailed modeling of the effect of HfO$_2$ coating thickness on QE. For IR detectors, Zinc sulfide ($SZn$) can be used as an AR coating ($n_{SZn} \sim 2.27$, $n_{HgCdTe} \sim 3.7$).

A purchaser of scientific grade detectors needs to give careful attention to the selection and tuning of AR coatings. The AR coating is one of the most important specifications in a detector procurement.

*Table 3.*   Energy (in eV) of optical and IR photons

| Wavelength ($\mu$m) | Energy (eV) |
|:---:|:---:|
| 0.3 | 4.13 |
| 0.5 | 2.48 |
| 0.7 | 1.77 |
| 1.0 | 1.24 |
| 2.5 | 0.50 |
| 5.0 | 0.25 |
| 10.0 | 0.12 |
| 20.0 | 0.06 |

## 6.      Charge generation

After the light has entered the material, it can be "detected". The energy of the electromagnetic field is extracted in quanta called *photons* with energy given by *Photon energy* $= h\nu$, where $h$ is the Planck constant ($6.626 \cdot 10^{-34}$ Js) and the frequency $\nu$ of light is related to the wavelength by $\nu = c/\lambda$. Since a Joule (J) is a relatively large unit of energy, the electron-volt (1 $eV =$ $1.602 \cdot 10^{-19}$ J) is typically used to express the energy of optical and infrared photons, as shown in Table 3.

### 6.1      The photovoltaic effect

The detector materials discussed in this chapter are crystalline structures. A crystal lattice is a regular, repetitive, three-dimensional pattern of the constituent atoms, as shown in Fig. 7 (left) for silicon. A crystal structure is tightly bound together by covalent bonds, which is the sharing of the outer electrons of adjacent atoms. These *valence band* electrons can be excited so that they escape from the hold of the crystal lattice and are free to move through the material in the *conduction band*. When a "free electron" is created, a corresponding *hole* (vacancy of electron, or positive charge site) is also created. An electron-hole pair can be created from the absorption of the energy of a photon or from thermal energy. When the electron-hole pair is created by light, the process is called the *photovoltaic effect*. When the electron-hole pair is created by thermal energy, the resulting flow of charge is called *dark current*.

The amount of energy that is required to excite an electron from the valence band to the conduction band is the bandgap ($\mathcal{E}_g$) of the material; see Fig. 7 (right). The wavelength (in microns) of the lowest energy photon that can be detected by a material is given by the energy of a 1 $\mu$m photon divided by the bandgap of the material (in $\mathcal{E}_g$) – this is called the cutoff wavelength, $\lambda_c$. To detect longer wavelength light, the bandgap must be smaller. A smaller bandgap also makes it easier for thermal energy to excite electrons, so materials sensi-

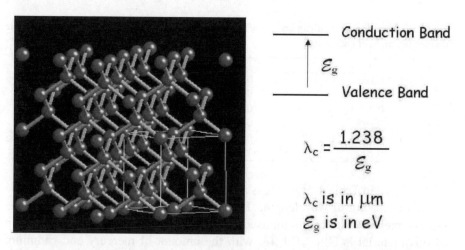

*Figure 7.* Silicon lattice and the bandgap of crystalline structures. Figure 7 (left) shows the regular pattern of the crystalline structure of silicon. Figure 7 (right) is an energy diagram that shows the bandgap ($\mathcal{E}_g$) that separates the valence band from the conduction band. Photons with wavelengths shorter than the cutoff wavelength, $\lambda_c$, have enough energy to excite electrons across the bandgap and thus be detected by the material.

tive to infrared light must be cooled to very low temperature to minimize dark current. The dark current depends on the maximum wavelength sensitivity, the temperature of the device, the size of a pixel, and quality of materials. CCDs, which are made of silicon, are typically cooled to -110 or -120 C, reducing the dark current to $<2$ e$^-$/pix/hr, low enough to be negligible. For HgCdTe IR detectors that are sensitive over the range of 0.9-2.4 $\mu$m, cooling to liquid nitrogen temperature (-196 C) is required, which reduces the dark current to as low as 15 e$^-$/pix/hr. Table 4 summarizes the properties of the most popular optical and infrared detector materials.

*Table 4.* Properties of the most popular optical/IR detector materials. $\mathcal{E}_g$ is the bandgap, of the material (in eV) – the minimal energy required to excite an electron from the valence band to the conduction band. The cutoff wavelength, $\lambda_c$, is the wavelength corresponding to the bandgap – photons of lower energy, or longer wavelength, will not be detected by the material. Note that HgCdTe has a range of cutoff wavelengths – it is a "tunable" material, as explained in the remainder of this section. The operating temperature (Op. Temp.) must be decreased for smaller bandgap materials in order to minimize dark current.

| Material | Symbol | $\mathcal{E}_g$ (eV) | $\lambda_c$ ($\mu$m) | Op. Temp. (°K) |
|---|---|---|---|---|
| Silicon | Si | 1.12 | 1.1 | 163 - 300 |
| Mer-Cad-Tel | HgCdTe | 1.00 - 0.09 | 1.24 - 14 | 20 - 80 |
| Indium Antimonide | InSb | 0.23 | 5.5 | 30 |
| Arsenic doped silicon | Si:As | 0.05 | 25 | 4 |

*Table 5.* Variation of bandgap and cutoff wavelength of $Hg_{1-x}Cd_x Te$ as a function of the concentration, x, of cadmium.

| x | $\mathcal{E}_g$ (eV) | $\lambda_c$ ($\mu$m) |
|---|---|---|
| 0.196 | 0.09 | 14 |
| 0.21 | 0.12 | 10 |
| 0.295 | 0.25 | 5 |
| 0.395 | 0.41 | 3 |
| 0.55 | 0.73 | 1.7 |
| 0.70 | 1.00 | 1.24 |

As shown in Table 4, HgCdTe is a "tunable" material for which it is possible to modify the cutoff wavelength. The energy gap of HgCdTe depends on the ratio of mercury and cadmium used in the material. A more exact expression of this material is $Hg_{1-x}Cd_x Te$, with the amount of mercury and cadmium equaling the total amount of telluride. A higher concentration of cadmium leads to a larger bandgap and correspondingly shorter cutoff wavelength. The variation of bandgap and cutoff wavelength as function of the concentration of cadmium is presented in Table 5. There is great value in being able to tune the bandgap of HgCdTe. For instance, if an instrument's science is limited to $\lambda < 1.6 \mu$m, it is much easier to build the instrument if longer wavelength photons cannot be sensed by the detector. The instrument can be operated warmer, and light shielding is easier. On the other hand, there are applications where it may be desirable to use HgCdTe to detect photons with $\lambda > 5.9 \mu$m, the cutoff wavelength of InSb.

---

**How small is an electron-volt?** *An electron-volt is a very small unit of energy. As an example, consider the DEIMOS multi-object spectrograph which was installed on the Keck II telescope in 2002. The CCD array in DEIMOS is the largest array on any optical spectrograph in the world, an 8K×8K mosaic of eight 2K×4K detectors, 67 million pixels total. On a busy night, the array will be read out 100 times – 25 science frames and 75 calibration frames. If DEIMOS is used one-third of the time and all nights are clear, DEIMOS will produce 1.6 Terabytes of data each year. On average, each pixel will contain ~5000 photoelectrons (primarily from calibrations, not science), resulting in $4 \cdot 10^{15}$ photons per year. While that may be a lot of photons, it is not much energy. After four years of heavy DEIMOS usage, the detector will have absorbed only $1.8 \cdot 10^{16}$ eV of photon energy – equivalent to the gravitational energy of dropping a peanut M&M° R candy (2 grams) a distance of 6 inches (15 cm) – not much energy at all!*

---

**One photoelectron per detected photon.** *A very useful feature of optical and infrared detectors is that, in most cases, there is a one-to-one correspondence between photoelectrons and detected photons. For every photon that is detected, one electron is moved from the valence band to the conduction band. This property provides a very linear response between incident energy and detected signal. The one-to-one correspondence is not valid when the photon energy becomes so large that multiple electrons are produced by a single photon. The onset of multiple electron production occurs for photons with energy $\sim 3.5$ times the bandgap; $\lambda < 320$ nm for silicon and $\lambda < 800$ nm for HgCdTe (if $\lambda_c = 2.5 \mu m$). For ground-based telescopes, the atmosphere is nearly opaque for $\lambda < 320 nm$, and IR detectors are seldom used for $\lambda < 800$ nm – thus the one electron per photon rule generally applies.*

## 6.2    Absorption depth

A very important attribute of light sensitive materials is their absorption depth. The absorption depth is the length of material that will absorb 63.2% of the radiation ($1/e$ of the energy is not absorbed). After two absorption depths, 87% of the light has been absorbed, and after three absorption depths, 95% has been absorbed. To make an efficient detector of light, the material thickness should be several times the absorption depth.

There is a very significant difference in the absorption depths of optical and infrared detectors. The difference is due to the fact that silicon is an *indirect* bandgap material, whereas HgCdTe and InSb are *direct* bandgap materials. An electron transition from the valence band to the conduction band in an indirect bandgap material is less likely to occur because the electron must undergo a momentum change in addition to an energy change. Thus, an indirect bandgap material is a less efficient absorber (and emitter) of light.

Since IR detector materials are direct bandgap materials (with no change in electron momentum required), they are very efficient absorbers (and emitters) of light – all IR photons are absorbed within the first few $\mu m$ of material. The reason that infrared detectors are 10 to 15 $\mu m$ thick is for structural and fabrication reasons, not for light absorption reasons.

Silicon, on the other hand, presents a very special challenge, since it is an indirect bandgap material with an absorption depth that changes tremendously over the wavelength region of interest, as shown in Fig. 8. Note that the horizontal (wavelength) scale is linear and the vertical (absorption depth) scale is logarithmic. The absorption depth varies from 8 nm in the UV ($\lambda \sim 0.3 \mu m$) to over 100 $\mu m$ in the far-red ($\lambda > 1 \mu m$) - more than a factor of 10,000 change in absorption depth! The absorption depth plot is very useful for understanding the quantum efficiency response of silicon-based optical detectors, and we will return to it after we discuss the quantum efficiency of infrared detectors.

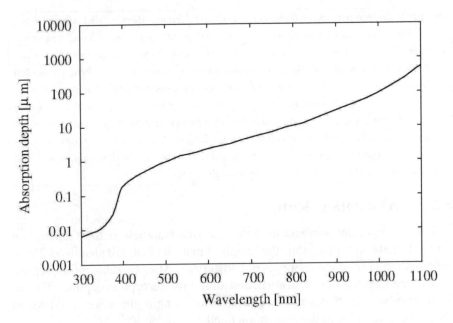

*Figure 8.* Absorption depth of photons in silicon. Notice the tremendous variation in penetration depth, from 8 nm in the UV to more than 100 $\mu$m in the far-red.

## 6.3     Infrared detector quantum efficiency

As shown in Table 2, an IR detector without an AR coating will reflect over 30% of the incident light. Several of the first generation IR focal plane arrays were fabricated on a glass substrate and it was not possible to put an AR coating between the glass and the detector material. There are many IR detectors being used today in astronomy that are without AR coating and have about 60% peak QE. If a single layer AR coating is used, infrared detectors typically achieve 80% QE, as shown in Fig. 9 for a Rockwell Scientific 2K×2K detector.

Infrared detectors can be made with several layer AR coatings, and the detectors for the James Webb Space Telescope, awarded to Rockwell Scientific, specify 4- or 5-layer coatings to achieve QE of 95% over a broad bandpass.

## 6.4     Optical detector quantum efficiency

In comparison to infrared detectors, it is much more difficult for silicon-based optical detectors to achieve high QE over a wide bandpass. The main challenge is the tremendous variation of absorption depth shown in Fig. 8. In addition, the index of refraction varies significantly for $\lambda = 0.32$–$1.1$ $\mu$m, as shown in Fig. 10, making it difficult to optimize anti-reflection coatings for broad bandpass.

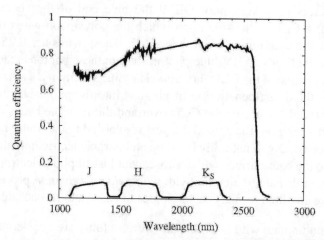

*Figure 9.* Quantum efficiency of a Rockwell 2K×2K HAWAII array. The atmospheric windows of J, H, and K are shown. Note the relatively constant QE across the 1-2.5 $\mu$m wavelength region, with peak QE of 84% in the K-band (centered at 2.2 $\mu$m). Figure courtesy of J. Garnett, Rockwell Scientific.

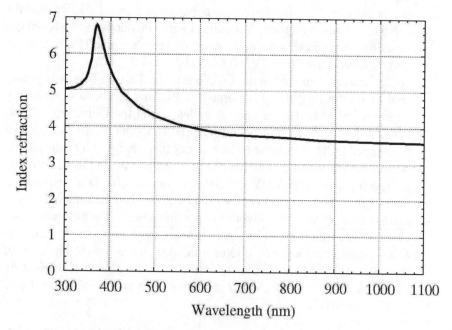

*Figure 10.* Index of refraction of silicon. The factor of $\sim$2 difference of the index of refraction for 0.32-1.1 $\mu$m makes broad bandpass AR coatings difficult to optimize.

The short penetration depth of UV/blue photons is the reason that "frontside" CCD detectors have very poor QE at the blue end of the spectrum. The frontside of a CCD is the side upon which the polysilicon wires that control charge collection and transfer are deposited. These wires are 0.25 to 0.5 $\mu$m thick and will absorb all UV/blue photons before these photons reach the photosensitive volume of the CCD. For good UV/blue sensitivity, a silicon detector must allow the direct penetration of photons into the photosensitive volume. This is achieved by turning the CCD over and thinning the backside until the photosensitive region (the epitaxial layer) is exposed to incoming radiation.

A thinned CCD will naturally build up an internal electric field that will pull electrons to the back surface with a subsequent loss of photoelectrons. In order to overcome this natural electric field, a *backside passivation* process must be used. There are three primary technologies to passivate the backside:

1  Boron implant with laser anneal. Boron atoms are accelerated into the backside of the CCD, replacing about 1 of 10,000 silicon atoms with a boron atom. The boron atoms create a net negative charge that push photoelectrons to the front surface. However, the boron implant creates defects in the lattice structure, so a laser is used to melt a thin layer (100 nm) of the silicon. As the silicon resolidifies, the crystal structure returns with some boron atoms in place of silicon atoms. This works well, except for blue/UV photons whose penetration depth is shorter than the depth of the boron implant. Variations in implant depth cause spatial QE variations, which can be seen in narrow bandpass, blue/UV, flat fields. This process is used by E2V, MIT/LL and Sarnoff.

2  Special coatings applied to the CCD backside. This process was pioneered by the Jet Propulsion Laboratory (JPL) and has been significantly improved by Mike Lesser and his colleagues at the University of Arizona. A thin layer of transparent material is applied to the backside of the thinned CCD - this layer takes on a negative charge. Since the silicon is undisturbed and the negative charge layer is outside the silicon, this approach provides hiqh UV QE and very uniform flat field response in the blue/UV. This process is now used by SITe, Fairchild and MIT/LL.

3  MBE growth of very thin layer of boron and silicon. The problems associated with boron implant and laser anneal can be overcome by growing a very thin (5 nm) layer of silicon with boron atoms on the backside of the thinned CCD (1% boron, 99% silicon). The growth is applied by molecular beam epitaxy (MBE) machines. This process was developed by JPL and MIT/LL.

The backside passivation processes are most critical for the blue and UV wavelengths. For 0.5-0.7 $\mu$m light, high QE is more easily attained - the absorption depth is a few microns, and photons penetrate beyond the backside passivation layer, but do not pass all the way through the CCD. However, at the red end of the optical spectrum, the photon can pass completely through the CCD - the absorption depths at 800, 900 and 1000 nm are 11, 29 and 94 $\mu$m respectively.

Combining the effects of index of refraction and penetration depth of photons, the predicted CCD QE for a 15$\mu$m thick CCD with the "Lesser" backside passivation process and single layer HfO$_2$ AR coating is presented in Fig.13. The sensitivity to thickness of AR coating and transparency of the CCD to longer wavelengths is very evident.

To provide high QE for $\lambda = 0.75 - 1.1\mu$m, the light sensitive silicon layer must be made much thicker than the standard 15 $\mu$m thick CCD. The challenge to producing a thicker device is finding a way to produce electric fields in the material that penetrate through most of the depth of the material to produce good PSF. The two basic kinds of CCDs produced are *n-channel* and *p-channel*, depending on whether the CCD collects electrons or holes. Most CCDs are an n-channel devices.

To produce a very thick n-channel device, the resistivity of the silicon must be made relatively high, about 5,000 to 10,000 $\Omega$-cm, as opposed to the 20-100 $\Omega$-cm material used in "standard" n-channel CCDs. Higher resistivity is required for greater penetration depth of the fields produced by the frontside polysilicon wires (penetration depth is proportional to the square root of the resistivity). These "thick" high resistivity CCDs have been developed for detection of soft x-rays with space satellites and can be procured from E2V and MIT/LL.

An example of the QE that can be obtained with the high resistivity n-channel devices combined with the "Lesser" backside process is shown in Fig. 11.

The p-channel CCD has been developed by Steve Holland and his colleagues at Lawrence Berkeley National Laboratory (LBNL) to provide very high QE in the far-red. *(A parallel effort on p-channel CCD development is underway at the Max-Planck Institut fur extraterrestische Physik under the leadership of Lothar Struder.)* The LBNL p-channel devices are 300 $\mu$m thick with 15 $\mu$m pixels, creating the interesting concept of a pixel "skyscraper". The backside of the CCD (where photons enter) is coated with a thin layer (60 nm) of a transparent, conductive material (indium tin oxide, ITO) that is charged to relatively high voltage (40 V). The backside charge produces a steep voltage gradient in the device that pushes holes to the frontside and minimizes charge diffusion. Amazingly, the PSF of these devices is as good as 15 $\mu$m thick n-channel devices. The ITO layer also acts as one layer of a two-layer AR coating. Typically, LBNL uses 100 nm of silicon dioxide (SiO$_2$) as the second layer to optimize far-red QE: 90% QE has been achieved at 700-900 nm and 60% QE at 1000 nm. However, the very thick p-channel devices do suffer

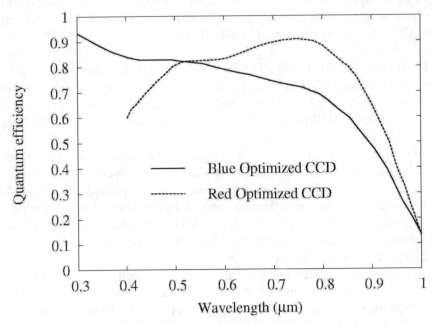

*Figure 11.*   Quantum efficiency of CCDs for the HIRES spectrograph on Keck I telescope. These CCDs will be highest QE devices ever mounted in an optical spectrograph. Three of these CCDs will be installed at the telescope in January 2004 and will make HIRES, once again, the most efficient high resolution spectrograph on the planet. The CCDs are all 45 $\mu$m thick, high resistivity n-channel devices produced by MIT/LL. The blue optimized CCDs (two of them) have undergone backside processing at the University of Arizona, followed by single layer $HfO_2$ AR coating tuned for maximum QE at 320 nm. The red optimized device has a boron implant / laser anneal treatment, with a two layer AR coating - $SiO_2$ over $HfO_2$. The QE curves were measured by Richard Stover and Mingzhi Wei of Lick Observatory.

from one significant drawback. Due to their 300 $\mu$m thickness, the devices are great detectors of many types of radiation, including cosmic rays and radiation created by terrestrial sources. In fact, a dark image can look like images from a particle physics experiment (see Fig. 12). Efforts are underway to find the best way to shield the LBNL CCD from terrestrial radiation and to most efficiently remove the unwanted radiation events from the data.

The challenges of achieving high QE over the 0.3-1.1 $\mu$m band is summarized in Fig. 14, which shows the optical absorption depth of photons in silicon with the range of thickness of different regions of a CCD. Figure14, which we like to call "the beautiful plot" captures the information needed for understanding the QE of silicon CCDs.

The challenges of achieving high QE over the 0.3-1.1 $\mu$m band is summarized in Fig. 14, which shows the optical absorption depth of photons in silicon with the range of thickness of different regions of a CCD. Figure14, which we

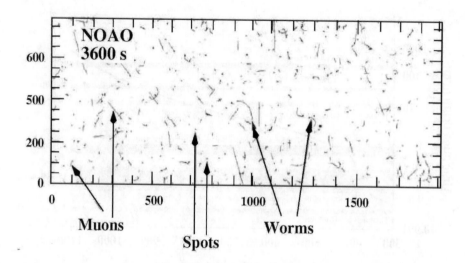

*Figure 12.* A one hour dark image from LBNL p-channel CCD with several types of radiation events. Figure provided by D. Groom (LBNL).

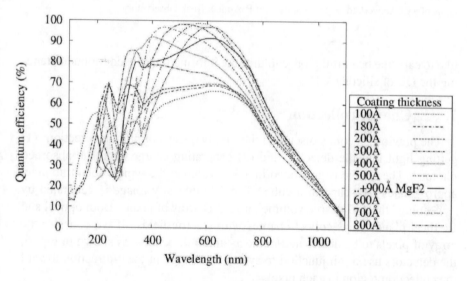

*Figure 13.* Theoretical quantum efficiency of single layer $HfO_2$ coating as function of coating thickness. Notice that the near UV QE (350-400 nm) can be optimized by making the AR coating 40 nm thick. Since there is a variance in the processing, some manufacturers do not attempt a 40 $\mu$m thick coating, since there is a signficant drop in QE for 30 nm thick coating. Thus, often the target thickness is 50 nm, which explains many of the QE curves measured for E2V 2Kx4K devices, which typically have a single layer $HfO_2$ coating. Figure courtesy of Mike Lesser and the Imaging Technology Laboratory, University of Arizona.

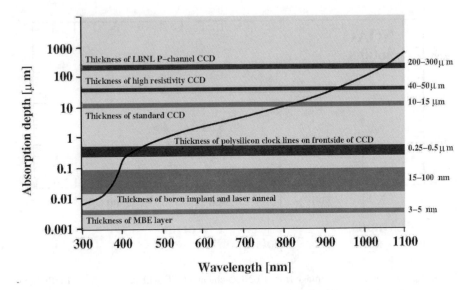

*Figure 14.* Optical absorption depth of photons in silicon with the thickness of different regions of a CCD overlaid. Figure courtesy of P. Amico, Keck Observatory.

like to call "the beautiful plot" captures the information needed for understanding the QE of silicon CCDs.

## 7.      Charge collection

The previous sections discussed the first two steps of photon detection: (1) getting light into the detector, and (2) generating charge from the absorbed photons. The charge is generated in a 3-D volume; the typical scientific pixel approximates the shape of a cube. The 2-D intensity image is generated by "sweeping" the 3-D charge volume into a 2-D array of pixels. Both optical and IR focal plane arrays collect charges with electric fields. If we consider the array of pixels to be distributed in the $x$- and $y$-directions, as shown in Fig. 5, the detectors use a p-n junction to sweep the charge in the z-direction toward the collection region in each pixel.

The photovoltaic detector potential well of a 15 $\mu$m thick n-channel CCD pixel is shown in Fig. 15. For silicon, the n-region is produced by doping the silicon with phosphorous, and the p-region has boron doping. The electric field of this p-n junction will attract electrons to the buried n-channel, which is 0.25-0.5 $\mu$m from the front surface of the detector. Keeping the channel away from the front surface of the CCD is very important for obtaining high charge transfer efficiency. The electric field separates the electron-hole pairs, with the holes flowing out of the detector via the substrate ground connection. Recall

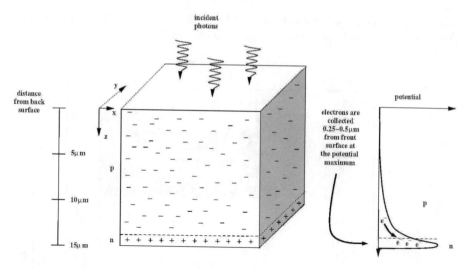

*Figure 15.* Photovoltaic detector potential well. The example in this figure is the p-n junction of a n-channel CCD. The x-y-z axes match the orientation shown in Fig. 5. The charge generated in the 3-D volume of a pixel is swept toward a 2-D layer, which is the buried channel that is 0.25-0.5 $\mu$m from the front surface of the detector. The z-direction potential is created by the p-n junction combined with the voltages on the polysilicon wires deposited on the frontside of the CCD (not shown in this figure).

that in a p-channel CCD, the internal electric fields collect holes instead of electrons.

The p-n junction collapses the 3-D charge distribution into a 2-D array. The pixel boundaries of the 2-D arrays are generated differently in optical and infrared detector arrays.

In CCDs, the boundaries in the column direction are defined permanently by doping the silicon. For an n-channel CCD, the *channel stops* are produced by boron doping; the 2-3 $\mu$m wide channel stop does not absorb photoelectrons since the boron is only implanted a small distance ($\sim$0.5 $\mu$m) into the front surface. In the row direction, the pixels are defined by the electric fields of polysilicon wires that are fabricated on the front surface of the CCD. There are three parallel wires per pixel in a "three phase" CCD. During integration, two phases of the CCD are kept at a positive voltage to attract electrons, with the third (barrier) phase biased at a negative voltage to provide a barrier between the rows of the CCD. Figure 16 shows a simplified schematic of the architecture of a CCD pixel.

In the infrared, the light sensitive material is bump bonded at each pixel to a silicon multiplexer that amplifies the charge and multiplexes the outputs to external electronics, as shown in Fig. 17. The column and row boundaries are defined by the electric field that attracts the charge to the silicon multiplexer. The geometry of an InSb detector pixel is shown in Fig. 18.

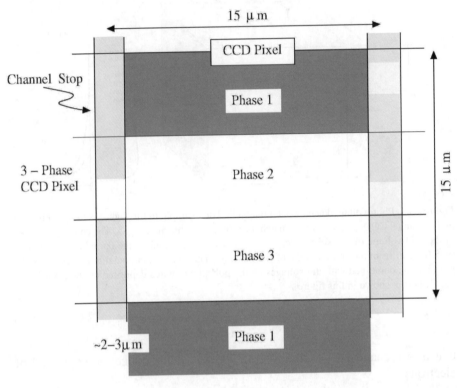

*Figure 16.* Simplified schematic of the architecture of a single CCD pixel. The channel stops permanently define the pixel boundaries in the column direction. The pixel rows are defined by the electric fields applied to the three pixel phases. For a 15 $\mu$m pixel CCD, channel stops are 2-3 $\mu$m wide and the phases are each $\sim$5 $\mu$m wide.

## 8.    Charge transfer

Infrared arrays do not have any charge transfer, since the charge is amplified at each pixel by the silicon multiplexer. Only CCDs have charge transfer and thus this section only discusses CCD array architecture.

The polysilicon wires that define the pixel geometry in the vertical (row) direction are also used to move charge across the array. For a detailed explanation of charge transfer, please see Janesick, 2001. Figure 19 shows the parallel and serial movement of charge to the output amplifier. Since the parallel phases for all pixels are wired together, the movement of all charge packets is "coupled" - this is the origin of the term *charge coupled device* (CCD).

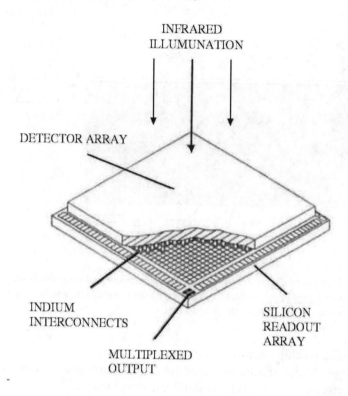

INFRARED
ILLUMUNATION

DETECTOR ARRAY

INDIUM
INTERCONNECTS

SILICON
READOUT
ARRAY

MULTIPLEXED
OUTPUT

*Figure 17.* Infrared array geometry. The light sensitive material is connected at each pixel by an indium bump bond to the silicon multiplexer that reads out the charge generated by the incident light. Figure courtesy of I. McLean (UCLA).

Charge transfer has both good and bad features. The negative aspects of charge transfer include:

⊗ takes time,
⊗ can blur the image if a shutter is not used,
⊗ can lose or blur the charge during the transfer (charge transfer inefficiency),
⊗ allows a saturated pixel to bleed charge up and down a column,
⊗ a trap in a pixel can block an entire column,
⊗ a "hot" pixel defect can release charge into all pixels that are moved past it during readout.

With proper design, fabrication and clocking, most of these negative aspects can be overcome, but there are usually some bad columns due to blocked columns or hot pixels. There are many positive aspects of charge transfer, giving the CCD some very good and unique attributes vis-a-vis an infrared detector:

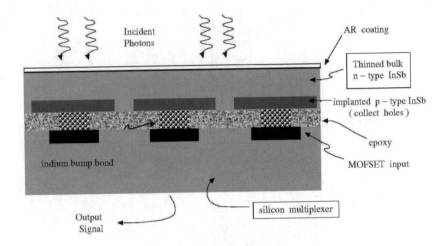

*Figure 18.* Cross section of an InSb pixel. This detector collects holes and amplifies the signal. The charge is collected in the vertical ($z$) direction by the p-n junction, and is separated in the $x$- and $y$-directions by the fields of the p-n junction.

- ⊙ binning on chip - a noiseless process,
- ⊙ charge shifting during exposure for tip/tilt correction or to reduce systematic errors (*e.g.*, nod and shuffle, drift scanning),
- ⊙ enables special purpose designs that are optimized for a particular application,
- ⊙ moving the charge to the edge of the array provides space for designing and building a very low noise amplifier.

With large detectors such as the 2K×4K, there must be very high charge transfer efficiency (CTE) for the signal to retain good fidelity. Devices with CTE of 0.999 99 are readily attainable and CTE of 0.999 999 5 has been achieved. Note that charge that is inefficiently transferred is not lost, but simply trails 1 or 2 pixels behind the original pixel that lost the charge. Also note that CTE is cumulative - a charge packet that moves through 4K pixels with 0.999 99 CTE will retain 96% of the original charge packet ($0.999\ 99^{4K} = 0.96$).

## 9.    Amplification

Although astronomy is accustomed to the detection of a few photons per pixel, the electric charge of a few electrons is extremely small. A critical part of the design of a focal plane array is the amplifier which converts the small amount of charge in each pixel into a signal that can be transmitted off the detector. The amplifier in an optical or infrared detectors is typically a *field effect transistor* (FET), a solid state structure which allows a very small amount

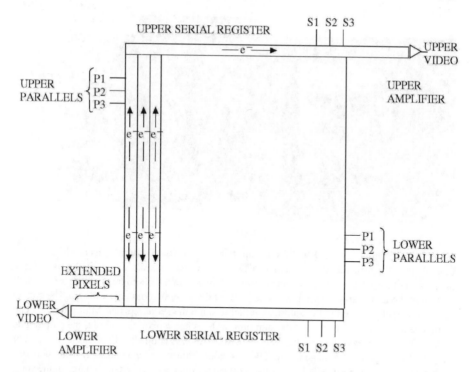

*Figure 19.* CCD array architecture, showing the parallel and serial transfers required to move the pixel charge to the output amplifier.

of charge to change the current flow of millions of electrons per second. In an IR array, there is an amplifier in each pixel. For optical CCDs, there are a small number of amplifiers at the edge of the detector, typically 2 or 4 amplifiers on a large scientific array.

Figure 20 shows the amplifier for a CCD developed for the Craf/Cassini space mission. This is a MOSFET (metal-oxide-semiconductor field effect transistor), which has a gate, source and drain. Charge on the gate will affect the flow of current between the source and drain. The amount of amplification is tremendous – a single electron added to the gate will reduce the current flow by 300 million electrons per second! The current is measured by connecting the amplifier output to ground via a resistor - a change in the current through the resistor will produce a change in the voltage at the output node. The sensitivity of an amplifier, which is called the *responsivity*, is typically stated in units of microvolts per electron ($\mu$V/e$^-$).

Amplifiers are not perfect. There is always random fluctuation in the current through an amplifier, even if there is no change in the charge on the gate of the amplifier. This random fluctuation of the current produces a false signal that looks like a real signal. The *readout noise* is the rms variation of the signal, and

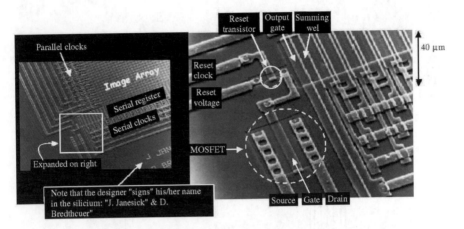

*Figure 20.* Example of CCD MOSFET amplifier - Craf/Cassini design. This is a high resolution image of the fronside of a n-channel CCD - note the 40 $\mu$m scale shown on the right. Only the portion of the CCD near the output amplifier is shown. In the photograph on the left, a small portion of the image array is seen. The operation of the pixel readout is as follows. The charge from each pixel is moved to the output amplifier through the serial register. Multiple pixels can be binned in the serial direction at the summing well. When a pixel is read out, the output gate is first set to the reset voltage by the reset transistor which is controlled by the reset clock. For a typical CCD, the reset voltage is about 9 volts, which "turns on" the current flowing between source and drain in the MOSFET. After the reset gate is open and closed, the charge in the summing well is pushed over the "wall" of the output gate by clocking the summing well positive (high enough voltage to push charge over the output gate, but not so high as to push charge back up the serial register). The charge on the output gate affects the current from source to drain - with more electron charge on the gate there is less current between source and drain. The linear dynamic range of a typical CCD MOSFET is 2 volts - a high responsivity (20 $\mu$V/e$^-$) amplifier will saturate at about 100,000 electrons. Figure courtesy of J. Janesick.

is stated in units of electrons at the gate of the amplifier. The noise spectrum of a FET, shown in Fig. 21, has two components, a white noise component and a $1/f$ noise component. By reading out slower, the random white noise fluctuations can be averaged, and lower noise achieved, until the $1/f$ noise becomes dominant. Since a CCD places the amplifiers at the edge of the array, there is room to design and place a very low noise amplifier. The amplifiers in IR arrays have less room and more signals in close vicinity, and thus IR amplifiers typically have higher noise.

The CCD MOSFET is a destructive readout device - there is only one measurement per charge packet. However, an infrared amplifier can be read out several times, with averaging and corresponding reduction in the effective readout noise (16 reads can reduce the noise by a factor of $\sqrt{16}$ or 4). In theory, multiple readout of an infrared amplifier could achieve extremely low noise, but in practice, due to other complications, the noise reduction usually reaches a limit of 4-5 improvement (achieved after 16-32 reads).

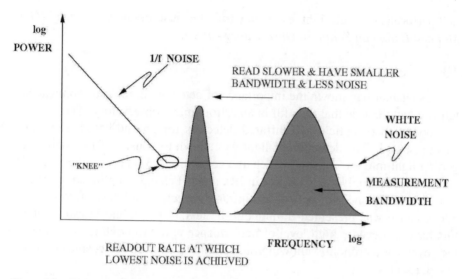

*Figure 21.* Noise spectrum of detector amplifiers. Note that both axes have logarithmic scale. There are two main components of noise - the white noise which is present at all frequencies, and the $1/f$ noise that is dominant at low frequencies. $1/f$ noise has a fractal structure and is seen in many physical systems. The bandpass of a measurement decreases for slower readout, and the readout noise will correspondingly decrease. A limit to reduction in readout noise is reached at the "knee" of the noise spectrum (where white noise equals $1/f$ noise) - reading slower than the frequency knee will not decrease readout noise.

CCD detector designers try to increase the signal-to-noise ratio of an amplifier in two ways: (1) increase the responsivity, or (2) decrease the random current fluctuation between source and drain. The responsivity can be increased by decreasing the amplifier size. Decreasing the amplifier size decreases the capacitance of the MOSFET. The responsivity of a MOSFET obeys the capacitor equation which relates voltage, $V$, to the charge $Q$ on a capacitance $C$ : $V = Q/C$.

For example, the amplifier shown in Fig. 20 has a gate dimension of 5 by 20 $\mu$m with a capacitance of 100 fF; this amplifer has a responsivity of 1.6 $\mu$V/e$^-$. New designs by MIT Lincoln Laboratory have a gate size of 1 by 10 $\mu$m, reducing the capacitance by more than a factor of ten, for a responsivity of 10-20 $\mu$V/e$^-$.

The best CCD amplifiers that have been produced achieve read noise of 1.5 e$^-$ at 50,000 pixels per second and 4 e$^-$ at 1 million pixels per second. *(These are rates per readout port. Lower noise can be achieved for a given frame rate by adding more readout ports.)* The best readout noise achieved with infrared detectors is 15 e$^-$ for a single read and 3-5 e$^-$ after many samples.

There will be new developments in the area of amplifier design during the next few years that may break the 1 electron noise barrier. Photon-noise limited performance at all light levels may truly be "just around the corner" *(Guy Monnet, European Southern Observatory - 1999).*

## 10.   Summary

This chapter has shown the importance of detectors in the astronomical enterprise and the role that they fill in an astronomical observatory. The physics and operation of optical and infrared detectors for ground-based telescopes was presented. The detection of light was shown to consist of five steps: (1) getting light into the detector, (2) charge generation, (3) charge collection, (4) charge transfer (only in CCDs, not in IR), and (5) charge amplification. These steps were explained for optical and infrared detectors and the state-of-the-art performance was presented throughout the text. Optical and infrared detectors have achieved a high level of performance and will continue to serve as the "eyes" of astronomy, allowing humankind to explore the cosmos on ever grander scales.

## References

Every three years, representatives of all of the major observatories and every detector manufacturer gather to exchange information on the state-of-the art. The proceedings from the past two workshops, held in 2002 (Hawaii) and 1999 (Germany), capture the most recent developments in optical and infrared detectors:

P. Amico and J. W. Beletic eds., 2003, *Scientific Detectors for Astronomy 2002, the beginning of a new era,* ASSL Series, Kluwer Academic Publisher, Dordrecht

P. Amico and J. W. Beletic eds., *Optical Detectors for Astronomy II, state-of-the-art at the turn of the millenium,* ASSL Series, Kluwer Academic Publisher, Dordrecht, 2000.

James Janesick has been the most influential person in CCD community for many years, not only making significant technological developments, but also playing the role of lead evangelist. His classic "CCD course", held bi-annually at UCLA, is a pilgrimage many of us in the field have made. He has finally captured all of his lecture notes into one tomb.

J. R. Janesick, 2001, *Scientific Charge-Coupled Devices,* SPIE Press, Bellingham

For an excellent overview of all aspects of astronomical instrumentation, please see the classic textbook by Ian McLean. This book contains a very good pedantic explanation of optical and infrared detectors.

I. S. McLean, 1997, *Electronic Imaging in Astronomy: Detectors and Instrumentation,* John Wiley and Sons, Chichester

All of the optical and infrared focal plane arrays are solid state electronic devices, and to fully understand their physics and operation, one should have a solid foundation in the solid state electronics. An excellent reference is:

B. G. Streetman, 1990, *Solid State Electronic Devices,* Prentice-Hall, Englewood Cliffs

# SPECTROSCOPY

Jeremy Allington-Smith

*University of Durham*
*Astronomical Instrumentation Group, Physics Dept.*
*South Rd, Durham DH1 3LE, UK*

j.r.allington-smith@durham.ac.uk

**Abstract**     The basic principles of astronomical spectroscopy are introduced and the main types of dispersing element surveyed. The principles behind two modern spectroscopic techniques, multiple object and integral field spectroscopy, are also discussed.

**Keywords:**     Astronomical spectroscopy, multiobject spectroscopy, integral field spectroscopy, dispersing elements, diffraction gratings, fibres, image slicers

## 1.     Principles of dispersion

## 1.1     Basic equations

Light incident on a diffraction grating is diffracted according to the *grating equation*:

$$m\rho\lambda = n_1 \sin\alpha + n_2 \sin\beta \qquad (1)$$

where $\rho = 1/a$ is the ruling density where $a$ is the centre-to-centre spacing of the rulings and $m$ is the spectral order. The geometry is shown in Fig. 1. The incident and diffracted light makes angles of $\alpha$ and $\beta$ to the normal in media of refractive index $n_1$ and $n_2$ respectively. Differentiating with respect to the diffracted angle gives the angular and linear dispersions respectively as:

$$\frac{d\lambda}{d\beta} = \frac{\cos\beta}{m\rho} \qquad \frac{d\lambda}{dx} = \frac{d\lambda}{d\beta}\frac{d\beta}{dx} = \frac{\cos\beta}{m\rho f_2} \qquad (2)$$

where, referring to Fig. 1 which shows a generic spectrograph, $f_2$ is the focal length of the camera and $x$ is the distance measured on the surface of the detector in the dispersion direction.

*R. Foy and F. C. Foy (eds.), Optics in Astrophysics, 155–179.*

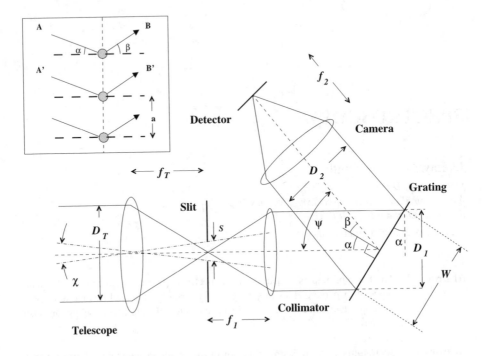

*Figure 1.* Layout of a generic spectrograph employing a reflection grating. Angles $\alpha$ and $\beta$ have the same sign if they are on the same side of the grating normal. Inset: the condition for interference of light by a diffraction grating.

The *spectral resolution*, defined as the width of the distribution of light projected on the detector, is

$$\delta\lambda = \frac{d\lambda}{dx}s' = \frac{\cos\beta}{m\rho f_2}s' \tag{3}$$

where $s'$ is the width of the image of the slit, whose actual width is $s$, on the detector. By conservation of Etendue, this is $s' = s(F_2/F_1)$ where $F_1$ and $F_2$ are the focal ratio of the collimator and camera respectively. Thus

$$\delta\lambda = \frac{s\cos\beta}{m\rho f_2}\frac{F_2}{F_1} = \frac{s}{m\rho F_1 W} \tag{4}$$

where $W = D_2/\cos\beta$ is the length of the intersection of the plane of the grating with the beam from the collimator and $D_2$ is the size of the camera aperture.

The dimensionless *resolving power* is

$$R \equiv \frac{\lambda}{\delta\lambda} = \frac{m\rho\lambda F_1 W}{s} \tag{5}$$

or, in terms of the parameters more relevant to an observer,

$$R = \frac{m\rho\lambda W}{\chi D_T} = (\sin\alpha + \sin\beta)\frac{W}{\chi D_T} \qquad (6)$$

where $\chi$ is the angular slitwidth and $D_T$ is the diameter of the telescope aperture.

From the second form of Eq. 6, it can be seen that the critical parameter for maximising $R$ is $W/(\chi D_T)$ since the geometric term $\leq 2$. Thus obtaining high resolving power depends on maximising the length of the grating-beam intersection and minimising the angular slitwidth. It is also necessary to ensure that $W$ scales in proportion to the telescope aperture (for constant $\chi$), i.e. extremely large telescopes require equally large instruments!

If the slit is very narrow, comparable to the diffraction limit of the telescope, $\chi \approx \lambda/D_T$, the resolving power is

$$R_* = m\rho W \qquad (7)$$

and so is determined only by the total number of illuminated rulings multiplied by the spectral order, *independent of the telescope aperture*.

Decreasing the slitwidth reduces throughput for extended objects or for point sources if the image is too broad. Thus a convenient figure of merit is the product of the throughput and the resolving power. One option for increasing this product is *image slicing* in which the image is divided into a number of thin slices which are rearranged into a long, narrow slit. If the slicing optics are efficient, the full benefit of high resolving power can be obtained without sacrificing throughput. A generalisation of this approach will be discussed in the section on *integral field spectroscopy*.

## 1.2    Reflection gratings in low order

This section refers to classically-ruled plane diffraction gratings where the grooves have a blazed profile as illustrated in Fig. 2a.

The intensity of light from a diffraction grating (e.g. Jenkins and White 1976) is given by:

$$I = \left(\frac{\sin^2 N\phi}{\sin^2 \phi}\right)\left(\frac{\sin^2 \theta}{\theta^2}\right) \qquad (8)$$

where $\phi$ is the phase difference between adjacent rulings and $\theta$ is the phase difference between the centre and edge of one ruling. The first term describes the *interference* between rulings while the second term describes the *diffraction* of light from a single ruling. This diffraction pattern modulates the interference pattern as shown in Fig. 2b. Unfortunately the peak of this envelope occurs at zero diffraction order ($m = 0$) whereas we would like it to peak at some useful order, such as $m = 1$. Shifting the envelope in this way is known as 'blazing'

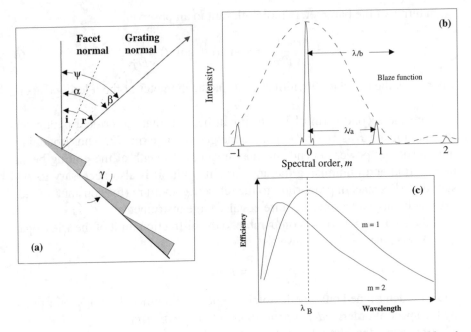

*Figure 2.* (a) Blaze condition for a reflection grating. (b) diffracted intensity for unblazed grating. (c) Typical relationship between efficiency and wavelength for a blazed grating.

and may be done simply by angling each ruling to form an inclined facet with angle $\gamma$ to the plane of the grating. The 'blaze condition' corresponds to specular reflection off each ruling as illustrated in Fig. 2a such that $\alpha + \beta = 2\gamma$. $\Psi = \alpha - \beta$ is the angle between the input and output rays which is fixed by the angle between the axes of the collimator and camera. Thus we can rewrite Eq. 1 for the blaze condition, at which $\lambda = \lambda_B$, as:

$$m\rho\lambda_B = \sin\alpha + \sin\beta = 2\sin\gamma\cos(\Psi/2) \qquad (9)$$

The variation of grating efficiency with wavelength shown in Fig. 2c is based on a simple model of diffraction which is valid for $a > \lambda$. For finer rulings, polarisation and resonance effects complicate the situation (Palmer 2000). In the simple case, the efficiency drops to 40% of the peak (blaze) value at the following wavelengths (Schroeder 2000).

$$\lambda_+ = \frac{2m\lambda_B}{2m-1} \qquad \lambda_- = \frac{2m\lambda_B}{2m+1} \qquad (10)$$

so the useful wavelength range is $\lambda_+ - \lambda_- \simeq \lambda_B/m$. The figure also shows the blaze profile for second order. In general, the peak of the blaze profile in order $m$ occurs at $\lambda_B(m = 1)/m$. As the expression for $\lambda_+ - \lambda_-$ shows, the width of the profile also scales with $1/m$.

An unusual feature of the geometry employed with reflection gratings is that the beam entering the camera may be squashed or expanded in the dispersion direction so that the beam size parallel to dispersion and along the slit respectively is

$$D_\lambda = W \cos \beta \qquad D_S = W \cos \alpha \qquad (11)$$

so there is an anamorphic factor

$$A = \frac{D_\lambda}{D_S} = \frac{\cos \beta}{\cos \alpha} \qquad (12)$$

which results in an elliptical beam at the camera, assuming that the incident beam is circular. If $A > 1$, it is necessary to oversize the camera aperture. In this case the image of the slit is also demagnified so that: (a) the number of pixels to which the slitwidth projects is decreased, thereby reducing the oversampling of the slit by the detector and increasing the wavelength range that fits on the detector; and (b) the separation of slits which are displaced in the dispersion direction (as is usual in *multiobject spectroscopy*) is reduced on the detector. Since the anamorphism is a function of the tilt of the grating, a spectrograph which allows the grating tilt to be varied to bring different wavelengths of interest onto the detector must account for variation in this magnification. The two main configurations for a spectrograph with $\Psi \neq 0$ are shown in Fig. 3.

Apart from using the grating in zero order so that it works like a mirror, maintaining a (compact) round beam at the camera ($A = 1$) requires that the spectrograph adopt the *Littrow configuration* for which $\Psi = 0$ and the incident and diffracted rays are parallel. From equations 6 and 9, the resolving power at blaze in the Littrow configuration is

$$R_B^L = \frac{2D_1 \tan \gamma}{\chi D_T} \qquad (13)$$

since $W = D_2 / \cos \alpha$ and $\alpha = \gamma$, $D_1 = D_2$ in Littrow. The blaze wavelength in a non-Littrow configuration is related to the Littrow blaze wavelength via

$$\lambda_B = \lambda_B^L \cos \left( \frac{\Psi}{2} \right) \qquad (14)$$

where $\lambda_B^L = 2 \sin \gamma / m\rho$ from Eq. 9.

Fig. 4 shows the efficiency of the set of reflection gratings delivered with the Gemini Multiobject Spectrograph (GMOS) for Gemini-North (Section 2.3).

## 1.3 Grisms

A grism consists of a prism with a grating optically interfaced to one side. The usual configuration is shown in Fig. 5. The prism deflects the zero-order

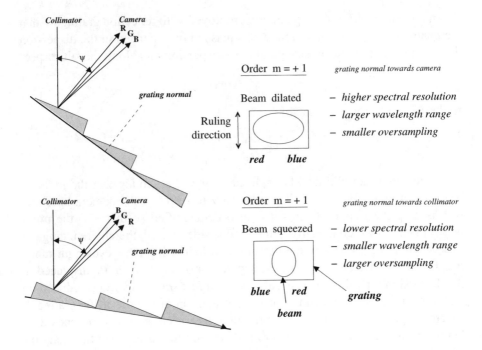

*Figure 3.* Illustration of the use of the same grating in *normal to camera* (top) and *normal to collimator* (bottom) configurations.

light to one side so that first order light can propagate directly along the optical axis. This allows the collimator and camera to be in line, so $\Psi = \pi \Rightarrow |A| = 1$, an anamorphic, Littrow configuration.

The grating equation (1) is modified such that $n_1 \equiv n$ where $n = n_G = n_R$, the refractive indices of the glass and resin respectively, and $n_2 = 1$. The condition for rays to pass undeviated through the grism is

$$m\rho\lambda_U = (n-1)\sin\phi \tag{15}$$

and the blaze condition (when $\theta = 0$, Eq. 8) corresponds to $\lambda_B = \lambda_U$. Although this is only true for the special case shown in the figure, it is the favoured geometry since it ensures that the blaze and undeviated wavelengths are equal.

Following the same arguments as before, it is simple to show that the resolving power is given by Eq. 6 as before, or at blaze, by substituting from Eq. 15:

$$R_B = \frac{(n-1)\tan\phi D_1}{\chi D_T} \tag{16}$$

since $W = D_1/\cos\phi$.

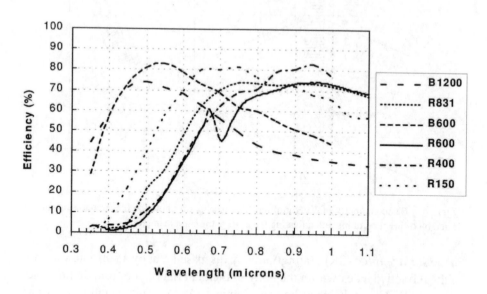

*Figure 4.* The efficiency of gratings used in the Gemini Multiple Object Spectrograph (GMOS) from measurements by the vendor in Littrow configuration.

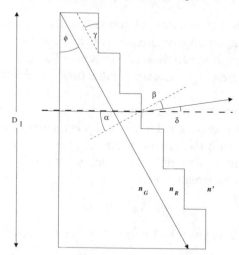

*Figure 5.* A grism with the facet size much exaggerated. The blaze condition is when $\delta = 0$.

## 1.4 Volume phase holographic gratings

A full description is given by Barden et al. (2000). In a VPH grating, the interference condition is provided not by structuring the surface into facets, as

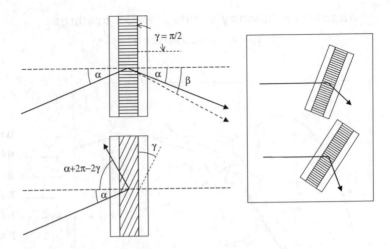

*Figure 6.*    Basic geometry of VPH grating. The inset shows how the blaze condition may be varied by tilting the grating and adjusting the collimator-camera angle.

in the *surface relief* (SR) gratings already discussed, but by a periodic variation in the refractive index within the body of the material (made from dichromated gelatine, DCG, sandwiched between optical glass). This usually has a harmonic form:

$$n(x, z) = n_g + \Delta n_g \cos\left[2\pi\rho_g(x \sin\gamma + z \cos\gamma)\right] \qquad (17)$$

where $\gamma$ is the angle between lines of constant $n$ (the 'fringes') and the grating surface, and $z, x$ are coordinate directions parallel to the optical axis and the orthogonal direction in which dispersion occurs. The geometry can be varied so that the fringes are perpendicular, parallel or at some arbitrary angle to the surface (Fig. 6).

VPH gratings obey the grating equation (1) with $n_1 = n_2 = 1$ if the incident and diffracted angles are defined in air, and $\rho$ is the frequency of the intersection of the fringes with the grating surface. Thus $\rho = \rho_g \sin\gamma$ where $\rho_g$ is the true fringe density. The diffracted energy pattern is governed by Bragg diffraction giving a maximum when

$$m\rho\lambda_B = 2 \sin\alpha \qquad (18)$$

The resolving power is given by Eq. 6 with $W = D_1 / \cos\alpha$, or at blaze,

$$R_B = \frac{2 \tan\alpha D_1}{\chi D_T} \qquad (19)$$

which is identical to the expression for a SL reflection grating with $\gamma = \alpha$.

Kogelnik's analysis of the efficiency (see Barden et al., 2000 is applicable when $2\pi\lambda d\rho_g^2/n_g > 10$ where $d$ is the thickness of the DCG layer. The width

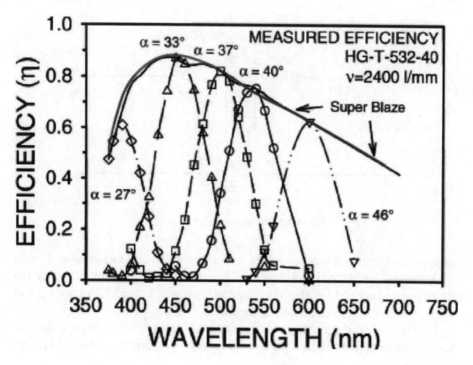

*Figure 7.* Efficiency of a VPH grating with $\rho = 1200$/mm ($=\nu$). The superblaze envelope is indicated (from Barden et al., 2000).

of the 'blaze' peak is given at FWHM is terms of the *Bragg envelopes* in terms of angle at fixed wavelength as

$$\Delta\alpha \propto \frac{1}{\rho_g d} \tag{20}$$

and wavelength at fixed angle as

$$\Delta\lambda \propto \frac{1}{\rho_g \tan\alpha_g}\Delta n_g = \frac{1}{\rho_g \tan\alpha_g}\frac{\lambda}{d} \tag{21}$$

Figure 7 shows the latter, *spectral* envelope for different incident angles and illustrates the *superblaze* envelope obtained by simultaneously optimising both wavelength and incident angle.

The advantage of VPH gratings is that the peak efficiency can be very high, e.g. ~90% in Fig. 7 compared with 70–80% for SL gratings (e.g. Fig. 4). However the width of the blaze peak is generally narrower than for SL gratings restricting their use in applications requiring a large simultaneous wavelength range. However, from equations 20 and 21, it is expected that the situation will

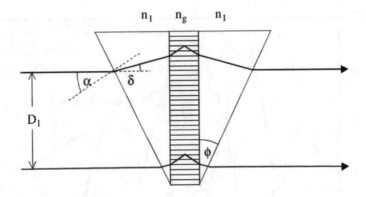

*Figure 8.*    Example of a VPH 'vrism'.

improve as manufacturers learn how to maximise $\Delta n$ and minimise $d$. Unlike SL gratings, the blaze (i.e. Bragg) condition may be altered by changing the tilt of the grating ($\alpha$) but this also requires a change in the collimator-camera angle, $\Psi$, which is mechanically undesirable.

To avoid the bent geometry required for utilisation of VPH gratings, they are sometimes combined with a prism analogously to a grism ('vrisms'; Fig. 8). For these,

$$R = \frac{m\rho\lambda W}{\chi D_T} = \frac{m\rho\lambda}{\chi D_T} D_1 (1 + \tan\delta\tan\phi) \tag{22}$$

where, for the configuration shown in the figure, $\delta = \phi - \arcsin(\sin\phi/n_1)$.

## 1.5    Limits to resolving power: immersed gratings

For the dispersion options examined so far, it is plain from Eq. 6, that the resolving power increases with ruling density and spectral order. However the illuminated grating length, $W$, imposes a limit. If this becomes too big the diffracted beam will overfill the camera, and, although $R$ will continue to increase, the throughput will fall. This is shown in Fig. 9 where the resolving power obtainable from a set of gratings intended for GMOS is plotted against wavelength for a range of tilt angles $\alpha$. The traces are terminated at the point where the camera is overfilled. It can be seen that $R$ cannot exceed $\sim$5000 for $\chi = 0.5''$. Increasing the spectral order will not help since the product of this and the ruling density is fixed for a given wavelength since $m\rho\lambda = (\sin\alpha + \sin\beta) \leq 2$.

Overcoming this limit is possible by optically-interfacing a prism to the reflection grating using an appropriate gel or cement. As shown in Fig. 10, the beam exiting the grating is squeezed in dispersion allowing it to fit within the camera aperture so that no light is lost. Lee & Allington-Smith (2000) show that practical immersed gratings can be built which effectively double the maximum resolving power. This is also demonstrated in Fig. 9 which includes traces for immersed gratings.

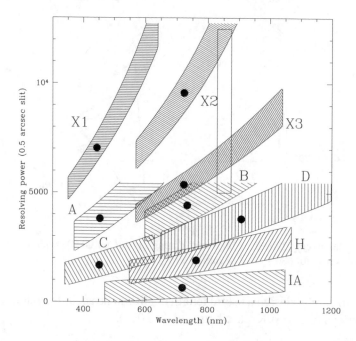

*Figure 9.* Predicted resolving power for GMOS equipped with the gratings described in Table 1 and slit width 0.5 arcsec. Those prefixed by 'X' are hypothetical immersed gratings with prism vertex angle, $\phi$, given in the Table 1. The dots mark the nominal blaze condition. The box marks the requirements for measuring the velocity dispersion of dSph galaxies using the Calcium triplet.

## 1.6    Limits to resolving power: echelle gratings

Another solution is to use unimmersed gratings in very high order. From Eq. 13, it can be seen that increasing the groove angle, $\gamma$, increases the resolving power at blaze but this requires large $m\rho$ since this $\propto \sin\gamma$ (from Eq. 9). Instead of increasing the ruling density, which is subject to a fabrication limit for large gratings, another option is to increase $m$. An echelle is a coarse grating with large groove angle which is used in high order. They are often characterised by $\tan\gamma$ so, for example, 'R2' implies $\gamma = 63.5$. Fig. 11 shows the general configuration. The disadvantage of working in high order is that the spectra will overlap unless a passband filter is used to isolate the order of interest. Since the free spectral range is $\simeq \lambda/m$, the useful wavelength range is small, so a better option is to introduce dispersion perpendicular to the primary dispersion direction to spread the orders out over a 2-D detector (Fig. 11). However, this makes it impossible to use for multiple-object spectroscopy, al-

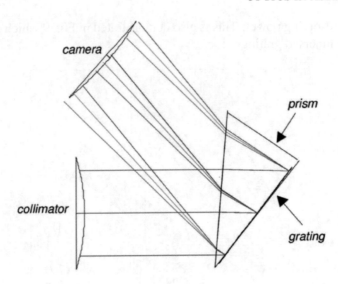

*Figure 10.* Ray-tracing through an immersed grating suitable for GMOS with $\rho = 1800$/mm, with a 35-deg prism for $630 < \lambda < 682$nm.

though with careful design a useful length of slit can be accommodated without order overlaps (e.g. ESI/Keck, Sheinis et al. 2002).

## 1.7 Prisms

Finally, we need to consider the use of prisms as dispersing elements. Although capable only of low dispersion, they have the advantage that, because they do not rely on interference effects, there are no multiple orders to overlap

*Table 1.* Details of gratings appearing in Fig. 9. The names in parenthesis refer to the actual gratings delivered with GMOS-N (except for R150 which is slightly different) $\phi$ is the vertex angle of the immersed prism if present.

| Name | $\rho$ (mm$^{-1}$) | $\gamma$ (deg) | $\phi$ (deg) |
|---|---|---|---|
| A (B1200) | 1200 | 17.5 | 0.0 |
| B (R831) | 831 | 19.7 | 0.0 |
| C (B600) | 600 | 8.6 | 0.0 |
| D (R600) | 600 | 17.5 | 0.0 |
| H (R400) | 400 | 9.7 | 0.0 |
| IA (R150) | 158 | 3.6 | 0.0 |
| X1 | 2400 | 21.0 | 30.0 |
| X2 | 1800 | 26.7 | 30.0 |
| X3 | 1200 | 17.5 | 30.0 |

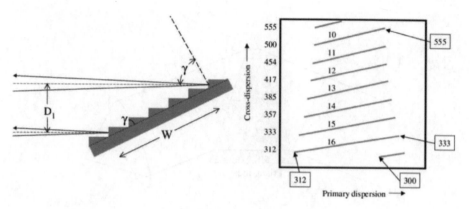

*Figure 11.* Left: Littrow configuration for an echelle grating. Right: example of the layout of orders (labelled by $m$) on the detector showing the wavelength ranges covered.

inconveniently. For this reason, they are sometimes used for multiple-object spectroscopy of faint objects where only low spectral resolution is required without the penalty of order overlap, or for cross-dispersion in echelle spectrographs. ESI (Sheinis et al. 2002) uses them in both these ways.

The equivalent of the grating equation for a prism is the application of Fermat's principle to the configuration shown in Fig. 12.

$$tn = 2L \cos \alpha \tag{23}$$

where $t$ is the baselength of the prism. Since $\beta = \pi - \phi - 2\alpha$, the angular dispersion can be written as

$$\frac{d\lambda}{d\beta} = \frac{D_1}{t} \frac{d\lambda}{dn} \tag{24}$$

in terms of the variation of refractive index with wavelength. Unfortunately, almost every suitable optical material has a highly non-linear $dn/d\lambda$ leading to large variations in dispersion with wavelength. However this problem can be ameliorated through the use of composites of different materials. The resolving power is derived in the same way as for grating spectrographs (i.e. by considering the projection of the slit on the detector and conserving Etendue) to get

$$R = \frac{\lambda}{\chi D_T} t \frac{dn}{d\lambda} \tag{25}$$

By comparison with Eq. 6, it can be seen that the analogues of $m\rho$ and $W$ are $dn/d\lambda$ and $t$, respectively.

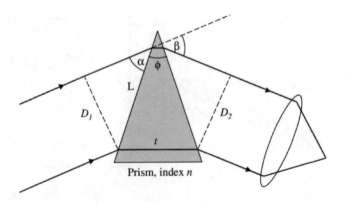

*Figure 12.* Prism configuration considered in the text.

## 2. Multiple object spectroscopy

### 2.1 Introduction

The purpose of multiobject spectroscopy (MOS) is to obtain spectra of many objects within a wide field of view. Modern spectrographs have fields of ∼5–200 arcmin within which several tens to hundreds of objects may be observed simultaneously. With *microslit* techniques, the *multiplex advantage* may be increased still further. The main technological options are (Fig. 13):

*Multi-slit systems* in which the field is mapped onto the detector via a mask. This admits only the light of pre-selected objects so that a spectrum is produced at the location of each slitlet. The slit mask is custom-made for each observation. Recent examples include GMOS (Hook et al. 2003; Table 2), VIMOS (Le Fevre et al. 2003) and DEIMOS (Faber et al. 2003).

*Multi-fibre systems* in which optical fibres are positioned on each target to guide the light to a fixed pseudo-slit at the spectrograph entrance. The fibres are positioned at the telescope focus typically by means of a *pick-and-place robot*. Because the fibres can be very long, this configuration is suitable for coupling gravitationally-invariant instruments to the telescope. Recent examples include 2DF (Lewis et al. 2002) and FLAMES (Pasquini et al. 2003).

### 2.2 Issues affecting MOS

**2.2.1 Overlaps between spectra.** If the distribution of targets in the field maps directly to the detector, as in a multislit system, the spectrum of one object may overlap that of another, leading to a restriction in the maximum surface density of targets which may be observed simultaneously. It is also

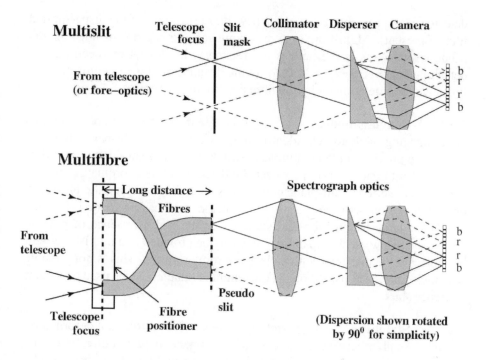

*Figure 13.* Illustration of the difference between multislit and multifibre spectrographs.

*Table 2.* Details of the Gemini Multiobject Spectrographs

| | |
|---|---|
| Image scale | 0.072 arcsec/pixel |
| Field of view | $5.5 \times 5.5$ arcmin |
| Detector | CCD: $3 \times (2048 \times 4608)$ pixels |
| wavelength range | $0.4 - 1 \ \mu$m |
| Slit width | 0.2 arcsec minimum |
| Resolving power | 5000 maximum with 0.5 arcsec slit |
| Flexure rate | 9(3) mas/hour in dispersion(spatial) direction |
| Velocity accuracy | 2km/s (projected) |
| Throughput of optics | ~80% |
| Modes | Single-slit, multislit and integral field spectroscopy |

possible for light of different orders from different objects to overlap. For example, at low dispersion where the spectrum are short, it may be possible to use multiple banks of slits (separated in dispersion) to increase the multiplex advantage. But this can lead to overlaps between the zero order of some objects and the first order spectra of other objects. The solution is careful mask design using software that accurately predicts the layout of the spectra on the detector taking account of the efficiency (blaze function) of the disperser (and filter if present). Multi-fibre systems do not suffer this problem since the distribution of targets in the field is not related to the layout of spectra on the detector. However they are limited by the distance of closest approach of the fibres which is determined by the fibre positioning system.

**2.2.2    Imaging mode.**    It is a good idea if the spectrograph includes a direct imaging mode to help identify targets (e.g. from multi-band photometry) and to provide contemporaneous astrometry (important due to the proper motion of reference stars). However the detector-frame coordinates must be carefully converted to slit-plane coordinates taking into account the inevitable optical distortion between the slit and detector. An imaging mode is also essential for target acquisition and station-keeping in the presence of guiding errors (see below). This capability can easily be provided in a multislit system by replacing the disperser with a clear aperture or mirror. This is not possible with multifibre systems, although coherent fibre bundles can be used to acquire reference stars.

**2.2.3    Stability.**    All spectrographs are subject to structural deformation due to thermal effects or, in the case of instruments mounted directly on the telescope, variation in the gravity vector. This can be divided into two parts.

*Before the slit.* Motion of the image delivered by the telescope with respect to the slit causes both a loss of throughput and an error in the barycentre of the spectral lines recorded on the detector, unless the object uniformly fills the slit (which implies low throughput). This can cause errors in measurement of radial velocities. For MOS, there is the particular problem of variations in the image scale or rotations of the mask. These can cause errors which depend on position in the field resulting in spurious radial trends in the data. Fibre systems are almost immune to this problem because the fibres scramble positional information.

*After the slit.* Flexure may cause the image of the slit to move on the detector. This can cause loss of signal/noise and errors in radial velocity determinations. The solution is to make sure that wavelength calibration observations are repeated more frequently than the timescale of the image motion although it is obviously better to minimise flexure to reduce calibration overheads. Active control of flexure is often required for larger instruments (e.g. GMOS: Murowinski et al., 2003). Thermally-induced errors can be reduced by eliminating heat sources within the instrument enclosure and insulating it to buffer external temperature changes.

**2.2.4    Calibration.**    *Wavelength calibration* to measure the *dispersion relation* between wavelength and position on the detector requires illumination by a lamp whose spectrum contains numerous well- defined spectral features. For MOS, this must be done using the same fibre or slitmask configuration as for the science observation since the dispersion relation changes slightly over the field. *Sensitivity calibration* requires illumination which is spatially-uniform but spectrally unstructured (e.g. from a quartz lamp). The instrumental response is a function not only of the detector but also of the specific dispersion and field configuration so the calibration exposure must be taken using the same setup as for the science observations. For both types of observation, the calibration light must be delivered to the instrument by a system which mimics the exit pupil of the telescope (e.g. GCAL on Gemini: Ramsay-Howat et al. 2000). Exposures of the twilight sky can be used to correct for large-scale inhomogeneities in the flatfield illumination provided by the calibration system. A common problem is detector *fringing* resulting from interference effects in the active layer of CCDs. Since the fringing pattern is sensitive to flexure, it is important to repeat calibrations *while tracking the target*. This is the big advantage of a calibration system which can deliver light at any time to the instrument without requiring a slew to acquire a particular calibration field.

**2.2.5    Atmospheric refraction.**    The apparent position of an object may be shifted vertically with respect to the true position due to refraction in the atmosphere. The shift is $A(\lambda) \tan \phi$ where $\phi$ is the zenith distance and $A \simeq 50$ arcsec. This has two effects.

**The chromatic component** , *atmospheric dispersion*, due to the wavelength-dependency of $A$, can be several arcsec over the wavelength range $0.4 - 1 \mu$m with most of the variation at the blue end. Unless the slit is placed vertically (at the *parallactic angle*), light at some wavelengths may miss the slit leading to spectrophotometric errors at wavelengths different from that used for guiding. The requirement for a vertical slit is inconvenient for MOS and, in any case, can only be satisfied at one time of the night, so a better option is to remove the dispersion using an *atmospheric dispersion compensator* placed before the instrument (Wynne and Worswick 1988).

**The non-chromatic component** produces a distortion $\simeq A\Psi \sec^2 \phi$ in the field, where $\Psi$ is half the field of view. In some parts of the sky the dominant effect is a rotation which can be corrected by the telescope control system. In other parts the effect is a squashing or stretching of the field which cannot be corrected by software. The effect is significant for very large fields or for smaller fields if high spatial resolution is also required. For MOS, the distortion can induce measurement errors in radial velocities which depend on both position within the field and the range of zenith distances at which the observations are made. For GMOS, to achieve its goal of measuring radial velocities to an accuracy of 2km/s *over a multiobject field*, requires a separate mask for each range of airmass, with the slit positions fine-tuned to match the expected distortion.

**2.2.6    Sky subtraction.**    With multislit systems, the use of slitlets allows the sky spectrum to be measured immediately adjacent to the object providing the slitlet is long enough. For multifibre systems, the sky must be estimated from the spectrum obtained from a different part of the field leading to possible errors in subtraction especially near to bright sky emission lines. An alternative is to alternately sample the sky and object with the same fibre and subtract the spectra. This must be done more frequently than the timescale for intrinsic variations in the background, especially at longer wavelength $(0.7 - 2\mu m)$ where OH emission from the atmosphere produces a forest of lines whose relative intensity varies on the timescale of minutes (Ramsay et al. 1992). These problems mean that multislit systems are often preferred to multifibre systems for observing very faint objects where quality of subtraction is critical. Various strategies have been proposed to alleviate this problem (e.g. Watson et al. 1998, see also Allington-Smith and Content 1998).

*Nod and shuffle* (see Glazebrook and Bland-Hawthorn, 2001 and references therein) makes use of the ability of some CCD controllers to move (shuffle) photoelectrons accumulated by a CCD from one part of the detector to another. If synchronised with a movement of the telescope (nod), the same detector pixels can be used to sample alternatively object and sky rapidly *without the overhead and loss of signal/noise involved in reading the detector before each nod.* This allows a reduction in the length of the slitlets since they no longer provide sky estimation at their ends. Thus the multiplex gain can be much increased with only a modest increase in overall exposure time.

# 3.    Integral field spectroscopy

## 3.1    Introduction

Integral field spectroscopy (IFS; Courtes 1982) is a technique to produce a spectrum for each point in a contiguous two-dimensional field, resulting in a datacube with axes given by the two spatial coordinates in the field and the wavelength. The advantages of IFS are as follows.

- Spectra are obtained for the full field simultaneously without the need for repeated pointings.
- There are no slit losses, unlike slit spectroscopy where the slit cannot be opened up without degrading spectral resolution.
- Target acquisition is easy: there is no need to position the object carefully on the slit.

- The position of the object in the field can be unambiguously determined *post facto* by summing the datacube in wavelength to produce a white-light image. In contrast, the position of the target with respect to the slit is difficult to determine in slit spectroscopy.
- By making use of the spatial information, the velocity field of an extended, structured object can be obtained unambiguously without errors caused by uncertainty in the position of a feature within the slit.
- Even in poor seeing, IFS acts as an image slicer to maximise the product of spectral resolution and throughput.

IFS is a specific example of a *3D* technique, which all produce datacubes. These include *Imaging Fourier Transform Spectroscopy* (IFTS) and *Fabry-Perot Spectroscopy* (FPS), tunable filters and stepped exposures with a longslit (see Bennett, 2000 for a general review). Unlike IFS, they all require multiple exposures to generate the datacube, involving scanning of a physical parameter of the instrument such as the Etalon spacing in an FPS. However, it can be shown that, *to first order*, all 3D techniques are of equivalent speed for the same number of detector pixels, $N_x N_y$. For example, compare IFTS and IFS. For IFTS, every pixel receives a signal obtained in time $t$ with a single setting of the scanning mirror position. After scanning through $N_s$ steps, the data series recorded in each pixel is transformed into a spectrum containing $n_\lambda$ samples, where for simplicity we assume $n_\lambda = N_s$. Thus the datacube obtained in time $N_s t$ contains $n_x n_y n_\lambda = N_x N_y N_s$ samples where $n_x = N_x$ and $n_y = N_y$ are the numbers of spatial samples in the field. For IFS, the same $N_x N_y$ pixels must record *all* the spectral *and* spatial information in one go, so $n'_x n'_y n_\lambda = N_x N_y$ where $n'_x, n'_y$ are the dimensions of the IFS field, which is much smaller than that of the IFTS since $n'_x n'_y = n_x n_y / N_s$. However in the same total time, a larger IFS field can be mosaiced from $N_s$ separate pointings to generate the same total number of samples, $N_x N_y N_s$. Thus the same number of (spatial×spectral) samples are produced in the same time.

From this, it can be seen that the optimum choice of 3D technique depends on the observing strategy appropriate to the particular scientific investigation. However there are other, second-order, considerations to be considered which violate this *datacube theorem*. These include whether the individual exposures are dominated by photon noise from the background or the detector, and the stability of the instrument and background over a long period of stepped exposures.

## 3.2    Techniques of IFS

There are many different techniques of IFS, summarised in Fig. 14. The first (e.g. Bacon et al., 2001) is the use of lenslet arrays to produce an image of the telescope pupil for every spatial sample. Because each pupil image is smaller than the extent of each spatial sample, this pattern of spots can be dispersed to produce a spectrum of each spatial sample. By tilting the angle of dispersion, the overlap between spectra can be minimised, although the number of spectral samples in each spectrum is necessarily limited. The second technique uses optical fibres to segment the image in the telescope focal plane and reformat the samples into a pseudo-slit at the entrance to the spectrograph (e.g. Allington-Smith et al., 2002). Lenslets (e.g. Lee and Allington-Smith, 2000) almost always have to be used to improve the coupling of light into (and sometimes out of) the fibres since the phenomenon of *focal ratio degradation* (e.g. Carrasco and Parry, 1994), a non-conservation of Étendue, is minimised at faster input focal ratios than that typically provided by the telescope. This technique allows the spectrum to cover many pixels since spectral overlaps are eliminated. This also reduces the number of wasted pixels between spectra so that the detector surface is used more efficiently than in fibreless techniques. The ultimate in efficient detector utilisation is the third technique in which the image is split into slices which are rearranged end to end to form a pseudo-slit (e.g. Thatte et al., 1997; Content, 1997; Dubbeldam et al., 2000). Because there are no unused pixels between samples in the spatial direction (except between slices), the detector utilisation can approach 100%. This improvement in performance is bought at the expense of a complex optical system requiring multifaceted mirrors made by techniques such as diamond-turning in metal. However the avoidance of fibres brings the benefit of easier use in cold environments which makes slicer systems a better choice for cooled infrared instruments.

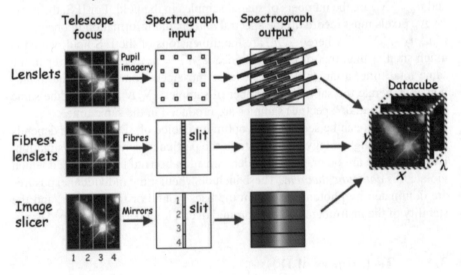

*Figure 14.*   The main techniques of integral field spectroscopy.

## 3.3 Sampling with fibre-lenslet systems

For *image slicers*, contiguous slices of the sky are re-arranged end-to-end to form the pseudo-slit. In that case it is obvious that the sky can be correctly sampled (according to the Nyquist sampling theorem) by the detector pixels on which the slices are projected in the same way as required for direct imaging.

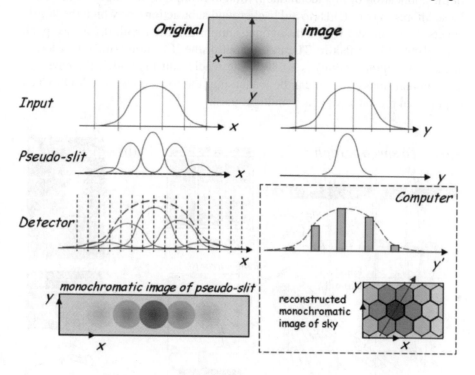

*Figure 15.* Illustration of the way in which the image is sampled in two dimensions by a fibre-lenslet IFU. In the direction in which the fibres are ordered at the slit ($x$), the distribution of light in the field, pseudo-slit and detector are shown in turn. At the slit, the finite width of the images produced by the fibres contribute only marginally to the overall width of the reconstructed image.

It is less obvious that this principle can also be applied to *fibre-lenslet* systems. At first sight, it might seem preferable to keep the output of each fibre independent of its neighbour by preventing the images produced by each fibre at the slit from overlapping. However this would severely limit the number of fibres which could address a given detector, so some overlapping must be tolerated. The situation is shown in Fig. 15 where it can be seen that the over-sampling is determined by the way that the image produced by the telescope is sampled by the lenslets at the fibre *input*. Subsequently, it is permissible for the images produced by each fibre at the pseudoslit to overlap *provided that* fibres which are adjacent at the slit are also adjacent on the sky and that the fibre-to-fibre response is reasonably uniform. Allington-Smith and Content, 1998 show that the only effect is a slight degradation in spatial resolution in the dimension parallel to the slit.

The *integral field unit* (IFU) built for GMOS serves as a successful example of the application of this technique (Allington-Smith et al., 2002). It has 1500 lensed fibres in two fields (to aid background subtraction), of which the largest covers $7 \times 5$ arcsec, with a hexagonal sampling pattern with 0.2 arcsec pitch and $\sim$100% filling factor. The throughput of the IFU alone (including losses at the spectrograph pupil) is $\sim$65%. The whole unit fits within the envelope of a multislit mask and so can be loaded into the focal plane of GMOS when required, in the same way as a multislit mask.

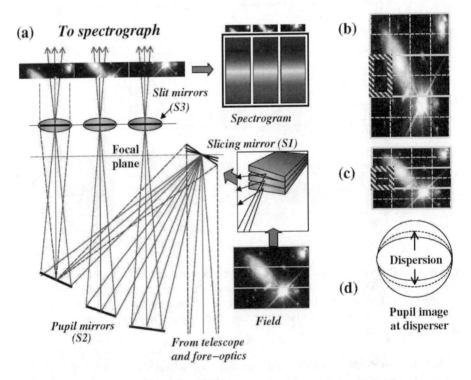

*Figure 16.* (a) Illustration of the principle of the *Advanced Image Slicer* (Content 1997) showing just 3 slices. For clarity the S3 mirrors are represented as transmission optics. (b) shows how the sky image may be anamorphically magnified before the IFU so that each slice can project to a width of two pixels (detector pixel boundaries shown as dashed lines) so that the sampling element (hatched outline) is square when projected on the sky as in (c). (d) shows how this squeezes the beam (ellipse) at the spectrograph pupil (circle) in such a way as to give extra room for the effect of diffraction (dashed ellipse).

## 3.4    Advanced Image slicers

The optical principle of an image slicer is shown in Fig. 16. This is an example of the *advanced* type proposed by Content (1997). The sky is imaged onto a set of inclined *slicing mirrors* (S1) which reflect the light striking each slice to a set of *pupil mirrors* (S2). The slicing mirrors are designed to form an image of the telescope pupil on the S2 mirrors. This means that the S1 mirror facets must be curved instead of flat as in early designs, but greatly reduces the lateral extent of the S2 mirrors leading to a reduction in the overall size of the instrument. The S2 mirrors are also figured so as to reimage the field onto the *slit mirrors*, S3, to form the pseudo-slit which consists of images of each slice arranged end to end. Finally, the S3 mirrors are also powered so as to reimage the telescope pupil onto the stop of the spectrograph (normally at the dispersing element).

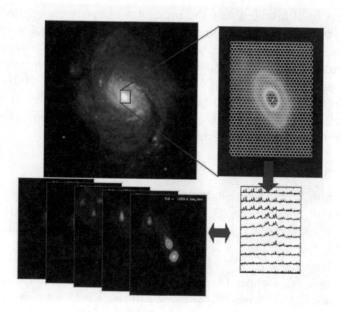

*Figure 17.*    Steps in the construction of a datacube of the nucleus of NGC1068 from observations with the integral field unit of the Gemini Multiobject Spectrograph installed on the Gemini-north telescope. The datacube is illustrated by a few spectra distributed over the field and equivalently by a few slices at a given radial velocity in the light of the [OIII]5007 emission line. Only a few percent of the total data content is shown.

An advantage of this arrangement is that the original image is resampled only in one dimension rather than the two of the other techniques. This means that diffraction (which is significant at wavelengths where the angular sampling is $\sim \lambda/d$ where $d$ is the physical width of the slice) broadens the beam emerging from the IFU only in the dispersion direction. This helps to improve throughput, particularly in the infrared. By introducing anamorphic magnification (e.g. $2\times$) before the IFU, the sky image is stretched in the dispersion direction so that the width of each slice projects onto 2 detector pixels. This allows the slit to be correctly oversampled in the dispersion direction, while allowing the spatial element, defined by one detector pixel in the spatial direction and the width of the slice in the dispersion direction, to be conveniently square when projected on the sky. This also brings the benefit of reducing the width of the beam at the spectrograph pupil by a factor of 2 so that the beam spreading produced by diffraction can be easily accommodated within the circular stop of a conventional design of spectrograph.

## 4.    Conclusion

This paper has illustrated how the principles of optics are applied to astronomical spectroscopy. I can think of no better way to conclude than by showing just one example of the astrophysical product. Fig. 17 shows an example of the construction of a datacube for the nucleus of the active galaxy NGC1068 using the IFU of GMOS-N (Allington-Smith et al., 2002). With proper exploitation of the aperture of current telescopes and the construction of 30-100m telescopes, data of this quality will be available routinely for much fainter and more distant objects than this.

## References

Allington-Smith, J., and Content, R. 1998, PASP **110**, 1216

Allington-Smith, J.R., Murray, G., Content, C., Dodsworth, G., Davies, R.L., Miller, B.W., Jorgensen, I., Hook, I., Crampton, D. & Murowinski, R., 2002. PASP, **114**, 892

Barden, S., Arns, J., Colburn, W. and Williams, J., 2000. PASP, **112**, 809

Bacon. R., Copin, Y., Monnet, G., Miller, B.M., Allington-Smith, J.R., Bureau, M., Carollo, C.M., Davies, R.L., Emsellem, E., Kuntschner, H., Peletier, R.F., Verolme, E.K. & de Zeeuw, T., 2001, MNRAS **326**, 23

Bennett, C., 2000, in *Imaging the universe in three dimensions*, ASP Conference Series, **195**, W. van Breugel and J. Bland-Hawthorn, eds.

Carrasco, E. and Parry, I., 1994. MNRAS **271**, 1

Content, R., 1997, SPIE, **2871**, 1295

Courtes, G. 1982, in *Instrumentation for Astronomy with Large Optical Telescope*, Proc. IAU Colloq 67, Astrophysics and Space Science Library **92**, p123. ed. Humphries, C.M., Dordrecht, Reidel

Dubbeldam, C.M., et al., 2000, SPIE **4008**, 1181

Faber, S. et al. 2003, SPIE **4841**, 1657

Glazebrook, K. and Bland-Hawthorn, J., 2001. PASP **113**, 197

Hook, I. et al. 2003, SPIE **4841**, 1645

Jenkins, F. and White, H., 1976, Fundamentals of Optics, 4th Edn, McGraw Hill

Lee, D. and Allington-Smith, J.R., 2000, MNRAS, **312**, 57

Lee, D., Haynes, R., Ren, D., and Allington-Smith, J. R., 2001, PASP **113**, 1406

Le Fèvre, O. et al. 2003, SPIE **4841**, 1670

Lewis, I. et al., 2002. MNRAS **333**, 279

Murowinski, R. et al. 2003, SPIE **4841**, 1189

Palmer, C., 2000, *Diffraction grating handbook* 4th edn., Richardson Grating Laboratory, Chap. 9

Pasquini, L., et al. 2003, SPIE **4841**, 1682

Ramsay, S.K., Mountain, C.M. and Geballe, T.R., 1992, MNRAS **259**, 751

Ramsay-Howatt, S., Harris, J., Gostick, D., Laidlaw, K., Kidd, N., Strachan, M., & Wilson, K., 2000, SPIE **4008**, 1351

Schroeder, D., 2000, *Astronomical Optics*, 2nd edn, Academic Press

Sheinis, A. et al. 2002. PASP **114**, 851

Thatte, N., Genzel, N., Kroker, H., Krabbe, A., Tacconi-Garman, L., Maiolino, R. & Tecza, M., 1997, ApSS, 248, 225

Watson, F. et al., 1998, in *Fiber optics in astronomy*, ASP Conference Series **152**, 50

Wynne, C. and Worswick, S., 1986. MNRAS **220**, 657

# ADAPTIVE OPTICS; PRINCIPLES, PERFORMANCE AND CHALLENGES

## Adaptive Optics

Nicholas Devaney

*GRANTECAN Project*
*Instituto de Astrofisica de Canarias 38200 La Laguna, Tenerife, SPAIN*
Nicholas.Devaney@iac.es

**Abstract**     There are several good reviews of adaptive optics (AO) as well as a few text-books (Hardy(1988), Roddier(1999), Tyson(2000)). The proceedings of earlier Cargese Summer schools also serve as a good introduction. While it is necessary to briefly explain the principles of AO, I will try to emphasize some particular aspects in more detail than the existing reviews. The more recent developments in AO will also be examined in some detail. These include Multi-Conjugate AO (MCAO) and the development of AO for the next generation of Extremely Large Telescopes (ELTs).

**Keywords:**     adaptive optics, multiconjugate adaptive optics, extremely large telescopes

## 1.     INTRODUCTION

One of the basic tasks of observational astronomy is to make images (or 'maps') of distant sources. The angular (and hence spatial) resolution of these images will depend on the optical quality of the telescope and camera we use and at best it will be limited by diffraction at the entrance aperture. This diffraction-limited resolution is given approximately by $\lambda/D$ radians, where $\lambda$ is the wavelength of observation and $D$ is the aperture diameter. Using the fact that 1 arcsecond corresponds to approximately 4.85 $\mu$rad, we could expect a 10m telescope to have a resolution of 0.01 arcsecond when observing at 0.5 $\mu$m. However, no matter how carefully we make the telescope we never obtain resolution which is this good. In fact, the resolution will be in the range of approximately 0.3-3 arcseconds and will not depend on the telescope diameter if the diameter is larger than about 10cm. This is due to the effects of the Earth's

181

*R. Foy and F. C. Foy (eds.), Optics in Astrophysics, 181–206.*

turbulent atmosphere on the amplitude and phase of electromagnetic waves passing through it. The effect of atmospheric turbulence on images is described in several texts and also in this book (Ch. 1). Here we repeat the definitions of the parameters of turbulence which are of most importance to adaptive optics. The most important parameter is $r_0$, the Fried parameter, which can be defined as the radius of a circle over which the mean square wavefront variance is 1 $rad^2$. A telescope of diameter less than $r_0$ will be diffraction-limited when observing through the turbulence, while a telescope of diameter larger than $r_0$ will be seeing-limited i.e. the resolution will be poorer than the diffraction limit. The value of $r_0$ depends on the integrated strength of turbulence along the line of sight. It also depends on the zenith angle of the observation, since the effective path through the turbulence is longer at higher zenith distance. Since it is defined in terms of radians, it will also depend on the wavelength of observation. For Kolmogorov turbulence the expression for $r_0$ is as follows:

$$r_0 = \left[ 0.423k^2 (\cos \gamma)^{-1} \int C_n^2(h) dh \right]^{-\frac{3}{5}}$$

where $k = 2\pi/\lambda$ is the wavenumber, $z$ is the zenith angle and $C_n^2$ is the turbulence structure function. The dependence of $r_0$ on wavelength and zenith angle are $r_0 \propto \lambda^{\frac{6}{5}}$ and $r_0 \propto (\cos \gamma)^{-\frac{3}{5}}$. As useful figures of thumb, the value of $r_0$ increases by a factor of three from the visible to the J band (1.25 $\mu$m) and a factor of 6 from the visible to the K band (2.2 $\mu$m). The full-width at half-maximum of a seeing-limited image is given by

$$fwhm_{seeing} \cong 0.97\lambda/r_0$$

which implies that the full-width at half maximum depends on wavelength as $\lambda^{-\frac{1}{5}}$ and is therefore slightly better in the near infrared than in the visible.

Atmospheric turbulence is a dynamic process and the wavefront aberrations are constantly changing. A characteristic timescale may be defined as the time over which the mean square change in wavefront error is less than 1 $rad^2$. If the turbulence were concentrated in a single layer with Fried parameter $r_0$ moving with a horizontal speed of $v$ ms$^{-1}$ then the characteristic time, $\tau_0$, is given by

$$\tau_0 = 0.314r_0/v$$

For example, a layer with $r_0 = 10$ cm at $0.5\mu$m moving at a speed of 10 m/s will give rise to wavefront errors having $\tau_0 = 3.1$ms at 0.5 $\mu$m. This will scale with wavelength in the same way as $r_0$, so that the corresponding value of $\tau_0$ at 2.2 $\mu$m is approximately 19 ms.

Another parameter of the turbulence which strongly influences the performance of an adaptive optics (AO) system is the isoplanatic angle. The light from stars in different parts of the telescope field of view will pass through different parts of the turbulence. At a given layer of turbulence, the degree of overlap of the beams corresponding to different stars depends on their separation and on the height of the layer. When the mean square difference in the wavefront errors corresponding to two stars is equal to 1 rad$^2$ then the stars are separated by the isoplanatic angle. For a single layer with a given value of $r_0$ at a height $h$, the isoplanatic angle is given by

$$\theta_0 = 0.314 r_0 \cos(\gamma)/h$$

For example, a layer of $r_0 = 10$ cm at 0.5 $\mu$m at an altitude of 1 km above a telescope observing at a zenith angle of 30° will give rise to an isoplanatic angle of 5.6 arcseconds. This would be considered a very small field of view for most astronomical applications. The isoplanatic angle has the same wavelength dependence as $r_0$ so that the corresponding value of $\theta_0$ at 2.2 $\mu$m would be $\approx$34 arcseconds, a much more useful value.

## 2. AO Principle

Since the image is degraded by rapidly varying phase and amplitude errors introduced by passing through turbulence, it should be possible to recover the degradation by measuring and correcting the phase and amplitude of the wavefronts before they form the final image. This is the basic idea of adaptive optics. It can be shown (Roddier, 1981) that the effects of amplitude errors on image quality should be negligible at good astronomical sites when observing in the infrared. Amplitude errors are introduced by the propagation of phase errors, and it is usually true that the telescope is in the near-field of the turbulence i.e. the propagation distance is short enough to ensure that the amplitude errors are small. For this reason, all AO systems built up to now measure and correct phase errors alone.

Figure 1 outlines the basic AO system. Wavefronts incoming from the telescope are shown to be corrugated implying that they have phase errors. Part of the light is extracted to a wavefront phase sensor (usually referred to as a wavefront sensor, WFS). The wavefront phase is estimated and a wavefront corrector is used to cancel the phase errors by introducing compensating optical paths. The most common wavefront compensator is a deformable mirror. The idea of adaptive optics was first published by Babcock (1953) and shortly after by Linnik (1957).

## 2.1 Modal Correction

The random phase error of a wavefront which has passed through turbulence may be expressed as a weighted sum of orthogonal polynomials. The usual set of polynomials for this expansion is the Zernike polynomials, which

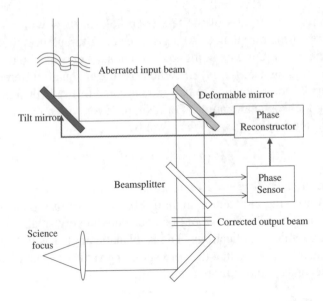

*Figure 1.*    Block Diagram of Adaptive Optics system.

are orthogonal on a circle of unit radius The Zernike polynomials are most concisely expressed in polar coordinates, whence they may be decomposed into radial and azimuthal terms;

$$Z_n^m = \sqrt{n+1} R_n^m(r) \left\{ \begin{array}{c} \sqrt{2}\cos(m\theta) \\ \sqrt{2}\sin(m\theta) \\ 1(m=0) \end{array} \right)$$

The index $n$ is the radial degree and the index $m$ is the azimuthal frequency. The function $R_n^m(r)$ is a radial function given by

$$R_n^m(r) = \sum_{s=0}^{(n-m)/2} \frac{(-1)^s (n-s)!}{s!\,[(n+m)/2 - s]!\,[(n-m)/2 - s]!} r^{n-2s}$$

The first Zernike polynomials correspond to familiar Seidel aberrations (piston, tip, tilt, defocus, astigmatism, coma) although modified to ensure orthogonality (e.g. Zernike coma includes some tip or tilt). We can consider that measurement and correction of Zernike polynomials become more difficult as one goes to higher orders; tip and tilt can be corrected by a simple tilting mirror, defocus could be corrected by a vibrating membrane while the wavefront corrector has to assume increasingly complicated shapes in order to correct higher orders. It is therefore of interest to determine how well a wavefront is corrected if we only correct the first $j$ polynomials of its Zernike expansion.

Noll (1976) determined the residual wavefront variance as a function of the number of corrected terms.

$$\Delta_j = \alpha_j \left(D/r_0\right)^{5/3}$$

and the coefficients $\alpha_j$ decrease as a function of $j$, the order of correction. In order to determine how image quality depends on the degree of correction, we consider first the structure function of the wavefront phase errors.

The phase structure function for a separation $r$ is defined as the mean value of the square difference of phase for all points with that separation. For wavefronts affected by Kolmogorov turbulence it is given by

$$D_\phi(r) = 6.88 \left(r/r_0\right)^{5/3}$$

and so it increases without limit as the distance $r$ increases. Using simulations it is possible to examine what happens to the structure function when the wavefront phase errors are corrected up to different orders of correction. The simulation generates phase screens corresponding to Kolmogorov turbulence as sums of Zernike polynomials. It is therefore simple to generate phase screens with Zernike polynomials removed. Figure 2 shows the phase structure function with no correction, and correction of 10, 20 and 40 Zernike polynomials for a case of $D/r_0 = 10$. The following behavior is observed; when there is correction of Zernike terms, the structure function no longer rises without limit, rather it saturates at a value for distance $r$ larger than a distance $l_c$. This distance is smaller as the degree of correction increases. From the definition of the structure function it can be seen that the value to which the structure function saturates is equal to twice the residual variance, $\Delta_j$. Finally, it can be seen that at distances shorter than $l_c$ the structure function has the same functional behavior as in the uncorrected case, but with a larger effective value of $r_0$. This behavior has been studied in detail by Canales and Cagigal (1996), and they define a 'Generalised Fried parameter', $\rho_0$, such that for $r < l_c$

$$D_\phi(r) = 6.88 \left(r/\rho_0\right)^{5/3}$$

The Optical Transfer Function (OTF) is related to the phase structure function as follows (Roddier,1981)

$$OTF = OTF_{Tel} \exp\left(-D_\phi/2\right)$$

where $OTF_{Tel}$ is the OTF of the diffraction-limited telescope. The Point Spread Function (PSF), which is the image of a point source, is related to the OTF by a Fourier transform. When the wavefront is partially compensated, the saturated region of the structure function leads to a plateau at high frequencies in the OTF which in turn gives rise to a sharp peak in the PSF. The amount of

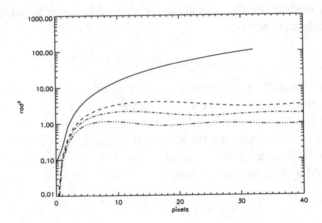

*Figure 2.*   Phase structure function with no correction (solid), 10 (dashed), 20 (dot-dash) and 40 (triple-dot-dash) Zernike polynomials corrected.

coherent energy in the peak is given by $\exp(-\Delta_j)$ times the total energy. The un-saturated region of the structure function ($r < l_c$) gives rise to a broad halo in the PSF. The energy in the halo is $(1 - \exp(-\Delta_j))$ times the total energy and the width of the halo is given by

$$\omega_h = 1.27\lambda/\rho_0$$

This result is very interesting; as we increase the degree of compensation, the amount of energy in the coherent diffraction-limited core increases and the size of the residual halo decreases. This behavior can be seen in Fig. 3.

Since the PSF of partially corrected images is not simple (like a Gaussian, for example) it cannot be completely characterized by a single parameter. However, a very useful parameter is the Strehl ratio, defined as the ratio of the peak intensity measured to that which would be provided by a perfect system. When the AO correction is good, then the Strehl ratio is given by $\exp(-\Delta_j)$, which ignores the contribution of the halo. When the correction is low order, then the coherent peak is negligible and the Strehl ratio is given by $(\rho_0/D)^2$ .The fact that AO correction leads to a concentration of energy in the core of the PSF means that much fainter objects can be detected when AO is used, providing that the AO system has high throughput. The calculation of the gain in sensitivity to point sources is straightforward and presented in other AO texts (e.g. Hardy,1998). In the case in which the main source of noise in the image is that due to the background (e.g emission from the sky and thermal emission from the telescope), the gain in signal-to-noise ratio for point source detection is given by

$$G = SRD/r_0.$$

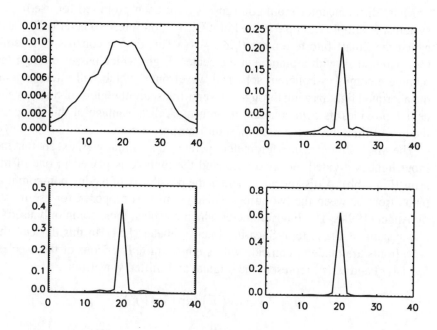

*Figure 3.* Point spread function section with no correction (top left), 0, 20 and 40 (bottom right) Zernike polynomials corrected for $D/r_0=10$. The pixel size is $\lambda/2D$ and Strehl ratio is given by the peak value.

For example, on a 10 m telescope with $r_0 = 1$ m a Strehl ratio of 0.6 corresponds to a gain of 6. On a 50 m telescope the corresponding gain would be 30. The limiting magnitude for point source detection is increased by $2.5 \log (G)$, so that the limiting magnitude of a 50 m telescope would be increased by 3.7. The time required to reach a given magnitude will be proportional to $1/G^2$, so that efficiency of a large telescope will be greatly enhanced by adaptive optics.

## 2.2 Wavefront Sensing

In the early days of Adaptive Optics, some systems controlled the deformable mirror using measurements made on the corrected image e.g. the shape of the deformable mirror is 'dithered' to maximize a measurement of image sharpness (Buffington et al., 1976). Nowadays, however, the vast majority of AO systems send part of the input light to a wavefront phase sensor. These wavefront sensors can be divided into different types according to different criteria. They can be divided into pupil plane and image plane sensors depending on whether the detector is placed at a pupil or image plane. We consider dividing wavefront sensors into Interferometric and Propagation type sensors.

Many different interferometric sensors have been proposed for use in AO systems. A simple example is the Point Diffraction Interferometer (PDI), also called the Smart Interferometer (Fig. 4ba). This device consists of a semi-transparent disk with a pinhole at the center. Light which passes through the pinhole emerges as coherent spherical wavefronts, which will interfere with the aberrated light passing through the semi-transparent region. In this way the device provides its own coherent reference. A disadvantage of this device is the loss of light due to the use of a semi-transparent plate.

The Mach-Zehnder Interferometer operates in a similar way, except that the input light is divided into two arms and the pinhole is placed in one of the arms (Fig. 4bb). A disadvantage here is the possibility of having non-common path errors between the two arms. This system was proposed for use in AO by Angel (1994). The Interferometer which has been most commonly used in AO systems is the Lateral Shearing Interferometer (LSI). In this device , the wavefronts are made to coincide with slightly shifted versions of themselves. Let $u1(r)$ and $u2(r)$ represent the original and shifted wavefronts.

$$u_1\left(\overrightarrow{r}\right) = a_1\left(\overrightarrow{r}\right)\exp\left(i\phi_1\left(\overrightarrow{r}\right)\right)$$

$$u_2\left(\overrightarrow{r}\right) = a_1\left(\overrightarrow{r} + \overrightarrow{d}\right)\exp\left(i\phi_1\left(\overrightarrow{r} + \overrightarrow{d}\right)\right)$$

When the shear distance is small, then the following Taylor expansion can be made

$$\phi_1\left(\overrightarrow{r} + \overrightarrow{d}\right) \approx \phi_1\left(\overrightarrow{r}\right) + \nabla\phi_1\left(\overrightarrow{r}\right)\cdot\overrightarrow{d} + \cdots$$

and the resulting intensity in the interference pattern is given by

$$I\left(\overrightarrow{r}\right) \approx \tag{1}$$

$$\left|a_1\left(\overrightarrow{r}\right)\right|^2 + \left|a_1\left(\overrightarrow{r} + \overrightarrow{d}\right)\right|^2 + \tag{2}$$

$$2\left|a_1\left(\overrightarrow{r}\right)\right| \times \left|a_1\left(\overrightarrow{r} + \overrightarrow{d}\right)\right|\cos\left(\nabla\phi_1\left(\overrightarrow{r}\right)\cdot\overrightarrow{d}\right) \tag{3}$$

This shows that the spacing of the fringes gives a measure of the gradient of the phase in the direction of the shear. It is necessary to also shear the wavefront in a perpendicular direction in order to recover the two-dimensional phase gradient. The sensitivity of the interferometer can be adjusted by changing the shear distance. An advantage of this device is that it does not require a coherent reference (since the wavefront is interfered with itself). However, unlike the PDI and Mach-Zehnder it measures the gradient of the phase errors rather than directly measuring the errors. In practical LSI, the shear is provided by a diffraction grating; the interference pattern is observed in the region of overlap of the zeroth and first order diffracted wavefronts.

Most WFSs in use in AO systems rely on propagation, i.e. converting phase errors into measurable intensities. A simple example of the underlying principle is the lens; the presence of a focused image implies a quadratic phase component in the wavefront. The intensity variation along the direction of propagation $(z)$ of a beam having phase $\phi(x, y)$ is given by the Irradiance Transport Equation, which is simply the continuity equation for light beams:

$$\partial I/\partial z = -\left(\nabla I \cdot \nabla \phi + I\nabla^2\phi\right)$$

the first term represents the transverse shift of the beam due to local wavefront tilt while the second term may be interpreted as the change in irradiance due to focusing or defocussing of the beam caused by local wavefront curvature. The intensity variations can be enhanced by placing a mask at one plane and measuring the intensity at another plane. An example of this is the Shack-Hartmann wavefront sensor which is a refinement of the Hartmann sensor. In the Hartmann sensor an array of apertures is placed in one plane and the resulting pattern of spots recorded (usually at the focal plane). The position of each spot is linearly related to the wavefront slope over the corresponding aperture. In Shack's modification the array of apertures is replaced by an array of lenses, which is more versatile and allows the whole wavefront to be measured. The spot positions are usually determined simply by measuring the centroides $x_n$

$$x_n \equiv I_n^{-1} \int \overrightarrow{x}\, I(\overrightarrow{x})d^2x = fI_n^{-1} \int a^2(\overrightarrow{r})\left[\nabla\phi(\overrightarrow{r})\right]d^2\overrightarrow{r}$$

where $I_n$ is the total intensity corresponding to the $n^{th}$ subaperture, $f$ is the microlens focal length, and $a(r)$ is the amplitude of the wavefront. This equation shows that amplitude variations will affect the centroide measurements. Voitsekhovich et al. (2001) have shown that the effect is small.

Practical Shack-Hartmann wavefront sensors employ microlens arrays, where the physical size of the microlenses is typically in the range 0.1-0.5mm. The spot images are usually formed on a CCD detector. The focal length of the microlenses is determined by the required image scale and the pupil demagnification factor i.e. the ratio of the size of the telescope entrance pupil to the size of the microlens array. Using the Lagrange Invariant, the following relation can be determined:

$$f = pD_{array}/D_{pupil}$$

where $D_{array}$ is the size of the microlens array, $D_{pupil}$ is the size of the telescope entrance pupil and $p$ is the required image scale (in m per radian). For example, if we require that pixels of side 13 $\mu$m correspond to 0.5 arcsecond when we map a 10m aperture onto a 3mm microlens array, then the microlens focal length should be 1.61mm. This length is very short and implies either that the microlens array has to be placed inside the detector housing or re-imaging optics has to be placed between the microlens array and the detector.

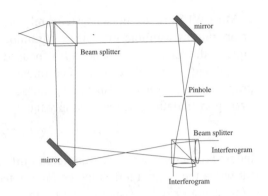

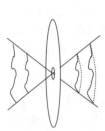

*Figure 4a.* Point Diffraction Interferome-          *Figure 4b.* Mach Zehnder Interferometer.
ter.

The idea of using propagation of intensity to measure wavefront errors is applied more directly in the curvature sensor. Conceptually, the irradiance is measured at two planes which are equally spaced about the telescope focus and the normalized difference in intensity obtained.

$$\Delta I \equiv (I_2 - I_1)/(I_2 + I_1)$$

It can be derived from the transport of intensity equation that this signal has two terms; at the edges of the pupil it is proportional to the local wavefront gradient in the direction normal to the pupil edge (i.e. radial in the case of a circular pupil) and elsewhere in the pupil it is proportional to the local wavefront curvature (Roddier, 1988). The signal is more intense when the planes are nearer the telescope focus, but diffraction will limit the spatial resolution more. Thus there is a trade-off between resolution and signal-to-noise (see Ch. 24).

In real curvature sensors, a vibrating membrane mirror is placed at the telescope focus, followed by a collimating lens, and a lens array. At the extremes of the membrane throw, the lens array is conjugate to the required planes. The defocus distance can be chosen by adjusting the vibration amplitude. The advantage of the collimated beam is that the beam size does not depend on the defocus distance. Optical fibers are attached to the individual lenses of the lens array, and each fiber leads to an avalanche photodiode (APD). These detectors are employed because they have zero readout noise. This wavefront sensor is practically insensitive to errors in the wavefront amplitude (by virtue of normalizing the intensity difference).

The pyramid wavefront sensor has recently been proposed (Ragazzoni, 1996). A four-sided pyramid is placed with its vertex at the telescope focus, and following the pyramid is some optics to image the pupil plane onto a detector.Since a four-sided prism is employed, four pupil images will be formed on the detector. In the absence of aberrations, the four pupil images will be uniformly illuminated. If we consider a local phase gradient (Fig. 5) then the aberrated ray will miss the vertex of the pyramid and illuminate only one of the pupil images. By comparing the pupil images it is therefore possible to see that at a particular position in the pupil there is a gradient in the $x$ (and/or) $y$ direction. However we do not have a measure of the gradient. This can be obtained by modulating the position of the input beam on the pyramid. If we consider an aberrated ray which strikes the pyramid as shown in Fig. 6 then circular modulation of either the pyramid or the beam will cause the ray to trace out a circle over the sides of the pyramid. The time spent on each face, and hence the intensity at the corresponding point in each of the four pupil images, will depend on the ray aberration. If we consider integrating four pupil images,$I_1, I_2, I_3$ and $I_4$ over at least one modulation cycle, then signals $S_x$ and $S_y$ can be defined as

$$S_x(x,y) = (I_1(x,y) + I_4(x,y)) - (I_2(x,y) + I_3(x,y)) / \sum I_i$$

$$S_y(x,y) = (I_1(x,y) + I_2(x,y)) - (I_3(x,y) + I_4(x,y)) / \sum I_i$$

It can be shown using geometrical optics that these signals are related to the wavefront gradient by

$$\partial W/\partial x \propto R\sin(S_x) \text{ and } \partial W/\partial y \propto R\sin(S_y)$$

Where $R$ is the modulation amplitude. It can be seen that the sensor response is approximately linear in the regime of small wavefront slopes. It can also be seen that the sensitivity of the wavefront sensor is inversely proportional to the modulation amplitude. It has been argued that in closed-loop AO the modulation amplitude can be reduced in order to increase the sensitivity of the sensor (Ragazzoni and Farinato, 1999), and that this is a fundamental advantage over the Shack-Hartmann sensor. However, in this regime the sensor will not be sensitive to large wavefront gradients which could be caused by non-compensated high spatial frequency errors.

## 2.3 Wavefront Correction

In order to compensate for the distortions in the wavefront due to the atmosphere we must introduce a phase correction device into the optical beam. These phase correction devices operate by producing an optical path difference in the beam by varying either the refractive index of the phase corrector (refractive devices) or by introducing a variable geometrical path difference (reflective devices, i.e. deformable mirrors). Almost all AO systems use deformable mirrors, although there has been considerable research about liquid crystal devices in which the refractive index is electrically controlled.

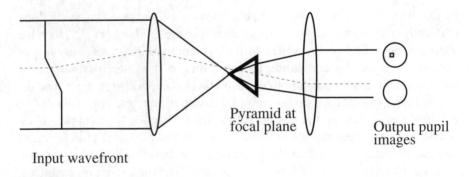

*Figure 5.* Pyramid wavefront sensor concept. Wavefront tilt causes the corresponding zone to be brighter in one pupil image and darker in the other.

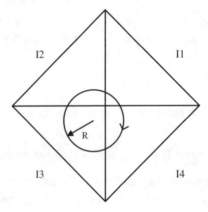

*Figure 6.* Path of aberrated ray on pyramid when circular modulation is applied. The offset of the circle from the pyramid vertex is proportional to the local wavefront slope.

The main characteristics which determine the performance of a wavefront corrector are the number of actuators, actuator stroke and the temporal response. The number of actuators will determine the maximum Strehl ratio which can be obtained with the AO system. The price of a deformable mirror is directly related to the number of actuators. The actuator stroke should be enough to compensate wavefront errors when the seeing is moderately poor. This can be derived from the Noll formula with $\alpha_0 = 1.03$. For example, on a 10m telescope with $r_0 = 0.05$m at 0.5 $\mu$m, the rms wavefront error is 6.7 $\mu$m. The deformable mirror stroke should be a factor of at least three times this. It should also include some margin for correction of errors introduced by the telescope itself. The required stroke is too large for most types of deformable mirror, and it is common practice to off-load the tip-tilt component of the wavefront error to a separate tip-tilt mirror. The Noll coefficient $\alpha_2 = 0.134$ and the rms wavefront stroke is then 2.4 $\mu$m. Note that the physical stroke required is independent of the wavelength being corrected.

Deformable mirrors basically consist of an array of electromechanical actuators coupled to a reflective faceplate. The actuators employed in deformable mirrors are usually either piezoelectric or electrostrictive. Piezoelectric materials expand when a voltage is applied. The most commonly used material is lead zirconate titanate, Pb(Zr,Ti)O3, usually referred to as PZT. The material is 'poled' by subjecting it to a strong field at 150°C and cooling. This process aligns the dipoles parallel to the applied field. The response of poled PZT to applied voltage is linear and bipolar (i.e. can be made to both expand and contract). The main disadvantage of PZT is that it exhibits 10-20% hysterisis. This means that when a voltage is applied and then removed, the actuator does not return to exactly the same length it had before the voltage was applied. The effect in a closed-loop AO system is to reduce the effective bandwidth. Electrostrictive materials exhibit a strain which is proportional to the square of the applied field. Lead magnesium niobate (PMN) is a commonly used electrostrictive material which shows a large electrostrictive effect. PMN actuators are generally composed of multiple thin layers in order to maximise the available stroke. A bias voltage has to be applied in order to allow bipolar operation. The optimal working temperature of PMN is around 25°C at which temperature its hysteresis is less than 1%. However the characteristics of PMN are strongly temperature dependent; the sensitivity and hysteresis of the material increases as its temperature is decreased. At about −10°C the hysteresis is ∼15 % and the sensitivity is twice the level of that at 25°C.

The faceplate is usually continuous, although some systems have employed segmented deformable mirrors in which groups of three actuators are attached to mirror segments. This has the disadvantage of requiring measurements to ensure that there are no piston errors between the segments. An alternative type of deformable mirror is the bimorph mirror, in which electrodes are placed between two sheets of oppositely poled actuator material. Applying a voltage to an electrode causes one sheet to contract and the other sheet to expand in the vicinity of the electrode. This results in a local curvature of the deformable mirror surface, and this type of mirror is usually (though not exclusively) employed in systems having a curvature sensor. A disadvantage of bimorph mirrors is that it is difficult to have a large number of actuators; if the electrode size is reduced then the bimorph thickness has also to be reduced in order to maintain stroke.

Deformable mirrors are usually placed at a greatly reduced image of the telescope entrance pupil - the typical diameter of PZT or PMN based deformable mirrors is in the range 10-20 cm. A completely different approach is to make one of the telescope mirrors deformable, and the best choice is the secondary mirror (relatively small and usually coincides with the aperture stop for infrared observations). The main advantage of this approach is that it does not increase the emissivity of the system and thereby allows detection of fainter objects in background limited operation. Equivalently, it allows a given limiting magnitude to be reached in less time. If the secondary mirror is adaptive then adaptive correction is available at any telescope focus (except prime) equipped with a wavefront sensor. The disadvantages are the increased cost and complexity when compared to standard deformable mirrors. Maintenance is also more difficult. Since the mirror has to be used when the telescope is not using adaptive optics (unless provision is made for interchange of secondary mirrors) the mirror has to maintain a good static figure when not in closed loop.

Correction with a deformable secondary mirror has recently been demonstrated at the MMT. The mirror consists of an aluminium cold plate, a solid Zerodur reference and a 1-2 mm thick meniscus shell with permanent magnets glued on the back. These are attracted by 336 voice coil actuators which pass through the reference. The position of the meniscus under each voice coil is controlled using a capacitive sensor which feeds back to the voice coil current through a local control loop. The spacing between the actuators is ~25 mm. The spacing between the shell and the reference is 40-50 $\mu$m. This space provides damping for the suppression of vibrations.

## 2.4    Performance Limitations

The performance of real adaptive optics systems are limited by several factors. There are two main approaches to estimating the performance of the system; numerical simulation and analytical modeling. In numerical simulation a model of the system is built in software and the response to a large number of simulated wavefronts is measured. The input wavefront errors should have the same spatial and temporal characteristics as expected from the theory of propagation through turbulence, and different techniques exist to do so. If the turbulence is distributed in layers and the near-field approximation is assumed, then wavefronts corresponding to each layer are simulated, shifted according to the wind-speed of the layer and summed. If near-field conditions are not assumed then the wavefronts should be propagated to the telescope entrance pupil using Fresnel transforms (Ellerbroek, 2001). Analytical approaches are generally faster since it is not required to simulate a large number of wavefronts. However models which aim to determine the optimal performance, which usually implies including a priori knowledge of the turbulent statistics in the AO control, can be quite involved. An approach which serves to build an error budget for the AO system is to use simple expressions for the error sources and combine them quadratically to give the final performance. The tip-tilt and higher-order error terms are usually considered separately. If the total tip-tilt error in radians of wavefront tilt is $\sigma_{tilt}$, then the corresponding

Strehl ratio is given by

$$SR_{tt} = \left(1 + \frac{\pi^2}{2} \left(\frac{\sigma_{tilt}}{\lambda_c/D}\right)^2\right)^{-1}$$

where $\lambda_c$ is the wavelength at which the AO correction is required. If $\sigma_{ho}$ is the total higher order wavefront error then the Strehl ratio is given by the Marechal expression.

$$SR_{ho} = \exp\left(-\sigma_{ho}^2\right)$$

The overall SR is given by the product of the tip-tilt and higher order Strehls.

The main error sources are noise in the wavefront sensor measurement, imperfect wavefront correction due to the finite number of actuators and bandwidth error due to the finite time required to measure and correct the wavefront error. Other errors include errors in the telescope optics which are not corrected by the AO system (e.g. high frequency vibrations, high spatial frequency errors), scintillation and non-common path errors. The latter are wavefront errors introduced in the corrected beam after light has been extracted to the wavefront sensor. Since the wavefront sensor does not sense these errors they will not be corrected. Since the non-common path errors are usually static, they can be measured off-line and taken into account in the wavefront correction.

The accuracy with which a wavefront sensor measures phase errors will be limited by noise in the measurement. The main sources of noise are photon noise, readout noise (see Ch. 11) and background noise. The general form of the phase measurement error (in square radians) on an aperture of size $d$ due to photon noise is

$$\sigma_{phot}^2 = \beta \left(\theta d/\lambda\right)^2 /n_{ph}$$

Where $n_{ph}$ is the number of photons used in the measurement, $\theta$ is the angular size of the guide image and $\lambda$ is the measurement wavelength. The factor $\beta$ depends on the detail of the phase measurement. This general expression will hold for all wavefront sensors (Rousset, 1993). The angular size of the guide source image depends on whether the source is extended or point-like, which is usually the case. For point-like sources, the image size depends on whether the aperture is larger or smaller than $r_0$; if it is larger, then the image is seeing-limited and the size is approximately $\lambda/r_0$. If $d$ is smaller than $r_0$, then the image is diffraction-limited and the size is approximately $\lambda/d$.

Increasing the diameter, $d$, increases the number of photons in the wavefront measurement, and therefore reduces the error due to photon noise. However, increasing the diameter also increases aliasing in the wavefront sensor measurement. If the deformable mirror actuator spacing is matched to the subaperture size, then the fitting error will also depend on the subaperture diameter. There is therefore an optimum subaperture diameter which depends on the

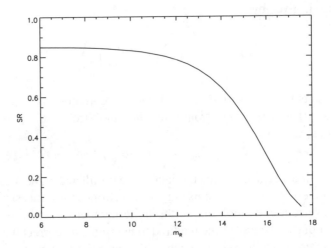

*Figure 7.* Strehl ratio at 2.2 $\mu$m as a function of guide star magnitude for typical system parameters (throughput = 0.4, read noise = 5 e$^-$ per pixel per frame, etc.) with $r_0 = 0.2$ m and $\tau_0 = 5$ ms at 0.5 $\mu$m.

guide star magnitude. In the same way, the exposure time can be chosen to minimize the sum of the errors due to photon noise and bandwidth. Figure 7 shows an example of K-band Strehl ratio predicted as a function of guide star magnitude when subaperture diameter and wavefront sensor exposure time are optimized at each magnitude. We can see that the Strehl ratio (in this system) can be greater than 0.2 for guide stars of magnitude $m_R < 16$. The probability of finding guide stars this bright within an isoplanatic angle of an arbitrary object is very small unless we are observing in the Galactic plane (where it is impossible to see interesting extragalactic objects). Laser guide stars (LGS) are necessary to increase the sky coverage (Ch. 15 and 14).

## 3.    Multiconjugae AO

### 3.1    Field anisoplanatism

One of the main limitations of 'conventional' AO is anisoplanatism; most objects of scientific interest are too faint to act as their own reference source for AO correction, and it would be useful to use bright nearby objects, preferably stars, which act as point sources. However, the light from the object of interest which is intercepted by the telescope passes through a part of the turbulence which is different from that traversed by the light from the reference star. This leads to a decorrelation between the wavefront measurements obtained with the reference star and that which should be applied to correct the wavefronts

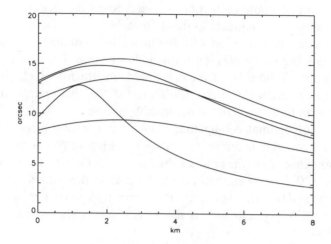

*Figure 8.* Isoplanatic angle at $2.2\mu$m as a function of conjugate altitude of the deformable mirror for different turbulence profiles obtained at the Observatorio del Roque de los Muchachos.

from the science object. The error is referred to as field anisoplanatism, and it is a function of the angular separation between the reference star and the science object. We have seen that typical values for the isoplanatic angle are $\sim 6$ arcseconds at visible wavelengths while the corresponding value at $2.2\mu$m would be $\sim 36$ arcseconds.

The main consequence of isoplanatism is to reduce the sky coverage of AO systems. In addition, the PSF is not constant inside the field of view, a fact which complicates the analysis of images obtained using AO. For example, astronomical photometry is usually performed by comparing objects in the field to a known point spread function which is considered constant over the field.

## 3.2    Optimal single-conjugate correction

If the turbulence really were concentrated in a single layer then placing the deformable mirror at an image of that layer would allow correction with an infinite isoplanatic angle. Even though the turbulence will never be completely in a single thin layer, there will be an optimal conjugate altitude for every turbulence distribution. Figure 8 shows the isoplanatic angle at $2.2\mu$m for different turbulence profiles obtained at the Observatorio del Roque de los Muchachoes (ORM) on La Palma in the Canary islands. It can be seen that a modest gain in isoplanatic angle can be obtained by placing the deformable mirror conjugate to $\sim$2.5 km.

Some practical complications are introduced when this technique is used; since the deformable mirror is not conjugate to the pupil it has to be made larger in order to allow correction of beams over the field of view. If the wavefront sensor is conjugate to the deformable mirror then it too has to be over-sized since the guide star will illuminate different areas of the wavefront sensor depending on its position in the field of view. For a given guide star position there will be parts of the deformable mirror for which no wavefront measurement is available. Veran (2000) has shown that extrapolation can be used to control those parts of the deformable mirror.

The idea of optimal conjugation can be extended by using multiple deformable mirrors-this is referred to as multi-conjugate AO or MCAO. Again, the question arises as to the optimal altitudes of the deformable mirrors. Tokovinin et al. (2000) showed that if the turbulence in the volume defined by the field of view and the telescope aperture is perfectly known, then the isoplanatic error in an MCAO system with $M$ deformable mirrors is given by

$$\sigma^2 = (\theta/\theta_M)^{5/3}$$

where $\theta_M$ depends on the number of deformable mirrors, their altitudes of conjugation and the turbulence profile. Using this expression with turbulence profiles obtained at the ORM we obtain that using two DMs the optimal altitudes are 660 m and 13 km, while for three DMs the optimal altitudes are 290 m, 3.9 km and 13.8 km.

The main difficulty in MCAO is how to obtain the information necessary to control multiple deformable mirrors. In single-conjugate AO one uses a wavefront measurement which corresponds to 'collapsing' or averaging the cylinder of light from the reference star onto a disk at the single plane of conjugation. In MCAO it is necessary to retain some 3-dimensional information, and for this reason wavefront sensing for MCAO is sometimes referred to as 'tomography' by analogy with medical tomography. However, it is not necessary to perform a full 3D reconstruction of the turbulent volume which we are correcting; rather it is necessary to obtain enough information to control the deformable mirrors in a way which will give a uniform (and high) Strehl ratio over the field of view. There are two basic approaches being developed; referred to as 'star-oriented' (SO) and 'layer-oriented' (LO) wavefront sensing. In both approaches multiple guide stars are used to 'illuminate' as much as possible of the turbulent volume. In a star-oriented scheme (Fig. 9) one implements a wavefront sensor for each reference star. The information from the wavefront sensors (e.g. the Shack-Hartmann centroid displacements) are combined into a single measurement vector which is multiplied by a pre-defined control matrix to give the required vector of actuator commands for the set of deformable mirrors. The control scheme is a simple generalization of single-conjugate AO.

In the layer-oriented approach each wavefront sensor 'sees' all the guide stars but each is coupled to a single deformable mirror. Each wavefront sensor

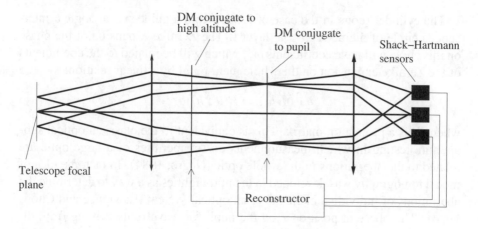

*Figure 9.* Schematic diagram of Star-Oriented MCAO; there is one wavefront sensor for each reference star.

is conjugate to the DM which it will control. In principle, each wavefront sensor-deformable mirror pair is independent of the others and has its own control loop. The LO approach potentially has some advantages over star-oriented; the light from several possibly faint guide stars can be added without requiring a wavefront sensor for each source. For example, if the low order WFS is a pyramid sensor then a pyramid is placed at the position of each guide star and the pupils are observed on a single detector. In addition, the spatial and temporal sampling of each detector may be optimized if there is knowledge of the temporal statistics as a function of altitude.

The control approach in both LO and SO can be either zonal or modal; most theoretical and simulation work is being done using the modal approach which may have some advantages including the rejection of waffle modes and avoiding problems due to the central obscuration.

## 3.3    Modal tomography

The modal analysis is a fairly straightforward generalization of the single conjugate case. In modal MCAO, Zernike polynomials are defined at layers conjugate to each DM. The unit circle on which the Zernikes are defined may be taken as the circle within which are inscribed the footprints from guidestars anywhere inside the field of view -this circle is usually referred to as the 'meta-pupil', for which at layer $l$, the Zernike expansion of the wavefront error is

$$\phi^{(l)}(r^{(l)}) = \Sigma_{i=1}^{n} a_n^{(l)} Z_n \left( \frac{2r^{(l)}}{D_m^{(l)}} \right)$$

The cylinder (cone in the case of LGSs) from a guide star at angle $\theta$ intercepts a circle of diameter $D^{(l)}$ at layer $l$. The Zernike expansion of the phase on this circle will have coefficients $b(l)$ which will be related to the coefficients of the Zernike expansion on the whole meta-pupil by a linear relation:

$$b^{(l)}(\theta, h_l) = P(\theta : h_l)a^{(l)}$$

where $P$ is a projection matrix. This is equivalent to the classical optical sub-aperture testing in which measurements on subapertures of a large optic are related to the aberrations in the whole optic (Negro, 1984). In fact, the idea of modal tomography was developed in the mid-eighties as a way to determine the aberrations of individual elements of an optical system (Lawrence and Chow, 1984). The phase at position $r$ on the pupil for wavefronts coming from direction $\theta$ is the sum of the phase from $L$ layers along that direction (assuming near-field conditions).

## 3.4     Limitations

There are (of course!) some limitations to the performance of MCAO. The first is a generalized aliasing whereby wavefront errors occurring at a certain altitude may be determined to have occurred at a different altitude. If we consider a sinusoidal wavefront error with a spatial frequency $f$ occurring at an altitude $H$, then it can be seen that for a guide star observing at angle $\theta$ it may appear that the wavefront error is at ground level if $\theta = 1/fH$. This aliasing then imposes a trade-off between the field of view and the maximum spatial frequency which can be corrected; it is generally found that increasing the field of view leads to reduced average Strehl ratio and greater variance in the Strehl ratio over the field. Another limitation on performance is due to the fact that the guide star footprints do not completely cover the meta-pupil, and this 'meta-pupil vignetting' increases as the guide star separation increases and also as the altitude of the turbulence being sensed increases. As with single-conjugate AO, a large effort has been put into numerical simulation of MCAO systems. Figure 10 shows an example result from such simulations (Femenia and Devaney, 2003). The Strehl ratio is mapped for fields of view of diameter 1 and 1.5 arcminute using a 10 m telescope (such as the GTC) and a triangle of natural guide stars. The average Strehl ratio is reduced on the larger field of view, and there are stronger variations in the Strehl ratio.

## 3.5     MCAO with LGS

We saw that the probability of finding bright guide stars is small outside the Galactic plane. It is clear that the probability of finding constellations of useful guide stars is very small, even if we make optimistic assumptions about the field of view and the limiting magnitude. The number density of stars on

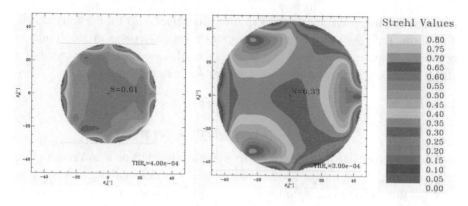

*Figure 10.* Results of numerical simulations to obtain the Strehl ratio at 2.2 $\mu$m over the field of view for a field of diameter 1 arcminute (left) and 1.5 arcminute (right) when using three natural guide stars.

the sky follows Poisson statistics; if the mean number of guide stars up to a certain magnitude in the field of view is $\mu$ then the probability of finding $n$ guide stars is given by

$$p(n) = \mu^n \exp\left(-\mu\right)/n!$$

The probability of finding at least 3 guide stars is therefore given by

$$p(n \geq 3) = 1 - (p(0) + p(1) + p(2))$$

$$= 1 - \exp(-\mu)\left[1 + \mu + \mu^2/2\right]$$

For example, at Galactic coordinates $(l,b) = (90,40)$ the average number of stars brighter than $m_R = 16$ in a field of view of diameter 2 arcminutes is 0.66. The probability of finding three such stars is therefore 3%. In addition, even if we have three stars which are bright enough it is unlikely that they will span the field of view. An obvious solution would be to employ LGSs. With multiple LGSs we can generate a guide star constellation around any field we wish to observe. The guide stars can always have the same spatial configuration and (approximately) the same flux, so that the system calibration is stable. This is an important advantage over natural guide star MCAO where it will be necessary to estimate new control matrices for each guide star configuration. The cone effect will not be a problem since 3D information is available from the guide star constellation. However, multiple LGS systems will be complex and expensive. In addition, the indeterminacy of tip-tilt (see Ch. 15) will lead to tip-tilt anisoplanatism which if uncorrected would practically nullify the benefit of MCAO. In the case of multiple LGS, the lack of information on their tip-tilt leads to indeterminacy in the altitude of quadratic phase errors (defocus and astigmatism) and this in turn gives rise to tip-tilt anisoplanatism. To see this imagine that the turbulence gives rise to a defocus of amplitude $a_1$ at altitude $h$ and a defocus $a_0$ at a layer at the telescope entrance pupil. For a star at direction $\theta$ the net phase error at position $x$ in the pupil is

$$\phi(x) = a_0 x^2 + a_1 (x + \theta h)^2$$

$$= a_0 x^2 + a_1 x^2 + a_1 (\theta h)^2 + 2 a_1 x \theta h$$

The last term is a tilt (phase linear in $x$) which depends on $\theta$ and is therefore anisoplanatic. A natural guide star constellation can detect all these terms (except the third which is a piston and does not affect image quality) and would completely correct the field of view if connected to two deformable mirrors, one conjugate to the pupil and the other to the layer at altitude $h$. However, LGSs would not detect the tilt term and tilt anisoplanatism will lead to a non-uniform Strehl ratio over the corrected field of view. Different solutions have been proposed to solve this problem. It is clear that it could be solved using a constellation of polychromatic guide stars. Another obvious approach is to measure tip-tilt on several (at least three) natural guide stars over the field of view. In principle, these guide stars could be very faint. This approach will be employed in the Gemini-South MCAO system. Finally, it has been proposed that combining guide stars which originate at different altitudes can solve the problem. Such hybrid approaches could involve combining sodium and natural guide stars or sodium and Rayleigh guide stars or Rayleigh guide stars at different ranges. To see the principle of these hybrid schemes, consider again the case of having defocus screens at pupil and altitude $h$. In the case of measuring with a LGS at direction $\theta$ generated at altitude $H$ the phase at position $x$ in the pupil is given by a modified version of the last equation:

$$\phi(x) = a_0 x^2 + a_1 \left(1 - h/H\right)^2 x^2 + a_1 (\theta h)^2 + 2 a_1 x \left(1 - h/H\right) \theta h$$

where the $1 - h/H$ term takes into account the cone shape of the laser beam (at $h = 0$ the beam covers the pupil, at $h = H$ the beam has zero extent). With the LGS constellation we cannot measure the third (no effect) or fourth terms. In fact, if $a_0 = a_1 \left(1 - h/H\right)^2$ then it would appear that there is no phase error ! If we employ a guide star generated at a different altitude $H'$ then this degeneracy cannot occur. There is information in the difference between the measurements obtained with the guide stars at $H$ and $H'$ about the altitudes at which the quadratic phase errors occur.

An extension of layer-oriented MCAO has been proposed as a way to get good sky coverage. In principle, LO can combine light from multiple faint guide stars inside the field of view. Even so, the sky coverage will be low for the fields of view over which good Strehl can be obtained ($\leq 2$ arcmin). The proposed 'multiple field of view layer-oriented' is based on the fact that layers near the telescope pupil are well sampled by stars over a very large field of view. The idea is that guide stars on an annular field of inner diameter 2 arcminutes and outer diameter $\geq$ 6 arcminutes would be sensed conjugate to the pupil, while stars on the inner 2 arcminutes would be sensed conjugate to high (i.e. $\sim$ 10 km) altitude. The stars in the inner field would control the deformable mirror conjugate to high altitude and the stars on the outer annular field would control the deformable mirror conjugate to the pupil. A complication arising with this approach is that the guide stars on the outer annulus are only partially corrected and so the wavefront measurements using them are not strictly in closed loop. The range and linearity of the wavefront sensors used has to be increased with respect to closed-loop sensors.

## 4. Measures of AO performance

The basic goal of AO systems is to provide diffraction-limited images in an efficient way. This goal is never fully achieved and we have seen that images obtained with AO systems typically have a diffraction-limited core on a halo which is somewhat smaller than the uncorrected image. The most important parameter characterizing AO-corrected images is the Strehl ratio. When an image is obtained with an AO system the usual procedure to determine the Strehl ratio is to compare it with an image obtained by Fourier transforming a mask corresponding to the telescope entrance pupil. Both images have to be normalized to contain the same total energy and the theoretical image has to be filtered to include the effect of the detector pixel size. It should also be taken into account that the real image will not be centered exactly in the center of a pixel. The Strehl ratio obtained from averaging the Strehl ratio of short exposure images will always be higher that obtained from an equivalent long-exposure image due to errors in the tip-tilt correction.

Astronomers are accustomed to describing images by their full-width at half-maximum (FWHM). If the Strehl ratio is high, then the fwhm should approximate to the theoretical value for a diffraction-limited image ($1.029\lambda/D$ for a circular aperture). For moderate values of Strehl ratio ($> 0.1$) there is a fairly monotonic relation between Strehl ratio and FWHM, while a low value of Strehl ratio may correspond to a wide range of FWHM. This is because in this regime the halo dominates the image, and its width depends on the current seeing conditions.

In order to evaluate system performance it is useful to plot SR as a function of $r_0$. This may be compared to simulations or model predictions and deviations indicate that there are problems. When this occurs, what are the diagnostics that can be examined ? There is a great deal of information in the wavefront sensor measurements and provision should be made to store them. Zernike decomposition of the residuals helps to identify if there are problems corresponding to some particular modes. Temporal spectra of open and closed-loop wavefront sensor data can be compared in order to show if the bandwidth of the system is as expected. This can also identify telescope vibrations, which can severely limit AO performance. The Strehl ratio should also be plotted as a function of guide star magnitude for observations taken in similar seeing conditions ($r_0$ and $\tau_0$) in order to determine whether the system is capable of providing correction up to the predicted limiting magnitude.

Finally a word about efficiency; observing time on large telescopes is a valuable asset, both in terms of cost and considering the ratio of observing time available to the time requested by astronomers. Marco et al. (2001) state that the observing efficiency defined as the ratio of 'science shutter time' to available dark time is 10-30% for the ADONIS AO system while the corresponding ratio for other instruments is 50-80%. Some of this difference is due to the fact that most AO exposures are of short duration and the readout time is significant. In addition, AO systems use time to close the loop and optimize performance. Observations may also be necessary to characterize the PSF.

Another indicator of the performance of AO systems is the number of refereed publications per year based on observations made using AO. This has been rising steadily over the last decade and is now >30 per year (see http://www2.keck.hawaii.edu:3636/realpublic/inst/ao/ao_sci_list.html).

# 5. Challenges for AO

## 5.1 Extreme Adaptive Optics

Since 1995 more than 100 planets have been detected around other stars. The detections have been made by indirect techniques, usually by detecting periodicities in radial velocity measurements, from which orbital parameters can be determined. Direct detection of light from extra-solar planets would greatly increase our knowledge of these objects allowing determination of parameters such as temperature and composition. However, the photons from extra-solar are lost among the relatively huge number of photons from the host star. The relative brightness of a mature (i.e. cool) planet to the host star is of order of $10^9$. If we are to detect these objects using AO then it is necessary to beat down the uncorrected halo of light from the central star so that the planet will be revealed. This implies obtaining very high Strehl ratios, so that a very large fraction of the stellar light is contained in the coherent core. Systems designed to do this are examples of 'extreme AO' (ExAO or XAO). On 8-10 m telescopes these systems will require several thousand actuators and have to run at a few KHz. They will use the central star as the source for wavefront sensing. In some designs the XAO system is a second stage corrector after a 'conventional' AO system. In addition to very high order correction all other sources of scattered light have to be controlled. These sources include stray light and ghost images, which are controlled by using very smooth optics, careful baffling, highly reflective coatings etc. Scattering of light at the edge of the telescope pupil is controlled by coronography. Extreme AO on an ELT such as OWL will require hundreds of thousands of actuators - the ultimate challenge for adaptive optics !

## 5.2    AO on Extremely Large Telescopes

The next generation of 'Extremely Large Telescopes' (ELTs) has been described in Ch. 7. We have seen that the detection power of these telescopes is enormously enhanced by AO. It is usually argued that AO is fundamental for ELTs, indeed some ELTs only consider operation with AO. How will AO scale to these telescopes of the future? This issue is being addressed by the different ELT project teams. We can get an idea of the complexity of AO systems on ELTs using simple expressions. In order to keep the fitting error constant as we increase the telescope diameter $D$ the number of actuators on the deformable mirror has to increase as $D^2$. The number of subapertures in the wavefront sensor has to increase accordingly as well as the number of pixels in the wavefront sensor detector. In order to maintain the bandwidth the pixel readout rate also has to increase as $D^2$ (while keeping the read-noise low !). If the control approach used on current AO systems is employed, then the required computing power scales as $D^4$. As an example, consider scaling the Keck AO system to a 30 m telescope (such as CELT) and a 100m telescope (such as OWL). The deformable mirror in the Keck AO system has 349 actuators. This would be increased to about 3000 actuators on a 30 m telescope. If the wavefront sensor employs quad cells then the detector would have $128 \times 128$ pixels, which corresponds to detectors which exist today. With 1 kHz sampling the computing power required is about 10 Gflop. The biggest challenge for a near-IR AO system on a 30 m telescope is therefore likely to be the deformable mirror. The same system scaled to OWL would have 35000 actuators and would need a $512 \times 512$ pixel detector. The computing power at 1kHz sampling would be of order $10^3$ Gflops. It has been pointed out (Gavel, 2001) that in the timeframe of the OWL project this computing power exceeds what will be possible even given Moore's law. It is therefore essential to develop alternative control strategies such as those employing sparse matrix techniques. It is clear that deformable mirrors having such a large number of actuators will require a major development effort. One possibility is to employ micromirrors (see Ch. 10) although the small actuator spacing implies a small mirror size, and mapping the (extremely) large entrance pupil onto a small mirror is optically difficult.

LGSs on ELTs also present severe challenges. The first problem is the perspective elongation which is a small problem on 8-10 m telescopes where the elongation is $\sim 0.73$ arcseconds (laser launched from behind secondary mirror) while it would be $\sim$ 2.2 arcseconds on a 30 m telescope and $\sim$ 7.3 arcseconds on a 100 m telescope. The laser power has to be significantly increased (without saturating the sodium layer) in order to compensate for this effect. Another problem arises from the fact that telescopes are designed to work at infinite conjugates while the LGS is at a finite distance. For example, in a telescope of focal length 170 m (e.g. the 10 m GTC) the sodium LGS focus is about 32 cm after the natural guide star one. In an ELT this distance will be greatly increased. The optical quality of the laser guide star image will be very bad in the case of OWL and will even depend on the telescope elevation.

# References

Angel, R., Nature, 1994, **368**, 203

Babcock, H.W., 1953, Pub. Astro. Soc. Pac. **65**, 229

Buffington, A., Crawford, F.S., Muller, A.J., Schwemin, A.J., Smits, R.G., 1976, SPIE **75**, 90

Canales, V.F., Cagigal, M.P., 1996, J. Opt. Soc. Am. A **16**, 2550

Ellerbroek, B., *Beyond Conventional Adaptive Optics*, 2001, ESO Conf. **58**, 239

Femenia, B., Devaney, N., 2003, Astron. Astrophys. **404**, 1165

Gavel, D.T., *Beyond Conventional Adaptive Optics*, 2001, ESO Conf. **58**, 47

Hardy, J.W., *Adaptive Optics for Astronomical Telescopes*, 1988, Oxford Univ. Press

Lawrence, G.N., Chow, W.W., 1984, Opt. Lett. **9**, 267

Linnik, V.P., 1957, Original article translated and reprinted in ESO Conference 48, ed. F. Merkle, Garching, 535

Negro, J.E., 1984, Appl. Optics **23**, 1921

Noll, R.J., 1976, J. Opt. Soc. Am. **66**, 207

Ragazzoni, R., J., 1996, Modern Optics **43**, 289

Ragazzoni, R., Farinato, 1999, J., Astron. Astrophy. **350**, L23

Roddier, F., 1981, Progress in Optics **19**, 281

Roddier, F., 1988, 1988, Appl. Optics, **27**, 289

Roddier, F. (ed.), 1991, *Adaptive Optics in Astronomy*, Cambridge Univ. Press

Rousset, 1992, *Adaptive Optics for Astronomy*, ed. Alloin, D.M., Mariotti, J.-M., NATO ASI **423**, 115, Kluwer Academic Publ.

Tokovinin, A., Le Louarn, M., Sarazin, M., 2000, J. Opt. Soc. Am., A 17, 1819

Tyson, R.K., 2000, *Introduction to Adaptive Optics*, SPIE press

Véran, 2000, J. Opt. Soc. Am.,A, 17, 1325

Voitsekhovich, V.V., Orlov, V.G., and Sanchez, L.,J., 2001, Astron. Astrophys., **368**, 1133

# LASER TECHNOLOGIES FOR LASER GUIDED ADAPTIVE OPTICS

Deanna M. Pennington

*Lawrence Livermore National Laboratory*
*P.O. Box 808, L-183*
*Livermore, CA 94550, USA*

pennington1@llnl.gov

**Abstract**     The principles of the laser guide star (LGS) for adaptive optics (AO) are described, as well as the physical processes involved in their generation. System
design issues are addressed, such as laser spot elongation and brightness. The
two main families of LGS, based either on Rayleigh or sodium resonant scattering are discussed in terms of performances and of laser technologies required to
match requirements. LGS aided AO systems built for astronomical telescopes
are described, with a special emphasis for the Lick and Keck systems which are
the only ones in routine operation for science. The future 30-100 m telescopes
need an intense R&D program for a new generation of robust turn key lasers.
These lasers must be compact, have nearly diffraction limited beam quality with
10 to 50 W output at 589 nm.

**Keywords:**     lasers, laser guide star, adaptice optics, sum-frequency lasers, fiber lasers, Raman lasers

## 1.     Introduction

The development of adaptive optic (AO) systems to correct for wavefront
distortions introduced by the atmosphere represents one of the major advances
in astronomical telescope technology of the 20th century. However, in spite of
the great progress in AO, sky coverage is limited to sources located near bright
stars that provide a measure of wavefront distortions.

In order to overcome the requirement of having a bright "natural guide
star" (NGS) near an astronomical source, a first generation of laser guide
stars (LGS), or laser beacons, has been deployed to produce bright artificial
stars from which wavefront distortions can be assessed. LGSs consist of the

*R. Foy and F. C. Foy (eds.), Optics in Astrophysics*, 207–248.
© *2006 Springer. Printed in the Netherlands.*

backscattered light generated by ground-based laser beams passing through the atmosphere. The beam is generated by a laser launched by a small telescope attached to the main telescope, transmitted into the atmosphere and pointed close to the source. The backscattered light returned from a column of a specific diameter, length, and altitude is observed as the artificial star.

The basic layout of a laser guided AO system is shown in Fig. 1. Implementation of LGS referencing requires the addition of a laser and launch telescope, plus one or more additional wavefront sensors (WFS), including a tip-tilt sensor. Multiple LGSs require additional lasers and launch systems, or a multiplexing scheme. Multi-conjugate AO (MCAO) requires additional deformable mirrors, operating in series, plus multiple WFSs.

This article gives a brief introduction into the formation of LGSs, either through Rayleigh backscattering in the lower atmosphere, or through resonant backscattering from the sodium layer at an altitude of 90 km. The most important properties of the sodium layer are summarized, and some technical issues relevant to the generation of sodium LGSs are discussed. The article concludes with a review of the laser technologies and their demonstration, as well as a description of systems under development.

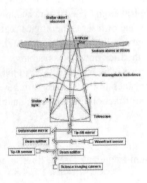

*Figure 1.* Schematic of a laser guide adaptive optics system. The laser is projected along or parallel to the telescope axis onto the science object. The deformable mirror can be a separate entity, or can be an adaptive secondary. The light is split among the cameras by dichroics.

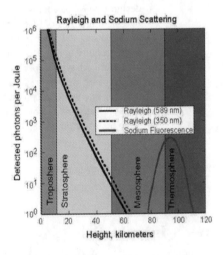

*Figure 2.* Magnitude of Rayleigh and sodium backscatter return as a function of atmospheric density.

## 1.1    History of laser guide stars

The idea of creating a LGS using Rayleigh scattering from a focused laser beam was proposed in 1981 by Feinleib and Hutchin, and also by Hunter. In 1982 Happer proposed using fluorescence in the mesospheric sodium layer for atmospheric compensation (Happer et al., 1994). In 1983 Fugate demonstrated that reliable wavefront measurements could be made from a Rayleigh LGS at 10km based on comparison with those from a natural guide star (NGS). In 1984 the first demonstration of a Na LGS was performed and compared with a nearby natural star to determine it's usefulness for atmospheric correction ( Humphreys et al., 1991). The first AO compensated stellar images were obtained with a Rayleigh beacon in 1988 (Zollars, 1992), followed by the first high pulse rate compensation demonstration in 1989 (Fugate et al., 1991) at the Starfire Optical Range (SOR). compensation was demonstrated in 1989 ( Fugate et al., 1991) at the Starfire Optical Range (SOR). The existence of the US Government-sponsored work on laser beacons described above was not publicly revealed until 1991 (Fugate et al., 1991; Primmerman et al., 1991).

In the meantime, similar ideas were published in the open literature by Foy and Labeyrie (1985). Their paper presented the concept of using Rayleigh or resonant backscatter from pulsed laser beams to measure wavefront distortion in AO systems. Based on this suggestion, Thompson and Gardner (1987) demonstrated a Na LGS using a pulsed dye laser on Mauna Kea. They published several papers covering the theory and practice of Rayleigh and Na LGSs, investigated the characteristics of the sodium layer, analyzed methods of implementing LGS/AO and expected performance. A comparison of Rayleigh LGS and NGS wavefront sensing was also carried outby Foy et al. (1989). (Foy et al., 1989). The first WFS measurements using a sodium LGS were performed at Lawrence Livermore National Laboratory (LLNL) in 1992 with a 1100 W dye laser, which had been developed for atomic vapor laser isotope separation (AVLIS)(Max et al., 1994; Avicola et al., 1994). They compared well with models (Morris, 1994). AO systems with LGSs are now included in the planning for nearly all major telescopes, and several prototype systems have been commissioned (Quirrenbach, 1997; Max et al., 1997; Pennington et al., 2002). Several good reviews of AO have been published in the past few years (Beckers, 1993; Hardy, 1998; Tyson, 1997; Dainty et al., 1998, see also Ch. 13). These contain a wealth of introductory material about LGSs. A number of detailed technical accounts have also been collected in Volume 11 of the JOSA. In addition, the last conferences and workshops on AO were organized by SPIE ( 4007, 4125, 4839). The present article is intended to give an overview of the basic principles of LGSs, and to set the stage for the later contributions in this volume, which will treat many aspects in more detail.

## 2. Laser Guide Stars

There are several factors to consider when using a LGS. First, beacons are formed within the earth's atmosphere, near the telescope aperture. As a result, they don't sample the full telescope aperture when focused on object at infinity. This effect, called the cone effect, leads to wavefront error called focal anisoplanatism (see Ch. 15). The laser light is also spread out by turbulence on the way up. This produces a finite spot size (0.5-2 arcsec), which increases the WFS measurement error. Even when using a LGS, a NGS is required to compensate for image motion because the laser beam is randomly displaced by turbulence (see Ch. 15). Use of LGSs complicates the telescope optical system more than conventional AO. Additional optical components are required for beam sharing with the laser. Unwanted scattered light, and laser light, must be rejected by the science-imaging path. Finally, lasers for LGSs have special requirements. The pulse repetition rate is the controlling factor in designing the AO feedback loop. It must be $\sim 10\times$ higher than the compensation bandwidth (several kHz). Sodium beacons must be accurately tuned to Na $D_2$ line (589.2 nm), with specific pulse shapes and spectral content.

There are two types of scattering used for LGSs. Rayleigh scattering is produced by scattering from air molecules, small aerosols, and is generated at altitudes up to $\sim 30$ km. Sodium resonance fluorescence occurs in a thin layer of sodium atoms deposited by meteors at an altitude of 90-100 km. To create sodium LGSs a laser, tuned to the 589 nm resonance line of Na, is used to resonantly excite the Na atoms, giving rise to fluorescence at the same wavelength. In addition, Mie scattering occurs in the lower atmosphere, producing backscatter that competes with the desired returns. This is radiation scattered from small particles and aerosols of a size comparable to the wavelength of the incident radiation, with no change in frequency.

The Lidar (light detection and ranging) equation defines the energy detected at a receiver of area $R_A$ because of the scattering process as the light beam propagates through the atmosphere. Assuming emission from a scattering volume at distance $R$ is uniform over $4\pi$ sr, the fraction of photons collected is $A_R/(4\pi R^2)$. Regardless of the type of scattering, the number of photons $N_\nu$ detected is the number of transmitted photons + the probability that a transmitted photon is scattered + the probability that a scattered photon is collected + the probability that a collected photon is detected + background photons (noise). With simplifying assumptions, described below, the Lidar equation may be expressed symbolically in the form:

$$N(H) = [\frac{E_L}{hc}](\sigma_B n(H)\Delta H)[\frac{A_R}{4\pi H^2}](T_{opt}T_{atm}^2\eta_d) + N_B \qquad (1)$$

where $N(H)$ is the expected number of photons detected in a range $\Delta H$, $E_L$ the laser energy (J), $\lambda$ the wavelength (m), $h$ the Planck's constant, c the speed of light,$\sigma_B$ the effective backscatter cross-section (m$^2$), $n(H)$ the number density of scatterers at range $H$, $\Delta H$ the receiver range gate length (m), $A_R$ the area of the receiving aperture (m$^2$), $T_{opt}$ the transmission of transmitter and receiver optics, $T_{atm}$ the one-way transmission of atmosphere between telescope and LGS, $\eta_d$ the quantum efficiency of the detector at $\lambda$, and $N_B$ the number of background and noise photoelectrons.

The assumptions are that $\Delta H \ll H$ , so that $n(H)$ and $T_{atm}$ are constant over $\Delta H$, and that the scattering process is linear so that $\sigma_b$ is constant. The first assumption is justified in the case of focused beams using a range gate in the detector optical system to select a small region around the focus. In the case of sodium beacons, there may be significant layering within the focal volume, which is accommodated by using an average value of $n(H)$. The Na resonance fluorescence process is easily saturated, so care must be taken to use an appropriate value of $\sigma_b$. The two factors $n(H)$ and $\sigma_b$ in the Lidar equation are specific to the type of scattering employed. For Rayleigh scattering,

$$\sigma_B^R = \pi^2(n^2 - 1)^2/(N^2\lambda^4) \tag{2}$$

where $n$ is the index of refraction and $N$ is the atom density. For Na, an effective cross-section is used, which is a function of the spectrum of the laser pulse (see Section 14.3.1.1).

## 2.1 Rayleigh Guide Stars

Rayleigh scattering results from interactions of the electromagnetic wave of the laser beam with atoms and molecules in the atmosphere. The laser electromagnetic fields induce dipole moments in the air molecules, which then emit radiation at same frequency as the exciting radiation (elastic scattering). Rayleigh scattering is not isotropic and the cross section depends on the polarization of the incident light. Along direction of propagation, radiation from both polarization components add. Perpendicularly, only the perpendicular component is scattered. Rayleigh scattering is much more effective at shorter wavelengths and depends strongly on the altitude. The product of the cross-section with the density of molecules at a given altitude is:

$$\sigma_B^R n(H) = \pi^2(n^2-1)^2\lambda^{-4}n(H) \cong 3.6{\times}10^{-31}\lambda^{-4.0117}P(H)/T(H)\mathrm{m}^{-1}\mathrm{sr}^{-1} \tag{3}$$

where $P(H)$ is the pressure in millibar, and $T(H)$ is temperature in degrees K at altitude $H$. Because $P(H)$ falls off exponentially with $H$, Rayleigh LGSs are generally limited to altitudes below 8-12 km. However, Rayleigh scattering dominates over sodium scattering for altitudes $< 75$ km (Fig. 2).

Pulsed lasers are used to generate a Rayleigh LGS, so that $H$ and $\Delta H$ can be selected through "range gating" for each pulse. A shutter in front of the WFS is opened when the light returning from the bottom end of the desired column reaches the detector, i.e., after a time $T = 2(H/cos(z) - \Delta H/2cos(z))/c$ has elapsed, where $z$ is the zenith distance. The shutter is closed again when it reaches its upper end at $T = 2(H/cos(z) + \Delta H/2cos(z))/c$. The angular size $\alpha$ of the beacon formed at the focus of a converging beam is determined by $\Delta H/cos(z)$. It is normally limited to 1-2 arcsec because of turbulence.

The power required for a Rayleigh LGS is $\approx 100$ W per beacon. The trade off between increasing $\Delta H$ to increase the return flux and decreasing it to decrease the spot size, which increases SNR, is given by:

$$\Delta H/H \leq 4.88(\lambda/D_{proj})(H/r_0) \qquad (4)$$

where $D_{proj}$ is the diameter of the laser projector and $r_0$ is the Fried parameter (see Ch. 1). Because of the strength of Rayleigh scattering, it is fairly easy to generate bright guide stars with this method. Furthermore, the exact wavelength of the laser does not matter. It is therefore possible to choose a laser technology that gives high power output at moderate cost. Many types of lasers have been used, including excimer (Thompson and Castle, 1992), copper vapor lasers (Fugate et al., 1994), frequency doubled or tripled neodymium-doped yttrium aluminum garnet (Nd:YAG) lasers (Fugate et al., 1991; Foy et al., 1989) and tunable dye lasers (Zollars, 1992). Rayleigh LGS have proved effective only for small telescope apertures ($< 2$m), because of cone effect. Multiple laser beams can be used to create a tomographic image of the turbulence,1 but this process makes an already complicated system even more complex.

## 3.     Sodium Guide Stars

Resonance scattering occurs when an incident laser is tuned to a specific atomic transition. An absorbed photon raises the atom to an excited state. The atom then emits a photon of the same wavelength via spontaneous or stimulated emission, returning to the initial lower state. This can lead to large absorption cross-sections, making the process much more efficient in generating backscattered light than molecular scattering. To create sodium LGSs, a laser, tuned to the 589 nm resonance line of Na, is used to resonantly excite Na atoms in the mesosphere at $\sim 100$ km altitude. This produces a bright artificial star.

See Table 1 for the values of important parameters of the sodium layer.

### 3.1     Physics of the Sodium Atom

Sodium has 1 valence electron, and 10 bound electrons. The first two excited states are the $3^2P_{1/2}$ and the $3^2P_{3/2}$ states. Transitions to these levels give rise to the $D_1$ and $D_2$ transitions respectively. There are two hyperfine levels in the $3^2S_{1/2}$ ground state, and four hyperfine levels in the $3^2P_{3/2}$ excited state (Fig. 3). There is no significant energy difference between the hyperfine levels in the $3^2P_{3/2}$ state. Thus, the six permitted lines appear in two groups, producing a double peaked spectral distribution, with the peaks separated by 1.772 GHz.

The 16 ns natural lifetime of excited Na is much shorter than the 140 $\mu$s mean time between collisions, thus the line broadening due to collision-induced

*Table 1.* Basic data about the mesospheric layer and the sodium atom.

| | | | |
|---|---|---|---|
| Layer attitude | 92 km | Oscillator strength of $D_2$ | 0.66 |
| Layer thickness | 10 km | $D_2$ wavelength | 589.2 nm |
| Temperature in laser | 215 ±15 K | Energy of 589 nm photon | $3.38\ 10^{-19}$J |
| Column density | $2 - 9\ 10^{13}$ m$^2$ | $D_2$ Doppler width at 215 K | 1.19 GHz |
| Peak sodium density | $4\ 10^9/m^3$ | $D_2$ Hyperfine splitting | 1.77 GHz |
| Molecular density | $7.1\ 10^{19}/m^3$ | $D_2$ hyperfine components ratio | 5:3 |
| Sodium-molecule collision time | 150 $\mu s$ | Peak $D_2$ cross section for natural linewidth | $1.1\ 10^{-13} m^2$ |
| Natural width of D lines | 10 MHz | Peak $D_2$ cross section for Doppler linewidth | $8.8\ 10^{-16} m^2$ |
| $D_2$ natural lifetime | 16 ns | Saturation intensity per velocity group | 64 W/m$^2$ |
| $D_1$ oscillator strength | 0.33 | Optical depth of Na layer | $\leq 0.1$ |

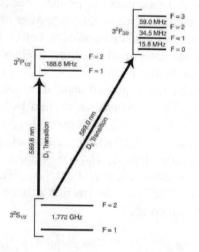

*Figure 3.* The ground and first two excited states of the Na valence electron. The $F$ values are the total angular momentum quantum number (Jeys et al., 1991).

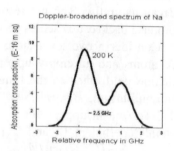

*Figure 4.* Absorption cross-section for the Na $D_2$ line. The center wavelength is 589.158 and the peak separation is 1.772 GHz.

transitions is negligible. The 1.07 GHz (at 200 K) Doppler broadened linewidth is $\sim 100$ times greater than the homogeneous one. Doppler broadening is therefore the dominant factor in determining the linewidth. The combined hyperfine energy separations and the Doppler broadening result in a 3 GHz full-width-half-maximum (FWHM) absorption profile. Sodium is normally in the ground state at 200 K. When a laser pulse passes through the medium, some of the atoms absorb a photon and are excited to the upper state. When you shine a laser on the sodium layer, the optical depth is only a few percent. Most of the light just keeps on going upwards. Atoms in the upper state may de-excite back to the lower state by either spontaneous or stimulated emission. With spontaneous decay, a photon of the same energy and wavelength is emitted incoherently and isotropically. At low levels of incident radiation, the mean time for an atom in the excited state to absorb a photon (provoking stimulated emission) is much longer than 16 ns mean lifetime, so most of the population remains in the excited state. In this regime the backscatter is proportional to the laser energy. The scattered fluxes are limited by three effects: saturation, optical pumping and radiation pressure, which will now be discussed.

### 3.1.1    Radiative Saturation.

Higher levels of radiation create a larger population in the excited state, allowing stimulated emission to become a competing process. In this process, atoms in the excited state absorb photons, which re-emit coherently; that is with the same frequency, phase and direction as the incident photon. Thus stimulated emission does not produce backscattered photons. As the incident energy increases, a greater proportion of the excited atoms absorb a photon and produce stimulated emission before they decay naturally. The net result is that the population of atoms available to produce backscatter decreases, i.e., the medium saturates.

Consider a two level atom that initially has a ground state $n$ containing $N$ atoms and an empty upper state $m$. The atom is then excited by a radiation field tuned to the transition $\nu = W_m - W_n/h$, where $h\nu \gg kT$. In equilibrium, $B_{nm}U(\nu)N_n = A_{mn}N_m + B_{mn}U(\nu)N_m$ , where $A_{mn} = 1/lifetime$ in upper state = Einstein's coefficient A, $B_{nm} = B_{mn}$ = Einstein's coefficient B, and $U(\nu)$ is the radiation density (J/cm$^3$Hz). Hence $N_m = N_n[B_{nm}U(\nu)]/[B_{nm}U(\nu) + A_{mn}]$. If $f$ is the fraction of atoms in level $m$ and $(1 - f)$ the fraction in level $n$ it comes

$$f = B_{mn}U(\nu)(1-f)/(B_{mn}U(\nu)+A_{mn}) = 1/(2+A_{mn}/[B_{mn}U(\nu)]) \quad (5)$$

At low $U(\nu)$, $f = B_{mn}U(\nu)/A_{mn}$, and the number of photons radiated in spontaneous emission/s is $NB_{mn}U(\nu)$. As the pump radiation increases,

$$\lim_{U(\nu)\to\infty} f = 1/2 \quad (6)$$

We can define the saturation level as the radiation field generating 1/2 this max: $U_{sat}(\nu) = A_{mn}/2B_{mn} = 8h\pi\nu^3/c^3$ J Hz /cm$^3$. The ratio $A_{mn}/B_{mn}$ is known from Planck's black body formula. The intensity of the radiation field $I(\nu)$ is related to $U(\nu)$ by $I(\nu) = U(\nu)c$ W Hz/cm$^2$ .

We can estimate the saturation intensity for a natural linewidth $\Delta\nu$ as: $I_{sat} = 4\pi(h\nu)\Delta\nu/\lambda^2$. Writing this in terms of the transition probability $A(= 2\pi\Delta\nu)$:

$I_{sat} = 2(h\nu)A/\lambda^2$ W/cm$^2$. Thus at 589 nm, $I_{sat} = 12.1$ mW/cm$^2$. More refined calculations, taking into account the hyperfine splitting and the polarization of the exciting radiation field, give:

$$I_{sat} = (\pi/2)(h\nu)A/\lambda^2 = 9.48 \text{ and } I_{sat} = (\pi/3)(h\nu)A/\lambda^2 = 6.32\text{mW}/\text{cm}^2 \tag{7}$$

Saturation can be controlled by limiting the peak pulse power of the laser, and by tailoring the pulse width, repetition rate and spectral bandwidth. From an altitude of 85-95 km, the density varies from $2.9 \times 10^{19}$ to $1.7 \times 10^{20}$ molecules/m$^3$ and the mean time between collisions of the Na atoms varies from 27 to 155 $\mu$s (U.S. Standard, 1976). With a single frequency laser tuned to the peak of D$_2$, only a few percent ($\approx$10 MHz/1 GHz) interact strongly with the field until they collide with another atom or molecule. With a multi-frequency laser one can excite many velocity groups at once, reducing saturation. The photon return per Watt of laser power depends on the spectral bandwidth (Fig. 5, d'Orgeville et al., 2000). The optimum laser linewidth is somewhat less than the Doppler width, with a value of 300 - 600 MHz being near optimum.

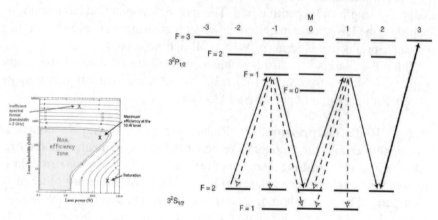

*Figure 5.* Dependence of photon return per Watt of laser power on the degree of spectral bandwidth added to the pulse.

*Figure 6.* The Zeeman sub-levels of the ground state and the second excited state, and the effects of optical pumping with right- hand circularly polarized light. The transition on the right represents the trapped state since decay to the lower ground state is forbidden. Solid lines: some of the allowed transitions that lead to trapped state. Dashed lines: other allowed transitions that either proceed away from the trapped state, or terminate in the $F = 1$ level and are lost to further pumping.

For a typical spot size of 1 arcsec the laser intensity is nominally 2 mW/cm$^2$, so the laser power must be $\lesssim 2.4$ W to avoid saturation for a single frequency

CW laser. By spreading the power over three spectral lines, the saturation fluence becomes $3\times$ larger, and 7.2 W can be used. Thus saturation is not a major concern for CW lasers with output powers of a few Watts. For pulsed lasers, however, the situation is very different. Depending on the pulse format, their peak power $P_{peak}$ can be many orders of magnitude higher than the average power $\bar{P}$. For example, a laser with a pulse duration of 20 ns, a repetition frequency of 10 kHz, and an $\bar{P} = 10$ W would have $P_{peak} \approx 50$ kW, which would badly saturate the $D_2$ line for any useful spot diameter. Peak power is defined as the $\bar{P}$/duty cycle, where the duty cycle is the fraction of time laser is "on". The pulsed dye laser at the Keck Observatory produces 120 ns pulses at 26 kHz. To achieve $P_L < 2.4$ W, the laser spectrum must be broadened over the full double peaked 3 GHz Doppler width at 10 MHz intervals, so the laser interacts with 300 velocity groups. Thus $P_L \leq 2.4 \times 3 \times 10^{-3} \times 300 = 2.2$ W. An analytical model can be used to predict saturation for high repetition rate lasers (Milonni et al., 1998). While this improves the photon return, CW lasers with only 500 MHz are more efficient because they match only the stronger of the two spectral peaks. Photon return saturates as $ln(1 + I_{peak}/I_{sat})$, where $I_{peak}$ is proportional to the power and inversely proportional to the LGS spot area, pulse length and repetition rate. The spot size assumption has a major influence on the laser power required; however increasing it to reduce saturation is counter-productive in terms of WFS SNR optimization.

With pulsed lasers it is also possible to redistribute the atoms between the magnetic sublevels (see Sect. 14.3.1.3). Nevertheless, saturation is a major concern for the design of pulsed Na LGS systems.

**3.1.2    Radiation pressure.**    Radiation pressure on the Na atoms can substantially change the amount of resonant fluorescence by altering their radial velocity and by pushing them out of resonance. This process is particularly important for single frequency laser, or laser that consists of a comb of frequencies separated by several homogeneous linewidths (Fugate et al., 1994). After an atom absorbs a photon, it radiatively decays by emitting a photon in an arbitrary direction. This process on average causes the atom to absorb one unit of photon linear momentum per absorption/emission cycle, which increases the absorption frequency of the atom by 50 kHz. At the saturation intensity, an atom absorbs a laser photon every 64 ns, thus the atom's resonant frequency shifts at a chirp rate of 50 kHz/64 ns = 0.78 MHz/$\mu$s. As the atom moves out of resonance with the radiation, the frequency chirp decreases.

Collisions between the Na atoms and molecules will redistribute the atomic velocities and fill the velocity hole created by radiation pressure. These collisions, however, are relatively infrequent. Therefore, ample time exists for a long laser pulse to push the atoms out of resonance before collisions can rethermalize the vapor. This loss in excitation efficiency can be overcome with radiation that has spectral content at intervals of the homogeneous linewidth. Then the Na atoms will be pushed merely from one homogeneous velocity group to another. This is the same spectral content requirement given earlier for complete excitation over the Doppler profile. It is also possible, however, to increase the Na excitation efficiency of a single-frequency laser, or of laser radiation that consists of a sparse comb of frequencies, by chirping the laser frequencies so they follow the resonant frequency of the Na atoms. This technique sweeps the Na atoms into velocity groups in resonance with the laser radiation, and increases the absorbance of the Na vapor.

### 3.1.3   Circular Polarization and Optical Pumping.

For laser illumination times that are long compared to the radiative lifetime, the effects of optical pumping must be considered. For the highest efficiency, the laser should be tuned to the transition between the $3^2S_{1/2}$ $F = 2$ state and the $3^2P_{3/2}$ level, because there are 5 angular momentum $M$ states in this level as opposed to 3 in the $F = 1$ one, with the corresponding increase in population. In the decay process, some of the electrons decay to the $F = 1$ level. These atoms are then effectively lost to further pumping (Jeys et al., 1991). However, there exists two Zeeman sublevels ($M = +3, M = -3$) of the $F = 3$ level for which there is no allowed transition to the $3S_{1/2}$ $F = 1$ level (Fig. 6). Due to conservation of angular momentum, absorption and emission of photons requires a shift of $\Delta M = \pm 1$ or 0 depending on whether the photon is right hand or left hand circularly or linearly polarized respectively. In Fig. 6 this would correspond to a shift of one column right or left or a transition straight down. Spontaneous photon emission is random in its polarization, therefore has no tendency to move the electrons toward either extreme. Similarly, linearly polarized causes transitions within the same column and therefore has no tendency to move the electrons toward either of the extreme Zeeman sublevels.

With right hand circularly polarized light, both ground states can be excited, thus all of the absorption transitions will cause $\Delta M = 1$. After several excitation/decay cycles the Na atoms will have absorbed angular momentum from the radiation field and will reside in the higher angular momentum hyperfine state ($F = 2$). This will cause the electrons to migrate to the $M = 3$ Zeeman sublevel. Once in this state the electron is "trapped" repeatedly transitioning between the $F = 3, M = 3$ excited state, and the $F = 2, M = 2$ ground state. The Na absorption profile will then consist of only the $F = 2$ peak. Left hand circularly polarized light causes the same effect but in the other "direction". There is no "trapped" state for the $F = 1$ level.

The earth's magnetic field does compete with the optical pumping process by causing a precession of the atomic magnetic moment at the Larmor frequency, $\nu_{Larmor} = 0.35$ MHz, thus allowing transitions that would otherwise be forbidden. In the absence of a perturbing magnetic field, about 20 excitation/decay cycles are required to pump most of the Na atoms into the $F = 2$ ground state with circular polarization. The cycle time of the Na atom should be short, preferably less than 500 ns and the intensity of the laser and the pulse format are important. For $\nu_{Larmor} > n_{min}/(20\delta t)$, the effects of circularly polarized pumping diminishes because fewer atoms are pumped into the $F = 2$ state. Here $n_{min}$ is the minimum number of times that each Na atom must spontaneously radiate a photon and $\delta t$ is the laser pulse length. Experimental observations have confirmed that the projection of circularly polarized light produces a $\sim 30\%$ increase in LGS brightness over linearly polarized light ( Ge et al., 1997; Morris, 1994).

## 3.2    Properties of the Sodium Layer

Na is likely deposited in the upper atmosphere by meteors along with other metals (Clemesha et al., 1981) and distributed by solar winds (Happer et al., 1994). This atomic layer is "eaten away" at its bottom by chemical reactions (e.g. molecule and aggregate formation). Fe, Al, Ca are more abundant than Na, but the $D_2$ transition is so strong that it provides the largest product of column density $c_{Na}$ and transition cross section, nominally $10^3 - 10^4$ atoms/cm$^3$. The layer has been studied mostly with Lidar technique (Blamont and Donahue, 1961; Albano et al., 1970; Bowman et al., 1969; Sarrazin, 2001).

There are considerable geographic, seasonal, and short-term variations in all parameters of the Na layer, that should be taken into account by the LGS system design. The average height of the layer is at $\approx 92$ km and is usually $\approx 10$ km thick. The variable height distribution is normally Gaussian in shape (Fig. 7), but it can become stratified, which can cause problems in the AO system. At mid-latitudes there are seasonal variations with a maximum in winter, which is $\approx 3$ times higher than the minimum in the summer (Fig. 8). The abundance profile also changes on time scales of hours (Fig. 9). All in all, there is a factor of at least 6 between the maximum and minimum abundance. The layer density also varies between 3 and $6 \times 10^9$/cm$^2$ on a seasonal basis ( Simonich et al., 1979), as well as daily and hourly (Papen and Gardner, 1996; Megie et al., 1978; O'Sullivan et al., 1999; Michaille et al., 2001). $c_{Na}$ can vary by a full order of magnitude in extreme cases. This alone gives a variation of the LGS by 2.5 magnitudes with serious effects on the performance of a LGS system, so efforts should be made to monitor it, and respond quickly to changes, in order to optimize the observing program.

## 3.3    Guide Star System Design Issues

The cone effect and image stabilization are addressed by Foy in Ch. 15.

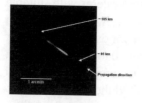

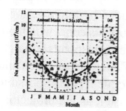

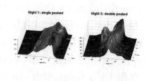

*Figure 7.* The vertical distribution of the sodium atoms can be seen by looking at the laser return from the side. This distribution varies dynamically on the time scale of a few minutes.

*Figure 8.* At University of Illinois: factor of three variation between December-January (high) and May-June

*Figure 9.* At La Palma: variation in the height (horizontal axis) and signal strength (right axis) of the Na layer as a function of time (left axis) for two nights.

### 3.3.1 Launch geometry and spot elongation.

The variations of the effective height, $H$, of the Na layer do not affect the performance of the system per se, but they have to be taken into account in the focusing procedures. For a large telescope, the Na layer is very far out of focus relative to the science object (see Ch. 7). The WFS must be mounted on a translation stage to follow the focus of the LGS, whose distance is $(H/cosz)$, where z is the zenith angle. This height must be calibrated regularly to account for motion of the Na laye. $H$ and thickness of the Na layer, $\Delta H$, give a limit on the maximum separation $d$ between the projector and the telescope. The Na "spot" is actually a long thin column, which appears elongated when viewed from a distance $d$ (Fig. 7). The elongation at zenith is given: $\varepsilon = d\Delta H/H^2 \approx 0.24d$ arcsec.

Therefore $d$ must not be larger than a few meters. Thus the beam projector has to be attached directly to the telescope structure. If it is mounted at the side of an 8-meter telescope the spot size changes by a factor of 3 across the pupil. This not only requires more laser power (to make up for the effect of the larger LGS), but optimum wavefront reconstruction is more difficult, since the noise is anisotropic the pupil. In addition possible stratification of the Na layer can introduce systematic errors into the wavefront determination. If the projector is located at the middle of the pupil, this effect causes a change of focus, which is not very important because the LGS already has to be automatically focused by its own control loop. If the laser is launched from the side, stratification introduces astigmatism into the wavefront measurement that must be corrected using data obtained from a NGS within the isoplanatic patch. This effect is independent on the spot size. Both these effects are minimized if the beacon is launched along the optical axis, a factor that must be traded against the added complexity and cost of placing the launch telescope in this position.

In many cases engineering drive the decision towards one or the other solution. The WFS must be able to separate the Na LGS from the much brighter

Rayleigh one from the lower atmosphere. In pulsed laser systems with suitable repetition rates this can be done by range gating. When viewed slightly from the side, the Rayleigh light forms a cone, which points towards the sodium spot (Fig. 24). The Rayleigh cone corresponds to heights from 0 to ≈40 km, the Na spot to ≈92 km. This gap of a few arc seconds allows us to reject the Rayleigh light by a field stop. In principle it is also possible to feed the laser beam into the the main telescope and use it as the projector. This "monostatic" configuration has severe drawbacks. The enormous amount of scattered light requires gating not only of the WFS, but also of the scientific camera. And phosphorescence of coatings or other materials in the telescope may produce a glow that persists long after the laser pulse, which is unacceptable for observations of faint objects. This approach is not suited for astronomical applications.

**3.3.2    Sodium Guide Star Brightness.**    From the properties of the Na atom and the Na layer (Table 1), we can estimate the power needed for the AO system. Assuming isotropic emission and no saturation, the photon flux $F$ of the LGS observed with the WFS is given by:

$$F = \eta.c_{Na}.\sigma_d P_L cos(z)/(4\pi H^2 E_{photon}) \tag{8}$$

where $\sigma_d$ is the peak $D_2$ cross-section, and $E_{photon}$ is the energy of a 589.2 nm photon. $\eta$ is an overall efficiency factor, which includes the instrumental and atmospheric transmission, and the quantum efficiency of the WFS detector. If we assume $\eta = 10\%, c_{Na} = 5 \times 10^13/m^2$ and require a minimum $F = 300000/m^2/s$ for AO correction in the IR, we find that an average laser power of 2.4 W is needed. This rough estimate demonstrates that LGSs are feasible in principle with currently available laser technology.

For any practical system, the variations of the LGS brightness due to $\partial c_{Na}/\partial t$ are a major concern. Furthermore, the LGS brightness depends strongly on $z$. The $cos(z)$ scaling results from the combination of a factor $cos^2(z)$ for the distance to the LGS, and a factor $1/cos(z)$ for the length of the Na column.

It should be noted that when we compare the brightness of a LGS to a NGS, the result depends on the spectral bandwidth, because the LGS is a line source, whereas the NGS is a continuum one. The magnitude scale is a logarithmic measure of flux per spectral interval (see Ch. 15). This means that a (flat) continuum source has a fixed magnitude, no matter how wide the filter is. In contrast, the magnitude of a line source is smaller for narrower bandpasses. It is therefore advisable to use the equivalent magnitude only for qualitative arguments. The photon flux should be used in careful system analyses.

Figure 9 shows that the Na LGS brightness can vary by several magnitudes over fairly short times. Clearly, these variations have a strong influence on the performance of the AO system. The performance degradation with increasing

$z$ is stronger than for NGSs because of the $cos(z)$ term in the LGS brightness, and the performance depends critically on the $c_{Na}$ and on the seeing.

**3.3.3 Spot size.** The size of the LGS is a critical issue, since it defines the saturation effects of the laser and the power needed to reach a given system performance, and also the quality of the wavefront sensing. There is an optimum diameter of the projector, because, if the diameter is too small, the beam will be spread out by diffraction and if it is too large it will be distorted due to atmospheric turbulence. The optimum diameter is about $3r_0$, thus existing systems use projection telescopes with diameters in the range of 30-50 cm.

**3.3.4 Safety Considerations.** High-power lasers raise a number of safety issues. There are the flammability and the toxicity of dye solutions. Most importantly, the eye hazards of laser radiation require careful shielding of the beam, and interlocks that restrict access to the laser room and to the dome. The laser could also dazzle aircraft pilots if they look directly down the beam. It is therefore necessary to close a shutter in the beam when a plane comes too close, either manually by human spotters, or automatically by use of radar, thermal IR or CCD cameras. Care must also be taken to avoid hitting overhead satellites in the case of pulsed or high power laser systems.

# 4. Laser Guide Star Technologies and Experiments

## 4.1 Rayleigh Guide Star Lasers

Because of the strength of Rayleigh scattering, it is fairly easy to generate bright LGSs with this method. Several commercial laser technologies can provide high power output at moderate cost. General requirements include a power output of $\approx 100$ W, pulse rate of >1 KHz and good beam quality to produce a small beacon diameter. Because the Rayleigh cross-section is proportional to $\lambda^{-4}$, scattering is most efficient at UV wavelengths and is useful up to $\approx 600$ nm. Historically the high power lasers needed for Rayleigh LGSs have had poor beam quality. This required the projection of the laser through the main telescope in order to create a seeing limited spot. Copper vapor lasers ($\lambda$ 510.6 + 578.2 nm), were among the first to be demonstrated. Excimer lasers have also been used to produce light at 351 nm. They work at best efficiency in the range of 300-350 Hz. Excimer laser technology is quite mature and is very reliable; however, they are fairly large and have a high power consumption.

Advances in laser technology now allow for solid-state lasers of high beam quality. These beams may be projected from a much smaller auxiliary telescope, which negates the need for optical switching and completely eliminates any main telescope fluorescence. Solid-state YAG lasers are the most common type of lasers commercially available. These lasers use a crystal as the lasing medium (solid-state), and can be reasonably compact and relatively efficient. The standard operating wavelength (1064 nm), can be frequency doubled in a nonlinear crystal to produce 532 nm or frequency tripled to produce 355 nm.

## 4.2    Rayleigh Beacon Demonstrations

### 4.2.1    Starfire Optical Range.
Rayleigh LGSs with AO at the SOR 1.5 m telescope (Fugate and Wild, 1994) have yielded near diffraction limited images (Fig.10). The LGS/AO experiments utilized a 75 W and later on 150 W copper vapor laser. The 546-actuators deformable mirror controled from a CCD Shack-Hartmann WFS had a 143 Hz closed loop control bandwidth. The system achieved a NGS vs. LGS Strehl for the same object of 0.59 vs. 0.48.

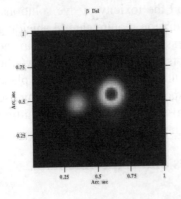

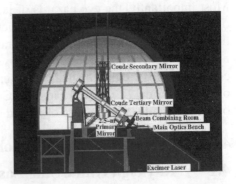

*Figure 10.* $\beta$ Del at $\lambda$ 850 nm observed at SOR with Rayleigh LGS/AO. Separation 0.275", magnitude difference 1.07. Without AO, the image would occupy the entire field of view. (Fugate and Wild, 1994).

*Figure 11.* (a) Schematic of the Rayleigh LGS system on the 2.5-m telescope at Mount Wilson.

### 4.2.2    Mt. Wilson Observatory.
The UnISIS excimer laser system is deployed on the 2.5 m telescope at Mt. Wilson Observatory (Thompson and Castle, 1992). A schematic of the system layout is shown in Fig. 11. The 30 W, 351 nm excimer laser is located in the coudé room. The laser has a 20 ns pulse length, with a repetition rate of 167 or 333 Hz. The laser light is projected from the 2.5 m mirror and focused at 18 km. A fast gating scheme isolates the focused waist. A NGS is needed to guide a tip-tilt mirror. Even with relatively poor seeing, UnISIS has been able to correct a star to the diffraction limit.

### 4.2.3    Advanced Rayleigh Guide Star Concepts.
The main disadvantage of Rayleigh is its low altitude and the cone effect. This problem is significantly reduced when Rayleigh LGSs are used in a constellation for multiconjugate adaptive optics (MCAO) (Beckers, 1987). Projection of multiple

seeing-limited Rayleigh beams from an auxiliary telescope with a low power solid-state laser has recently been reported (Lloyd-Hart et al., 2001). There are also two novel and attractive proposals for Rayleigh LGSs: first, the use of a dynamic refocusing mirror in the telescope's focal plane can make the LGS much brighter and/or higher (Angel and Lloyd-Hart, 2000), and second, the explicit measurement of the return from each laser at multiple heights which can improve the ability to discriminate between atmospheric layers.

In the past, Rayleigh performance has been limited by poor laser beam quality and by depth of field limitations. New near diffraction-limited YAG-based lasers remove the depth of field restriction for the projected beam. UV lasers are now available with sufficiently high quality beam profile to allow $\approx 20$ cm diameter projectors. If light returned by Rayleigh scattering could be collected by a WFS over a large range of height, then bright LGSs could be made using relatively cheap lasers of a few W. To be useable though, the returning light must be held in focus on a WFS. Lloyd-Hart et. al.(2001) have demonstrated that a sinusoidal motion of a mirror in the focal plane can maintain a well-focused image over heights from about 16 to $\gtrsim 30$ km. For an AO running at 1 kHz at 354 nm, the mean return flux integrated over this range is about 10 times the flux from a Na laser of the same power. Because of the low altitude multiple beams could be used to provide complete sampling of the turbulence, and this approach moves naturally towards an implementation of MCAO.

A preliminary demonstration of multiple Rayleigh LGSs was done at Steward Observatory's 1.5 m telescope (Fig. 12). The laser was a pulsed, frequency-tripled YAG, triggerable at rates from a few hundred Hz to 10 kHz with pulse powers in the range 3-6 J, depending on pulse rate. The beam had a Strehl $\sim 0.5$. The projected beam was 25 cm in diameter. The waist width was 2.3 arcsec FWHM, consistent with the (one way) seeing of 1.5 arcsec. Preliminary reductions indicated the system produced $N = 1.4\ 10^3$ photons/m$^2$/s/km per spot, assuming atmospheric transmission of 66% over the 14 km to the telescope at 2.5 km altitude. Thus high beam quality of the YAG laser allows projection from a small projector, greatly simplifying the system and eliminating fluorescence in the main telescope. 6W lasers and better beam quality are available, and should be powerful enough to provide beacons at 33 km altitude.

The use of dynamic refocusing optics is currently being tested at the Stewart 1.5-m telescope (Georges et al., 2002), and will eventually be deployed at the 6.5-m Multiple Mirror Telescope (MMT) (Lloyd-Hart et al., 2001). In addition, a similar scheme using phase diversity instead of dynamic focus has been proposed (Lloyd-Hart et al., 2001). With refocused beacons, implementing height diversity can be accomplished by multiply gating each beacon as the pulse flies away from the telescope. Wavefront sensor detectors are required that can either be read out or store an image on a timescale of $10\mu s$. For extremely large telescopes range gating or refocusing will be needed even for sodium LGSs,

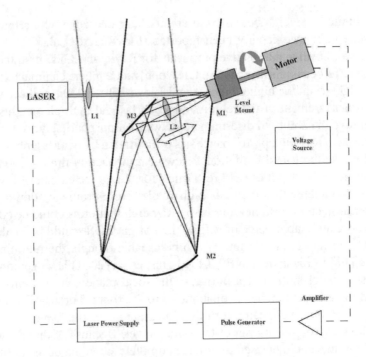

*Figure 12.* Schematic view of the projector used to launch 3 simultaneous Rayleigh beams into the sky. A spinning mirror M1 at an image of the exit pupil directed successive laser pulses into three separate beams (only 2 shown here) without causing beam wander on the primary mirror M2. Electronic synchronization was implemented between the motor and the laser.

which then lose the advantage of simpler focal plane optics. A combination of Rayleigh and sodium beacons may be desired for these systems.

## 4.3    Sodium Guide Star Lasers

For large telescope apertures, Na LGS offer improve sampling of the atmospheric turbulence due to their much higher altitude. Single beam systems are now being developed for and deployed on 8-10 m class telescopes. Since resonant backscattering from the mesospheric Na layer is the method chosen for most LGS projects, we will concentrate mostly on this technique.

Finding a suitable laser for the generation of a Na LGS is very non-trivial task because of the highly demanding and partially conflicting requirements. The need to operate at the $D_2$ wavelength limits the choices to tunable or specially designed lasers, which are difficult to obtain with high power output. Good beam quality is needed for the generation of seeing limited LGSs. Pulsed systems must have a pulse format and spectral bandwidth that minimize saturation. The pulse repetition frequency has to be at least as high as the desired read rate of the WFS. Furthermore, the ideal laser system should be affordable

and easy to maintain. No laser has been built so far that fulfills all these requirements simultaneously; thus compromises have to be made. Consequently,
there is no agreement yet about the most promising technology, and a number
of different approaches have been taken in the existing and planned Na LGS
systems. The $D_2$ line doesn't correspond to any common laser materials, so
clever approaches been necessary to produce the required wavelength. The
approaches that have been demonstrated or are on the horizon include:

1 Dye lasers, which can lase in either pulsed or CW formats. They may
   be pumped by flashlamps or by other lasers such as copper vapor, argon
   ion or by frequency doubled Nd:YAG lasers.

2 Sum-frequency mixing of two solid-state YAG lasers in a nonlinear crystal (see Ch. 20) to generate 589 nm in CW, CW mode-locked and macro-
   micro pulse formats. The Nd:YAG lasers can be pumped by flashlamps,
   but higher efficiency is obtained using diode lasers.

3 Sum-frequency mixing of two diode pumped fiber lasers (938 and 1583
   nm) in a nonlinear crystal. CW format has been demonstrated at low
   power levels; higher powers and pulsed formats are under development.

4 Frequency doubled/Raman shifted fiber laser to generate 589 nm. This
   is a diode pumped, CW format currently under development.

A laser tuned to $D_2$ produces Rayleigh scattering up to $\approx 40$ km. When projected coaxially with the telescope, the $D_2$ return must be separated from the
Rayleigh scattering from lower layers. For a pulsed format, this can be done
with range gating. Scattered light at the top of the Na layer should be received
at the telescope before the next pulse is transmitted. This requires a period of
$667\mu s$. The maximum pulse length is governed by the physical gap between
the Rayleigh and Na returns. Rayleigh becomes negligible above 40 km, and
the lowest altitude for Na generation is $\approx 80$ km. To avoid overlap between
these two signals the pulse length should not exceed the physical distance between them ($\approx 40$ km), which corresponds to a pulse length $< 133\mu s$. Thus in
a coaxial projection geometry, the maximum pulse repetition rate should not
exceed 1500 Hz and the duty cycle should not exceed 20%.

For projection from the side of the telescope, the Na and Rayleigh returns
are angularly separated. Thus a field stop in the focal plane of the WFS eliminates the pulse overlap problem, allowing higher pulse repetition rates with no
limitation on the duty cycle. The final projection geometry is from behind the
telescope secondary. Delivery method of the beam to the projector determines
the pulse format at present. Any temporal format can be used with free space
propagation across the secondary. A more desirable method is to transport the
beam in an optical fiber to reduce black body radiation and scattered light;
however, the current state-of-the-art fiber transport requires a CW beam ($<$
10W) with significant bandwidth. Rapid advances in fiber technology should
enable the transport of other pulse formats and higher powers in the future.

**4.3.1    Continuous-Wave Dye Lasers.**    Only one commercially available laser type is directly applicable to Na LGSs. This is a CW dye laser, pumped by an argon ion $Ar^+$ or a frequency doubled YAG laser (Jacobsen et al., 1994; Quirrenbach, 1997). One advantage of the single- frequency CW laser system is that it is always "on". This provides the lowest peak power, thereby reducing saturation. With atmospheric transmission of 0.6-0.8, the maximum CW power that can be used without adding spectral bandwidth is 4-5 W. A ring dye laser made by Coherent (CTI), produces $\sim 3$ W, when pumped by a 25 W $Ar^+$ laser (Georges et al., 2002). The dye molecule is Rhodamine 6G dissolved in water or ethylene glycol, and pumped through a nozzle under high pressure, so that a free dye jet is generated. The pump laser is focused into the dye jet, producing stimulated emission over a wide spectral range. The exact frequency and linewidth are then selected by tuning the cavity. The power output of CW lasers is a nearly linear function of the pump power, up to 4-6 Watts because of the tendency of the dye jet to become unstable when the pump deposits too much heat. In addition, a modeless, 3 GHz dye oscillator has recently been demonstrated (Pique and Farinotti, 2003). The continuous spectral profile provided by this design would be of assistance in reducing saturation, allowing increased return for lower power CW formats.

$Ar^+$ lasers are robust, but the efficiency is low, $\approx 0.05\%$. The heat generating additional turbulence near the telescope is a serious problem. They are now replaced with frequency doubled YAG lasers. These cannot yet deliver quite the same power at 532 nm, but they are much more efficient. CW dye laser systems have now been used to demonstrate Na LGSs at several observatories.

**ALFA.**    ALFA (Adaptive Optics with Laser guide star For Astronomy) was a joint project of the Max-Planck- Institutes for Astronomy and for Extraterrestrial Physics from 1997-2001 (Glindemann and Quirrenbach, 1997). The system was installed at the 3.5-m telescope in Calar Alto, Spain. ALFA used a commercial CW dye laser with an argon ion pump laser. The commercial dye laser was modified and optimized for high power output, producing 4-6 W single mode with good beam quality. This corresponds to a calculated $m_v$ of 9-12 magnitude depending on $z$ and $C_{Na}$. The best Strehl produced by the LGS/AO system was $\sim 0.23$.

**ESO VLT/Max Planck CW Dye Laser.**    The MPI is developing a CW dye laser for deployment on one ESO 8-m VLT telescope in 2004 (Fig. 13). The oscillator is a Coherent 899 ring dye laser, with a 2-5 W output, pumped by a 10 W, Coherent Verdi frequency-doubled Nd:YAG laser. The beam is amplified in a four-pass amplifier with 4 high velocity dye jets pumped with 4 10 W Verdi lasers. The system utilizes Rhodamine 6G in ethylene glycol; however, because of the high pump power, the dye degrades quickly , and must be changed weekly. The laser system, PARSEC, has now been demonstrated at 12.5W, with good beam quality. Phase modulation will be used to add ∼ 500 MHz of bandwidth to mitigate saturation, and to allow delivery of the beam via a photonic crystal fiber to a projector behind the telescope secondary mirror. Production of this modulator has required a significant R&D effort since the bandwidth is added at high power, requiring a large aperture to prevent optical damage. The large aperture requires a very efficient, high power RF driver, technically challenging to produce.

Another technical challenge is presented by the beam projection system behind the secondary mirror. The VLT will be the first telescope to utilize a fiber delivery system to the projector. This required the development of a special photonic crystal fiber to allow the transport of high power without inducing nonlinear optical effects in the fiber. The current fiber is limited to transport of ∼ 8 W, due to simulated Brillouin scattering in the fiber (SBS). SBS is caused by an electrostrictive interaction between light propagating in a material and the material itself, which creates an acoustic wave in the material. This in turn creates an index grating that reflects the beam back on itself, thus limiting the amount of power which can propagate in the fiber core. This effect scales as the product of the intensity of the beam in the fiber and the fiber length (Agrawal).

### 4.3.2 Pulsed Dye Lasers.

To generate Na LGSs pulsed dye lasers can be pumped by copper vapor lasers (Hogan and Web, 1995) or by pulsed Nd:YAG lasers (Friedman et al., 1994). They can deliver much higher output power than CW lasers, because the pulses are much shorter than the time needed for the development of thermal instabilities if the velocity of the dye is such that the dye is completely changed in the active region before the next pulse. Velocities of ≈ 10-30 m/sec are possible, so that pulse repetition rate can be of ≈ 10-30 KHz, with pulse lengths of ≈ 120 ns. The effectiveness is thus limited by the duty cycle (≪ 1%). It is always necessary to use a spectral format matched to the Doppler width of the $D_2$ line to avoid saturation.

### LLNL AVLIS Laser.

The first WFS measurements using a Na LGS were performed at LLNL (Max et al., 1994; Avicola et al., 1994). These experiments utilized an 1100 W dye laser, developed for atomic vapor laser isotope separation (AVLIS). The wavefront was better than 0.03 wave rms. The dye laser was pumped by 1500 W copper vapor lasers. They are not well suited as a pump for LGSs because of their 26 kHz pulse rate and 32 ns pulse length. The peak intensity at the Na layer, with an atmospheric transmission of 0.6 and a spot diameter of 2.0 m, is 25 W/cm$^2$, ≈ 4× the saturation. The laser linewidth and shape were tailored to match the $D_2$ line. The power was varied from 7 to 1100 W on Na layer to study saturation. The spot size was measured to be ∼ 7 arcsec FWHM at 1100 W. It reduced to 4.6 arcsec after accounting for saturation (diffraction limit = 3.1 arcsec). The spot centroid motion was $\sim 0.5$ arcsec rms. Measurements of the LGS intensity, size and motion compared well with models of Na LGS (Morris, 1994). Even in the highly saturated state, the LGS proved to be bright enough, at sample rates of 125 Hz, for an AO system.

**Lick Observatory.**     The success of the LLNL/AVLIS demonstration led to the deployment of a pulsed dye laser / AO system on the Lick Observatory 3-m telescope (Friedman et al., 1995). LGS system (Fig. 14). The dye cells are pumped by 4 70 W, frequency-doubled, flashlamp-pumped, solid-state Nd:YAG lasers. Each laser dissipates $\sim 8$ kW, which is removed by water-cooling. The YAG lasers, oscillator, dye pumps and control system are located in a room in the Observatory basement to isolate heat production and vibrations from the telescope. A grazing incidence dye master oscillator (DMO) provides a single frequency 589.2 nm pulse, 100-150 ns in length at an 11 kHz repetition rate. The pulse width is a compromise between the requirements for Na excitation and the need for efficient conversion in the dye, for which shorter pulses are optimum. The laser utilizes a custom designed laser dye, R-2 perchlorate, that lasts for 1-2 years of use before replacement is required.

The cavity is frequency locked to the $D_2$ emission from a Na cell. The seed pulse is broadened to 3 GHz using two phase modulators to mitigate saturation and to allow transport of the low power seed pulse through an optical fiber up to the telescope for additional amplification. The frequency-doubled Nd:YAG pump beams are delivered to the amplifier cabinet mounted on the side of the telescope via large mode optical fibers. It contains a dye preamplifier, spatial filter and dye amplifier, bringing the laser power up to $\sim 18$ W. The amplified beam is relay-imaged to a diagnostic package mounted $\sim 7$m above the final dye amplifier on the side of the telescope, prior to injection into the projector. A high-speed tip-tilt mirror, driven by the AO system, stabilizes the laser beam in the optical field of view. It also compensates for jitter in the outgoing beam plus the residual anisotropic wavefront tilt errors due to the physical separation between the telescope and the laser projector. There is also a WFS to measure the beam quality prior to projection. This table also contains the first element of the afocal beam expander and a fast shutter to cut off the laser beam in an emergency. Centering of the beam into the projection is accomplished using two adjustable mirrors in a dogleg configuration. Sensing of the beam position is implemented with 2 CCD cameras. The status of the beam is transmitted to the computer control in the laser room, and to the telescope control room. The objective of the laser projector is mounted near the top of the telescope tube. The projection aperture is 29 cm, $\approx 3 \times r_0$ at the Lick site. A radar dish adjacent detects approaching airplanes, and shutters the laser when triggered. The entire system is monitored by a Macintosh using Labview.

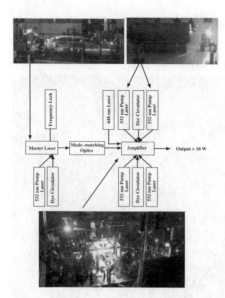

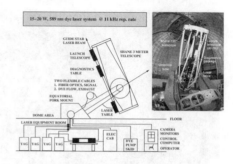

*Figure 14.* Layout of the Lick Observatory laser guide star system.

*Figure 13.* The Parsec CW dye laser system

The Lick AO system demonstrated the first wavefront correction using a Na LGS (Max et al., 1997), and engineering tests demonstrated LGS corrected Strehls of 0.5-0.6 at 2.2$\mu$m (Fig. 15, Gavel et al., 2003). In 2002, the LGS system was turned over to the Observatory staff for operation in science observing mode. It is used almost 100 nights per year. The first refereed science paper using a sodium LGS/AO system was published by Perrin et al. (2004).

**W.M. Keck Observatory.** A LGS system similar to the Lick Observatory design has been commissioned at the Keck Observatory (see Section 14.4.3.4).

### 4.3.3 Sum-Frequency Nd:YAG Lasers.

The sum- frequency mixed Nd:YAG laser (Jeys et al., 1989) is specifically developed for Na LGSs. Nd:YAG lasers normally operate at 1064 nm, but they can also lase at 1319 nm. By a fortunate coincidence, the sum-frequency of the two Nd:YAG wavelengths produce 589 nm, which is the Na $D_2$ wavelength: $1/1064 + 1/1319 = 1/589$ nm (Fig. 16). The outputs of two Nd:YAG master oscillators tuned to 1064 and 1319 nm are amplified and then superimposed with a dichroic mirror. The beams are then passed through a nonlinear crystal (lithium triborate (LBO)), which produces the sum-frequency. This laser type has the advantage of being all solid-state, avoiding the chemistry of dye solutions. Sum frequency generation can be very efficient, providing an inherently high quality wavefront and its spectral and temporal properties can be matched to the Na layer.

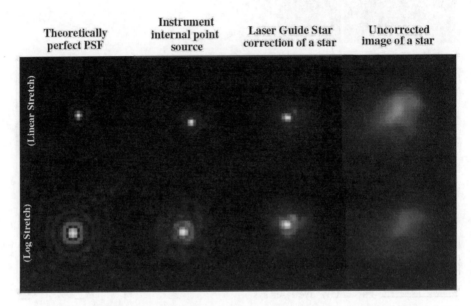

| Theoretically perfect PSF | Instrument internal point source | Laser Guide Star correction of a star | Uncorrected image of a star |

RX J0258.3+1947   10/20/00 2:45   Ks  V = 15   K= 13.32   20s   S = 0.53

*Figure 15.* Lick Observatory 3 m telescope results with their experimental LGS-AO.

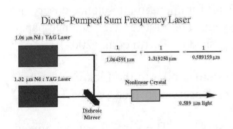

*Figure 16.* Sum-frequency mixing two Nd:YAG lasers at 1064 and 1319 nm in a nonlinear crystal produces 589 nm light.

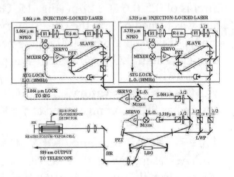

*Figure 17.* Schematic diagram of the AFRL CW sum-frequency laser.

On the other hand, the technologies for obtaining stable laser action at 1319 nm and for the nonlinear mixing of the two Nd:YAG wavelengths are far from trivial, and the robustness of this system for routine observations is only now being demonstrated. Conversion efficiency is proportional to the square of the intensity. Thus two beams must be overlapped and tightly focused into the LBO crystal. However, the angular content of the focused radiation must be kept less than the phase-matching angular acceptance of the crystal. The narrow angular acceptance of these crystals limits the degree of beams focusing. In addition, if the light is CW or pulsed with a high duty cycle, the tight focusing results in a high average intensity at the focus, resulting in local heating. The small temperature acceptance range, relatively high absorption coefficients and high susceptibility to photorefractive damage in some crystals severely limit the utility of these materials in high average intensity applications.

**Macro-Micro Pulse Sum Frequency Lasers.** The sum-frequency lasers, pumped by flashlamps, were originally demonstrated in a macro/micro pulse format at Lincoln Laboratory, giving 5 W at 10 Hz (Jeys et al., 1989) and then 10-20 W at an 840 Hz (Jelonek, 1992; Jeys, 1992). A duty cycle of 10-15% can be achieved allowing macro pulse lengths of $100 - 150\mu s$ at a repetition rate of 1 kHz. By mode-locking the master oscillators, the output is obtained as a train of sub-nanosecond micro pulses with a spectral width of 2-3 GHz, matching the $D_2$ line FWHM. These short pulses also produce efficient conversion in the nonlinear crystal. More recently, a pure CW format has been demonstrated at the 20 W level by the Air Force Research Laboratory (AFRL), for use on the SOR (Bienfang et al.). A mode-locked, CW pulse format is being developed by CTI for the Gemini North Observatory (McKinnie, 2003). Various theoretical studies have been completed comparing the efficiency of different pulse formats, including CW (Milonni et al., 1999). For average laser powers $\gtrsim 20$ W, diode lasers are more attractive than flashlamps for pumping. They are efficient, compact and produce little electrical noise. They also produce their output at near-IR wavelengths for which YAG lasers have strong absorption bands. Currently, diodes are expensive and have limited lifetime. However, their reliability is steadily improving, while the cost is decreasing.

A third generation macro-micro pulse system was developed in collaboration with the University of Chicago (Kibblewhite and Shi, 1999). Each YAG head consisted of double-pass zigzag slabs pumped by 40 diode lasers. Deformable mirrors were used in each cavity to maintain beam quality. The infrared light passed twice through an LBO sum frequency crystal. At a 100 MHz micro pulse frequency, the pulse format appears CW to the sodium atoms. The crystal, however, sees a higher peak power, leading to an increase in conversion efficiency of $\approx 5$: 30% conversion efficiency was achieved ($m_v = 9.2$ with $r_0 = 10$ cm). The laser produced 4-10 W average power with 1 GHz of bandwidth. The total power used by the laser and support electronics was $\sim 2300$ W. Efforts are now in progress to scale the slab technology to higher powers. A redesigned system that should be capable of scaling to higher powers has now been demonstrated at >4 W, and is intended to be installed on the Hale telescope at Palomar Observatory.

**AFRL CW Sum-Frequency Generation Laser.**    This laser (Fig. 17) uses a doubly-resonant external cavity to create the high field strength required to achieve highly efficient nonlinear sum-frequency generation with CW laser sources (Vance et al., 1998). Furthermore, high-power, linearly polarized, TEM$_{00}$, stable, single-frequency, injection-locked lasers are required to obtain significant circulating fields (Teehan et al., 2000). A high-power ring-laser oscillator is injection-locked to a low-power, nonplanar ring oscillator (NPRO), produced by Lightwave Electronics Corp. It provides a stable single-frequency reference to which the high-power ring oscillators are phase locked using the Pound-Drever-Hall (PDH) technique (Drever et al., 1983). A piezoelectric transducer-mounted mirror in the ring oscillator acts as the actuator for the phase- locked loop. Phase sidebands for the PDH technique are created using resonant electro-optic modulators in the low-power NPRO beam. These phase sidebands are also used in phase locking the sum-frequency generation cavity.

The high-power ring oscillators contain two diode-pumped Nd:YAG rod amplifiers built by Lightwave Electronics Corp. Each amplifier uses a 1.6-mm diameter, 63-mm long Nd:YAG rod. It uses three 20-W diode in the 1064 nm ring oscillator and three 25-W diode in the 1319 nm one. A $20\times5\times5$ mm$^3$, noncritically phase matched LBO crystal is used for sum-frequency generation (SFG). Nearly 60% optical-to-optical conversion efficiency was obtained through implementation of doubly-resonant sum-frequency mixing in LBO. The SFG cavity is made simultaneously resonant with single longitudinal modes of the pump and signal lasers by the PDH technique. A piezo actuator adjusts the position of the second flat cavity mirror to lock it to 1319 nm, while the frequency of the 1064 nm laser is tuned by adjusting the frequency of the NPRO to which this laser is locked to make it resonant with the SFG cavity by using the NPRO's internal PZT actuator. The output wavelength is fixed with respect to Doppler-free absorption at the $D_2$ line in a sodium cell.

"First light" on the sky was November, 2002. For 12 W of circularly/linearly polarized light, the LGS magnitude was $V = 7.2 / 7.9$, with a measured flux of 1050/650 photons/cm$^2$/s. The laser power has now been scaled the power to 50 W, prior to permanent installation on SOR in 2004.

**Gemini North Observatory/CTI Mode-locked SFG Laser.**    CTI is developing the first commercial solid-state Na LGS system. It will be installed on the center section of the 8-m Gemini North telescope, with the output beam relayed to a projector behind the secondary mirror. The projected beam is required to be 10-20 W power, with $M_2 < 1.5$. The architecture is based on sum-frequency mixing two mode-locked solid-state Nd:YAG lasers. The mode-locked format provides significantly higher peak intensity than CW, enabling more efficient SFG conversion. The laser is also free of the thermal and intensity transients that are inherent in the macro pulse format. The chosen

*Table 2.* Keck laser specifications

| Average power | 20 Watts |
|---|---|
| Pulse repetition rate | 25 kHz |
| Pulse duration | 125 ns |
| Beam quality | 2.5 × diffraction limit (DL) |
| Beam size | 50 cm |
| Wavelength | 589 nm (sodium $D_2$ line) |

format is also compatible with upgrade paths for the higher power level and modulated pulse formats required for MCAO.

Linearly polarized, near-diffraction-limited, mode-locked 1319 and 1064 nm pulse trains are generated in separate dual-head, diode-pumped resonators. Each 2-rod resonator incorporates fiber-coupled diode lasers to end-pump the rods, and features intracavity birefringence compensation. The pulses are stabilized to a 1 GHz bandwidth. Timing jitter is actively controlled to < 150 ps. Models indicate that for the mode-locked pulses, relative timing jitter of 200 ps between the lasers causes < 5% reduction in SFG conversion efficiency.

The SFG stage design calls for ≈ 50 W of IR light focused into a SFG crystal, to generate ≳ 20 W at 589 nm. To date, CTI has demonstrated 30 W of linearly polarized light at 1064 nm, and 18 W at 1319 nm, both with near-diffraction-limited beam. At the time of writing this article, >4 W at 589 nm has been demonstrated by SFG in a nonlinear periodically poled KTP crystal. Periodically poled materials are a recent advance in nonlinear materials that enable high efficiency frequency conversion of CW beams by overcoming the phase matching limitations of long crystal lengths. They have been used reliably in the IR; but they suffer from photorefractive damage in the visible. Solving this problem is an active area of materials research; as it has not solved in the required timeframe for Gemini deployment in late 2004, bulk LBO crystals and a resonant ring scheme will to be used to achieve the power required.

### 4.3.4 A Systems Design Example: The Keck Laser Guide Star.

**System Design.**    For the Keck AO, LGS requirements are 0.3 ph/cm$^2$/ms for the return flux and 0.6 arcsec for the spot diameter. The resulting laser system specifications are given in Table 2. The repetition rate was increased by a factor of 2 relative to the Lick system to improve the sodium return.

The overall conceptual layout of the pulsed dye laser LGS system is shown in Fig. 18. A thermally insulated room located on the dome floor houses much of the laser system to minimize vibrations on the telescope and the heat dissipated within the dome. The enclosure houses 6 frequency-doubled Nd:YAG pump lasers, the DMO, the associated laser electronics and diagnostics, the safety system and the computer systems that operate the laser system. The dye pumps are located in a flammable materials cabinet on the dome floor outside the laser room. The dye flow tubes, large mode fibers for pump beam delivery and single mode fiber to deliver the 589 nm seed beam to the laser table are routed to the side of the telescope elevation ring. The package there consists of three parts: the dye amplifier system and the diagnostic system on a common table, and the projector. The projector is mounted to the outside of the telescope frame. The input telescope lens is located on the laser table, and the final projection lens is mounted at the top of the secondary mount. An IR camera is mounted adjacent to the final lens for aircraft surveillance.

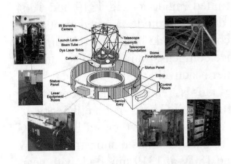

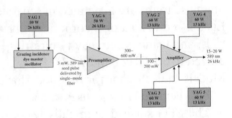

*Figure 19.* A set of 6 frequency-doubled, Q-switched Nd:YAG lasers are used to pump the DMO, preamplifier and power amplifier.

*Figure 18.* Layout of the pulsed dye laser guide star system at the Keck observatory.

The 6 Nd:YAG lasers pump the DMO, preamplifier and power amplifier (Fig. 19, Friedman et al., 1998). The YAG lasers are built from commercially available flashlamp/laser rod assemblies, acousto-optic Q-switches and frequency doubling crystals (LBO and KTP). Most of the mirror mounts and crystal holders are commercial. Nd:YAGs are frequency doubled to 532 nm using a nonlinear crystal. The Nd:YAG rod and nonlinear crystal are both in the pump laser cavity to provide efficient frequency conversion. The 532 nm light is coupled out through a dichroic and fed to multimode fibers which transport the light to the DMO and amplifier dye cells.

The pump lasers were designed and built at LLNL. Two laser cavity configurations are employed. Two "L" shaped cavities run at the full system repetition rate of 26 kHz, producing 40-50 W per laser. They pump the DMO and preamplifier dye cells. Four "Z" cavity lasers run at 13 kHz, each producing between 60-80 W. They are interleaved in the power amplifier dye cell to produce an effective 26 kHz repetition rate. Flashlamps were used to pump the frequency-doubled YAG lasers as diode-pumps were much more expensive at the time the Keck LGS was designed. In addition, high wavefront quality is not required from the pump lasers, since the light is used simply to illuminate the dye cells, and transport to the cells via multimode fiber.

The frequency-doubled YAG beams pump the custom dye, R-2 perchlorate, to get the 589.2 nm beam. The dye waveform generator (WFG) consists (Fig. 20) of the DMO, mode and wavelength diagnostics, 2 phase modulators and the single-mode polarization maintaining (SMPM) fiber launch. They are located in a thermal enclosure to maintain the 1 C temperature control for wavelength stability. DMO dye cell is pumped from the side by one YAG laser, delivered by 2 multimode fibers and coupled into the cell by mode matching optics. A single 90 ns long YAG output is divided between 2 fibers, differing in path length. The 2 fiber outputs are recombined at the cell, producing a longer pump duration (150 ns) for the oscillator. The DMO cavity is formed by 2 mirrors on either side of a dye cell. The cavity contains a grating to tune the wavelength, and an etalon to produce a narrow linewidth. The cavity length is adjusted by a piezoelectric mirror to keep the output beam single mode. Automatic control loops maintain the spectral purity and wavelength of the beam, and enable tuning on and off resonance for background subtraction.

The DMO output is diagnosed with a scanning interferometer to determine its mode content, then passed through 2 electro-optics modulators that generate spectral sidebands to match the $D_2$ line profile. Another scanning interferometer validates the spectral profile. Two diagnostics make up the wavelength acquisition and tracking system. A hollow cathode lamp containing a buffer gas of Ne and Na emits a spectrum containing the $D_1$ and $D_2$ sodium lines along with adjacent neon lines. A simple fixed spectrometer is used to tune the DMO output to the $D_2$ wavelength. Fine-tuning is obtained by observing the fluorescence from the vacuum Na cell using both a TV camera and a photodiode. Two control loops ensure that the laser stays single mode (before the modulators) and centered on wavelength. Once the beam has been properly formatted, it is injected into the SMPM fiber toward the dye laser table.

The SMPM fiber and the pump fibers exit the room and travel $\approx 80$ m through the azimuth and elevation bearings to reach the rest of the laser system on the telescope elevation ring. The laser package consists of a dye preamplifier and a power amplifier both pumped by fibers from the YAG pump system. Only 3.5 mW of 589 nm light can be delivered to the laser table due to the long fiber propagation path. SBS is not an issue because of the phase-modulated bandwidth that is added prior to the fiber. The main issue is stimulated Raman scattering (SRS), which becomes detectable at $\sim 3$ mW. By 5 mW, a significant portion of the light has been Raman shifted reducing the system efficiency. Once the seed beam exits the fiber, an integrated lens assembly produces a 300 $\mu$m beam waist in the preamplifier cell. The beam enters the dye cell at a grazing incidence angle and bounces off the side of the cell where the pump fiber is located to provide a more spatially uniform intensity profile.

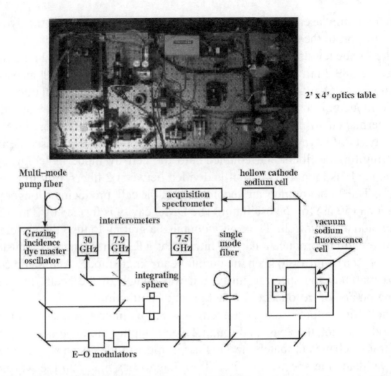

*Figure 20.*   Photograph and schematic diagram of the Keck dye laser waveform generator.

After the preamplifier, the beam is expanded to 2 mm, collimated and imaged onto a 1 mm aperture, producing a flat-top intensity profile. A 3-element telescope relays the aperture plane to the amplifier with a collimated 0.5-mm diameter. The telescope contains a spatial filter pinhole. The nominal power levels are 3 mW into the preamp, 500 mW out of the preamp and 200 mW out of the aperture. A 6° angle of incidence bounce beam geometry is utilized in the amplifier cell. The "bounce" footprint overlaps with the 4 pump beam fibers, arranged in 2 time sets of 13 kHz. The pump fibers have ≈50-60% transmission. The amplifier brings the power up to $\lesssim$ 20 W at 26 kHz.

Motorized mirror pairs, networked alignment cameras and energy monitors are located in between each major assembly described above to provide a means of identifying and correcting system misalignment and performance degradation via a remote computer interface. An automated pointing and centering loop keeps the beam centered in the projection optics, while a high-bandwidth, tip-tilt mirror stabilizes the laser beam in the optical field of view using a signal derived from the wavefront sensor of the adaptive optics system on the main telescope. A diagnostics suite directly below the projector provides an essential monitor of the parameters of the projected beam:

1  Beam position and angle.

2 Near field beam profile: to check the spatial and intensity profiles.
3 Total power.
4 Beam collimation and static aberrations.
5 Dynamic wavefront errors.
6 Polarization.
7 A fast shutter/calorimeter.
8 A camera to look down through the projector for alignment of L4.

Several launch concepts were investigated for the Keck including launching from behind the secondary with reflective designs. The off-axis configuration was chosen because although it does require about twice the laser power as the on-axis case, the implementation is much easier. For example, getting the beam behind the secondary requires transmitting the beam across the primary mirror in the shadow of one of the secondary support struts. If the beam is not baffled either because it is too large or not well controlled, scattering could disrupt the control loop. Encasing the beam in a tube will increase the IR emissivity of the background and trying to keep a small diameter beam behind a secondary strut would probably require a separate centering loop. Finally the transmission losses due to propagating from behind the secondary are higher than those of the off-axis beam train making the comparison more complicated. Reflective designs may be significantly cheaper than refractors but there is always the problem of aligning the reflective primary or inserting a flat to retro the beam. Also, tilt or jitter of the mirror will result directly in a tilt or jitter of the beam. Refractive designs can be made with a flat surface as the last surface and tilt of the lens does not tilt the beam, to first order. Placing the flat facing the sky does force one to use an aspheric surface for the curve, increasing the cost, but it also eliminates the need for a large flat and gimbaled mount to hold it.

The projector chosen for Keck is an afocal Gregorian telescope (see Ch. 3) which expands the beam to a full 50-cm round profile. The final lens is placed with the plane surface facing the sky and the aspheric surface ground to give a 1/10 wave (rms) transmitted wave front. A back reflection from the last surface is used to monitor the quality of the wave front as it leaves the dome.

The laser control system has multiple GUI interfaces, coordinated via EPICS. A state sequencer streamlines start-up and shutdown. Safety systems and interlocks prevent the operator from inadvertently damaging the system. An alarm handler detects faults and provides error and status messages. Real-time power diagnostics provide a means of identifying power drops due to alignment drift, and cameras provide references for repeatable alignment. Automation allows remote alignment and calibration of the laser. Remote controlled power adjustment allows the YAG powers to be independently controlled. Temperature monitoring is provided to sense misalignment of the beams into the fibers, and to control the YAG, DMO and dye temperatures. Counters on external YAG shutters monitor when to change dye, while internal shutter counters monitor the number of hours the flashlamps have operated. The wavelength control system provides detune capability to allow background subtraction.

**Safety Systems.**     A safety system is provided with built in safety shutdowns and emergency stop buttons. Crash buttons are located in the laser room, the laser table enclosure and the dome. The laser system is tied into the Observatory emergency stop system. Included both in the laser room and on the laser table are surveillance cameras, heat exchangers, alcohol sensors and fire detectors. In addition to personnel safety features, extensive interlocks have been installed in the laser to prevent the operator from inadvertently damaging it.

The Mauna Kea Laser Guide Star Policy's (Wizinowich et al., 1998) main requirements include a Laser Traffic Control system for the mountain, Federal Aviation Association (FAA) approval of an aircraft safety program and participation in the U.S. Space Command program to protect satellites. The need for a beam collision avoidance system was identified early in the effort to permit the use of LGS above Mauna Kea. A collision occurs when the projected laser beam enters the field of view of an instrument mounted on a secondary telescope, which would impact its observations. The implemented laser traffic control system includes the collection of participating telescope's pointing data, the determination of potential collision, and the automatic shuttering of the laser prior to a collision (Summers et al., 2002). The aircraft control system consists of an IR camera that automatically block the laser whenever an aircraft enters the area close to the beam. This system is augmented by two human spotters on opposites sides of the dome equipped with crash buttons. Keck participates in the Satellite Clearinghouse Program. All planned observing targets must be manually submitted 2 days in advance of planned use. A list of shutdown times per object is then faxed to the Observatory.

**Laser System Performance.**     Prior to installation on the telescope, a serie of three 40-hour performance runs were performed to demonstrate the required laser performance. The average energy for the final run was 17.4 W, with a daily variation of $\pm 0.2 - 0.5$ W (Fig. 21). The power drifted down $\sim 0.1$ W/hr due to temperature and alignment drift. Under observing conditions, this drift can be corrected on line by remote alignment. The average Strehl ratio of the amplified beam was $\sim 0.83$. The Strehl varied less than 1% over a day (Fig. 22). The laser had <1% down time during the final 40-hour run, in compliance with the specification for < 3% downtime during a run. The initial turn-on of the system each day is nominally <5 minutes, while system warm-up, alignment and calibration take $\sim 2$ hours/day prior to a run. This time frame should reduce as operation experience with the system increases.

First projection of the laser occurred on December 22 and 23, 2001 (Fig. 23). The projected power was $\approx 17$ W. The LGS magnitude was $m_v \approx 9.5$.

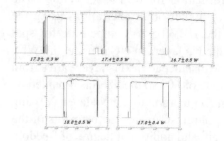

*Figure 21.* Plot of energy as a function of time during the final performance run prior to installation at the summit.

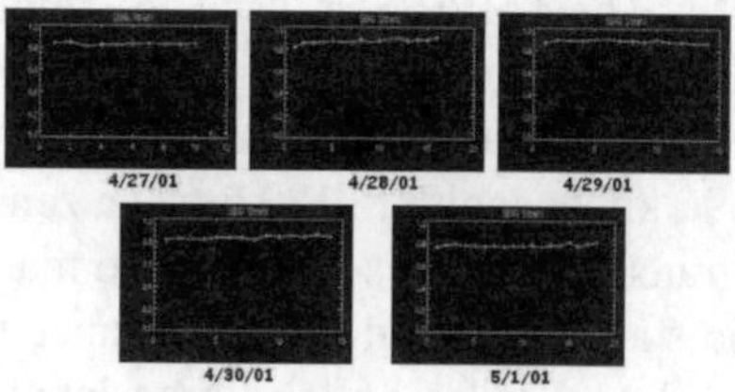

*Figure 22.* The Strehl ratio varies < 1% over the course of a day.

Under 1 arcsec seeing, the LGS FWHM was 1.4 arcsec. Images taken with the Keck primary mirror defocused (Fig. 24) demonstrated the level of distortion expected for an off-axis projection system. Low-level scattering was also minimal, consistent with expectations and seeing conditions.

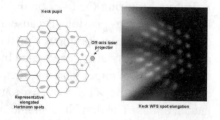

*Figure 23.* First light with the Keck guide star laser, December 2001. (Phototaken by J. MacDonald of the CFHT)

*Figure 24.* The acquisition camera image of the Keck LGS with the segmented mirror "unstacked." The brightening on the left is the Rayleigh scatter of light from the laser. The 36 spots show an elongation increasing with distance from the laser source, which is left of the camera.

**LGS/AO System Integration.** A number of modifications have been implemented in the Keck II AO system to accommodate LGS/AO (Contos et al., 2002): a tip-tilt sensor to provide NGS- based tip-tilt, a low bandwidth WFS (LBWFS) to measure long term focus using a NGS, focus tracking for the high bandwidth WFS (HBWFS) looking at the LGS, and a tip-tilt control loop for the laser pointing.

To overcome the effects associated with laser uplink, a quadrupole tip-tilt sensor on the AO bench is interfaced with the tip-tilt mirror on laser table to determine coordinate and gain transformation for tip-tilt correction. The HBWFS monitors the LGS to correct for higher-order aberrations using the DM. The LBWFS is used to calibrate the HBWFS to compensate for spot elongation due to the finite thickness of the Na layer. In NGS mode, the WFS focus is conjugate to infinity. However, the distance to the Na layer varies from $\approx 90$ km at zenith to 180 km at a zenith angle of $60°$ so the WFS is mounted on a translation stage (>250mm) to maintain conjugation to the Na layer. Moving the stage also changes the distance to the pupil, so one of the elements of the WFS pupil relay optics is translated as well, and additional degree of freedom are implemented for pupil singlet tracking. Na light is rejected from the science path by an IR transmissive dichroic.

To validate that aberrations obtained from the LGS are affecting the science object as well, a dim NGS ($\approx 19^{th}$ mag) is monitored with the LBWFS. The LBWFS has time-averaged focus information, while the HBWFS measured focus term also includes changes in the height of the Na layer. Since the LBWFS monitors a much dimmer star, it must integrate for tens of seconds to minutes to get the required signal level. The time averaged focus and wavefront information provided to the AO system is utilized by adjusting the stage position, which drives the DM focus, ultimately off-loading to the secondary piston.

Before acquiring the laser on the WFS it is important that the telescope be focused as well as possible, because the height of the sodium layers not known exactly. Next the laser is viewed on the TV-guider, which produces an excellent picture of the Rayleigh cone tipped by the LGS. The LSG is steered to overlap with the science target position, then the TV-guider is moved out so the LGS can be seen on the WFS. The positioning can then be fine-tuned using the laser launch optics. The dim NGS should be located within 60 arcsec of the science object. Both the launch telescope and WFS focus are adjusted to obtain the perfect focus positions so that the spots are as small and bright as possible. A reference fiber is then moved into the laser focus position in the AO bench and used to calibrate modes on the WFS. Once calibration is complete, the fiber is moved out and the system is ready to be used.

On September $20^{th}$,2003, all subsystems finally came together to reveal the unique capability of the Keck LGS/AO system. The system locked on HK Tau, a $15^{th}$ magnitude well-known T Tauri binary, revealing details of the circumstellar disk of its companion (Fig. 25). This was the first demonstration of a LGS/AO system on a large telescope. The FWHM is 50 milli-arcsec, compared to 183 milli- arcsec for the uncorrected image. While locked on a $14^{th}$ magnitude star, the LGS/AO system recorded Strehls $\sim 0.36$ (at $2.1\mu$m), 30-sec exposure time, compared to 0.04 for uncorrected images.

Of the above Na LGS/AO systems, only 4 have obtained sharpened images, namely those at Lick, MMT, Calar Alto and most recently Keck. They show that the technology is available today to deploy LGSs, but a substantial amount

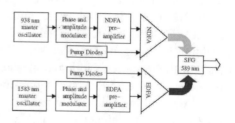

*Figure 26.* Block diagram of the proof of concept 589 nm CW fiber laser system.

*Figure 25.* First light image of the Keck LGS AO system. The lens-like nebula at upper left is a disk of dust and gas surrounding the young star HK Tau B. The star is hidden from direct view, seen only in light reflected off the upper and lower surfaces of the disk.

of system engineering is required before observations with LGS become a routine affair. Current Na LGSs are just sufficient for system development of the single laser case. It is clearly not feasible to deploy multiple laser systems of the technologies described in this article for next generation AO systems. It is also clear that an LGS/AO system requires multidisciplinary technologies, very well engineered systems and an overall tuning together with the science instrument, in order to perform as expected at a good astronomical site.

### 4.3.5 Next Generation Sodium Laser Technologies.

**LGS issues for ELTs.**

Next generation LGS/AO systems will employ multiple LGSs for MCAO to enlarge the corrected field of view up to $10\times$ those of current- generation AO systems (see Ch. 13 and 15). Furthermore, concepts for MCAO on 30m class telescopes rely strongly on the feasibility of Na LGSs. It is generally conceded that Rayleigh LGSs at ranges of 20-30 km, by themselves, are not a viable option for ELTs due to the severity of the cone effect. Detailed simulations and analytical AO performance models indicate that Na LGSs will enable MCAO systems with diffraction-limited image quality in the near IR over 1-2 arc minute fields of view, with performance similar to the 8-m class MCAO systems now under development. But new laser and beam projection concepts must be developed with performances well beyond the current generation. 30-m telescopes will be strongly affected by cone effect, spot elongation and Rayleigh backscatter from multiple LGSs. In addition, $\sim 50 - 100$ W power will be required for extending AO capability toward visible and MCAO applications. The elongation of a LGS varies as $hd/H^2$, where H is the range to the Na layer, $h$ is its thickness, and $d$ is the separation between the projector and a WFS subaperture. The value of $d$ reaches at least 15 m for a 30 m ELT, yielding an elongation $\gtrsim 3$ arcsec. Wavefront measurement error varies linearly with the source width and measurement accuracy can only be recovered by increasing the SNR. Power requirements will increase significantly for ELT LGS/AO systems unless new concepts are developed. Several methods have been proposed to defeat the elongation problem, including laser in the 10-50 W range, innovative pulse and spectral formats, and/or a multiplicity of projectors for each LGS location.

Although feasible in principle, increasing Na laser powers to $\sim 50$ W will require significant advances in the engineering of the components, and will increase the cost and complexity of the final systems as well.

Since elongation occurs in only one dimension on each WFS subaperture, multiple lasers with separate launch locations could be used to generate multiple, co-located LGSs with distinct orientations. WFS measurements of two or more elongated spots could then be used to accurately determine both components of the wavefront gradient vector. Nelson, et. al., have shown that this approach should be able to eliminate the impact of elongation on measurement accuracy if the total power is increased by a factor of 3 to generate the additional LGSs (Ellerbroek, priv. comm.). A complication is the increased number of projectors and beam transport systems, and custom-made detector arrays required. While feasible, this is also a costly and complicated solution.

Another complication is that multiple lasers create overlapping Rayleigh scatter regions, making it challenging to discriminate the LGS at 100 km altitude from low-altitude scattered light. The potential issues associated with Rayleigh laser backscatter in MCAO systems are not yet fully understood. Many options for LGS constellations and projector configurations require at least some of the subapertures in each WFS to observe their guide star through Rayleigh backscatter if CW or quasi-CW lasers are employed. Models indicate that the variability in the Rayleigh signal induced by photon arrival statistics should not prevent accurate wavefront sensing *if* the mean backscatter level can be accurately determined and subtracted off from the LGS signal. This last assumption needs to be verified empirically. Meanwhile, a pulsed laser format would allow individual beams to be gated on and off at different times and/or repetition rates to prevent fratricide between beams. This provides a strong motivation for employing a pulsed laser format.

An alternate approach to eliminate spot elongation is by dynamically focusing on a short laser pulse traveling up through the Na layer Beletic (priv. comm.). A $2\mu s$ pulse would correspond to a pulse length of 65 km, which would be acceptable for a 30-m class telescope. The laser duty cycle should be as high as possible to limit saturation, but a pulse-to-pulse separation of about $60\mu s$ is desirable to avoid multiple pulses in the Na layer at any one time. This separation would also eliminate low-altitude Rayleigh backscatter during the WFS measurement if the pulse-to-pulse timing were adjustable. This technique will require the development of specialized CCD design able to track the pulse in the correct direction and speed within each subaperture. Beletic, et.al., are developing such a CCD. It will utilize a novel pixel geometry to transfer photoelectric charges to accumulate the laser return within a small number of pixels. This is similar to "time delay and integration" or "nod-and-shuffle". The CCD design will enable shuttering out the Rayleigh backscatter and will allow the on-chip (noiseless) integration of return from multiple pulses. Combining a $2\mu s$ pulsed laser with the custom- designed CCD looks to be the most promising method for dealing with spot elongation.

One final area of necessary research is that of laser delivery. With multiple guide stars, it will be highly desirable to minimize the size and complexity of projectors systems. Fiber transport would also reduce the cost, complexity, and throughput losses associated with relaying the beam from the laser to the projector. ESO has been supporting the development of a large core fiber optics. Transport fibers are now available to deliver up to 10 W of CW power to launch sites on the telescope over distances < 25 m. To date, these fibers are limited to use with CW laser systems with significant bandwidth (>500 MHZ), due to nonlinear optical effects in the fiber. However, for ELTs, it will be necessary to develop fiber usable at the required powers and possibly a pulsed format without inducing SBS. Hollow core photonic crystal fibers should match these requirements. This is an important area of investment for ELTs.

The search for better laser technologies continues. A number of solid-state techniques to obtain 589 nm light have already been discussed above. Some of the more promising approaches employ fiber laser technologies developed for the telecommunications industry. Fiber lasers are very compact, efficient and rugged. They are alignment-free and provide built-in fiber delivery with diffraction-limited outputs, making these lasers ideal for LGS applications. One such laser that involves sum-frequency mixing of 938 nm Nd:doped fiber laser with a 1583 nm erbium:doped fiber laser to generate 589 nm, is being developed by LLNL (Payne et al., 2001). Another fiber laser concept that involves Raman-shifting an Ytterbium:doped fiber laser to 1178 nm and frequency doubling the output to 589 nm, is being developed at ESO (Hackenberg et al., 2003). A third concept, being pursued by CTI, sum-frequency mixes two YAG-based wave-guided lasers to generate 589 nm. It will be interesting to see which new ideas will come to fruition over the next few years. A more detailed description of the two fiber laser systems mentioned above will now follow.

**Fiber laser guide star systems.**     It is now widely appreciated that the heat- dissipation characteristics of fibers, coupled with the high efficiencies demonstrated (> 80%) and excellent spatial mode characteristics, make fiber lasers a preferred candidate for many high power applications. Based on these features, fiber laser technologies would provide a compact, efficient, robust, turnkey laser source, ideally suited for LGS applications.

Fiber lasers with single-frequency CW output powers of 500 W and good beam quality have been demonstrated (Limpert et al., 2003). In the telecom world, chains of 30+ fiber amplifiers are used in series in undersea cable applications where they must operate continuously without failure for >20 years. Optical-to-optical conversion efficiencies of >60% are regularly obtained in fiber lasers and amplifiers which, combined with diode laser wall plug efficiencies of 40% and non-linear conversion efficiencies of 40%, yield total wall plug efficiencies of close to 10%. This minimizes power consumption and cooling requirements to the lowest that are technically feasible in any high power laser system. Because the light in fiber lasers propagates within single mode, polarization maintaining waveguide cores, fiber lasers do not need routine alignment. Finally, these units are lightweight and are rugged with respect to vibration and thus can be easily mounted in convenient locations on the telescope with power and system operation controlled remotely.

**LLNL sum-frequency generation fiber laser.**     LLNL is developing a 589 nm CW fiber laser system, in collaboration with ESO. It is based on sum-frequency mixing two fiber lasers in a nonlinear crystal (Fig. 26). By combining a 1583 nm Er/Yb:doped fiber laser with a 938 nm Nd:silica fiber laser, one can generate 589 nm light via sum-frequency mixing in a periodically poled crystal. Erbium-doped fiber lasers (EDFA) are widely used in the telecom industry. A 1583 nm EDFA system, with 11 W of linearly polarized output, has now been demonstrated in an all-fiber form. The main technical challenge in the design is the Nd:fiber laser. Prior to the LLNL work, no 938 nm fiber laser with > 0.1 W output (Dragic and Papen, 1999) had been demonstrated. LLNL has now demonstrated an 8 W, 938 nm fiber laser, with linear polarization and near diffraction limited beam quality (Dawson et al., 2003). A Nd:doped fiber operating at 938 nm is a somewhat novel device, since the $Nd_{3+}$ ions must operate on the resonance transition (i.e. $^4F_{3/2} - ^4I_{9/2}$), while suppressing ASE losses at the 1088 nm transition. To circumvent this difficulty, a high power seed pulse is used to extract gain preferentially at 938 nm and custom-designed, low numerical aperture amplifier fiber is used to suppress the competing 1088 nm (Dawson et al.). This design should scale to >100 W with access to additional pump diodes.

The CW design utilizes a periodically poled material for the sum frequency mixing process; however, if these materials do not prove to be sufficiently robust, a resonant ring design will be employed with a bulk crystal to achieve the required conversion efficiency. Both the 1583 and 938 nm laser subsystems are currently operating at their design power and as of this writing, 0.5 W of 589 nm light has been demonstrated via sum-frequency mixing in PPKTP. LLNL expects to demonstrate 5-10 W at 589 nm by the end of 2004. The system is expected to fit in one standard size rack when field-hardened and will run on standard wall power.

The LLNL fiber system architecture also lends itself to producing a pulsed format. The primary challenge in making a pulsed fiber laser is in avoiding nonlinear effects in the fiber due to high peak power, in particular, SBS to be <1%. The SBS threshold for a given peak power is a function of the amplifier gain and length, the core size of the fiber and the bandwidth. SBS can be partially offset by broadening the bandwidth of the signal in the fiber. The SBS gain bandwidth is $\sim 50 MHz$ at 1583 nm and $\sim 300 MHz$ at 938 nm ( Dawson et al., 2003). As the bandwidth becomes increasingly larger than the SBS gain bandwidth, the power that can be transmitted through the same fiber increases as length and core size increases. Use of 1 GHz bandwidth increases the upper limit on the output power from the fibers by a factor of 10. For a 1% duty cycle a bandwidth of 1 GHz is required (also beneficial for reducing saturation in the Na layer). As the duty cycle increases it would be possible to operate the amplifier with less bandwidth. LLNL expects to demonstrate a pulsed version of its fiber laser within 3 years.

**ESO Raman fiber laser.**

ESO is developing a Raman fiber laser. A 40 W CW commercial ytterbium-doped fiber laser (YDFL) operating at 1121 nm is used to pump a germanosilicate single mode fiber. The pump laser photons excite the molecules of the fiber and are thus downshifted in frequency by an amount equal to the energy difference between the final and initial states of the molecule. This so-called "Stokes" shift depends on the material composition. It is 17 THz in silica, thus the pump photons are frequency shifted to exactly 1178 nm. The efficiency reaches 80-90% in single mode fibers. YDFLs are commercially available at output powers $\lesssim$100 W CW, which make them the ideal pump source for this application. In a final step, the 1178 nm output of the Raman fiber laser is frequency-doubled in a nonlinear crystal to produce 589 nm. Ideally the frequency doubling will happen in a single pass through a periodically poled crystal.

The two configurations of amplifier and resonator are being prototyped. In the Raman amplifier configuration, the Raman fiber is seeded with light from a tunable narrowband laser (recently commercially available) operating at twice the $D_2$ wavelength. The output is the seed light amplified by the SRS process in a single pass. Parasitic nonlinear optical effects like SBS are suppressed by the use of a sufficiently short Raman fiber. In the Raman resonator configuration, a narrowband resonator is created within a single mode fiber using a pair of fiber Bragg gratings tuned to 1178 nm. The Bragg gratings for the Raman-shifted light are designed such that they only reflect 1178 nm and not wavelengths that might be created by parasitic nonlinear processes such as SBS. Again, SBS is suppressed by use of a sufficiently short fiber This is the more attractive of the two designs, as high power narrowband Raman fiber lasers are not commercially available at this time.

With commercially available YDFL as pumps, powers $\gtrsim 40$ W at 1178 nm are feasible. This sets an upper limit to the conversion efficiency needed in the subsequent second harmonic generation. Numerical simulations for the amplifier and resonator Raman laser configuration indicate feasibility of the system with sufficient SBS suppression. ESO has assembled the amplifier configuration, and has demonstrated up to 4 W CW at 1178 nm. ESO's goal is to have compact and turnkey commercial fiber lasers for LGS/AO within 3 years.

# References

Agrawal, G.P., 1995, *Non-linear fiber optics 2nd Edition* Academic Press, Chapter 9

Albano, J., Blamont, J.E., Chanin, M.L., Petitdidier, M., 1970, Ann. Geophys. **26**, 151

Angel, R., Lloyd-Hart, M., 2000, SPIE **4007**, 270

Avicola, K., Brase, J.M., Morris, J.R., Bissenger, H.D., Duff, J.M., Friedman, H.W., Gavel, D.T., Max, C.E., Olivier, S.S., Presta, R.W., Rapp, D.A., Salmon, J.T., Waltjen, K.E., 1994, JOSA A **11**, 825

Azuma, Y., Shibata, N., Horiguchi, T., Tateda,M., 1988, *Wavelength dependence of Brillouin-gain spectra for single-mode fibers*, Electronics Letters, **24**, 250

Beckers, J.M., 1993, Annu. Rev. Astron. Astrophys., **31**, 13

Beckers, J.M., 1987, *Adaptive Optics in Solar Observations*, ed. F. Merkle, O. Engvold and R. Falomo, Lest Tech. Rep. **28**, 55

Bienfang, J.C., Denman, C.A., Grime, B.W., Hillman, P.D., Moore, G.T., Telle, J.M., 2003, Opt. Let. **28**, 2219

Blamont, J.E., Donahue, T.M., 1961, J. Geophys. Res. **66**, 1407

Bowman, M.R., Gibson, A.J., Sandford, M.C.W., 1969, Nature **221**, 456

Clemesha, B.R., Kirchhoff, V.W.J.H., Simonich, D.M., 1981, Geophys. Res. Lett. **8**, 1023

Contos, A.R., Wizinowich, P.L., Hartman, S.K., LeMignant, D., Neyman, C.R., Stomski, P.J., Summers, D., 2002, SPIE **4839**, 470

Dainty, C., Ageorges, N., 2000, *Laser Guide Star Adaptive Optics for Astronomy*, NATO/ASI **551**

Davies, R.I., Hackenberg, W., Ott, T., Eckard, A., Rabien, S., Anders, S., Hippler, S., Kasper, M., Kalas, P., Quirrenbach, A. , Glindemann, A., 1999, A&AS **138**, 345

Dawson, J., Beach, R., Drobshoff, A., Liao, Z., Pennington, D., Payne, S., Taylor, L., Hackenberg, W. , Bonaccini, D., 2003, SPIE **4974**, 75

Dawson, J., Beach, R., Drobshoff, A., Liao, Z., Pennington, D., Payne, S., Taylor, L., Hackenberg, W., Bonaccini, D., *High Power 938 nanometer fiber laser and amplifier*, Patent Application, (case number IL-11224)

Dragic, P. and Papen, G., 1999, *Efficient amplification using the 4F3/2-4I9/2 transition*, in *Nd-doped silica fiber*, IEEE Phot. Tech. Lett. **11**, 1593

Drever, R.W.P., Hall, J.L., Kowalski, F.V., Hough, J , Ford, G.M., Munley, A.J., Ward, H., 1983, Appl. Phys. B: Photophys. Laser Chem. **31**, 97

Foy, R. and Labeyrie, A., 1985, Astron. Astrophys. **152**, L29

Foy, R., Migus, A., Biraben, F., Grynberg, G., McCullough, P.R., Tallon, M., 1995, Astron. Astrophys. Suppl. Ser. **111** , 569

Foy, R., Tallon, M., Séchaud, M., Hubin, N., 1989, SPIE **1114**, 174

Friedman, H.W., Cooke, J.B., Danforth, P.M., Erbert, G.V., Feldman, M., Gavel, D.T., Jenkins, S.L., Jones, H.E., Kanz, V.K., Kuklo, T., Newman, M.J., Pierce, E.L., Presta, R.W., Salmon, J.T., Thompson, G.R., Wong, N.J., 1998, SPIE **3353**, 260

Friedman, H.G., Erbert, G., Kuklo, T., Salmon, T., Smauley, D., Thompson, G., Malik, J., Wong, N., Kanz, K., Neeb, K., 1995, SPIE **2534**, 150

Friedman, H.G., Erbert, G., Kuklo, T., Salmon, T., Smauley, D., Thompson, G., Wong, N., 1994, SPIE **2201**, 352

Fugate, R.Q., Ellerbroek, B.L., Higgins, C.H., Jelonek, M.P., Lange, W.J., Slavin, A.C., Wild, W.J., Winker, D.M., Wynia, J.M., Spinhire, J.M., Boeke, B.R., Ruane, R.E., Moroney, J.F., Oliker, M.D., Swindle, D.W., Cleis, R.A., 1994, JOSA A **11**, 310

Fugate, R.Q., Fried, D.L., Ameer, G.A., Boeke, B.R., Browne, S.L., Roberts, P.H., Ruane, R.E., Wopat, L.M., 1991, Nature **353**, 144

Fugate, R.Q., Wild, W.J., 1994, Sky & Tel. **87**, 25

Gavel, D.T., Gates, E., Max, C., Olivier, S., Bauman, B., Pennington, D., Macintosh, B., Patience, J., Brown, C., Danforth, P., Hurd, R., Severson, S., Lloyd, J., 2003, SPIE **4839**, 354

Ge, J. Angel, J.R.P. Jacobsen, B.P., Roberts, T., Martinez, T., Livingston, W., McLeod, B., Lloyd-Hart, M., McGuire, P. , Noyes, R., 1997, Proc. ESO Workshop *Laser Technology for Laser Guide Star Adaptive Optics Astronomy*

Georges III, J.A., Mallik, P., Stalcup, T., Angel, J.R.P., Sarlot, R., 2002, SPIE **4839**, 137

Glindemann, A., Quirrenbach, A., 1997, Su. W. **26**, 950

Hackenberg, W., Bonaccini, D., Werner, D., 2003, SPIE **4839**, 421

Happer, W., MacDonald, G.J., Max, C.E., Dyson, F.J., 1994, JOSA A **11**, 263

Hardy, J.W., 1998, *Adaptive Optics for Astronomical Telescopes*, Oxford University Press

Hogan, G.P., Web, C.E., 1995, in *Adaptive Optics*, ESO Conf. Proc **54**, 257

Humphreys, R., Primmerman, C.A., Bradley, L.C., Hermann, J., 1991, Opt. Lett. **16**,1367

Jacobsen , B.T., Martinez, R., Angel, R., Lloyd-Hart, M., Benda, S., Middleton, D., Friedman, H., Erbert, G., 1994, SPIE **2201**, 342

Jelonek, M.P., Fugate, R.Q., Lange, W.J., Slavin, A.C., Ruane, R.E., Cleis, R.A., 1992, *Laser Guide Star Adaptive Optics Workshop* Proceedings, ed. R.Q. Fugate, 213

Jeys, T., 1992, The Lincoln Laboratory Journal **4**, 132

Jeys, T.H., Brailove, A.A., Mooradian, A., 1989, Appl. Opt. **28**, 2588

Jeys, T.H., Brailove, A.A., Mooradian, A., 1991, Appl. Opt. **30**, 1011

Kibblewhite, E.J., Shi, F., 1999, SPIE **3353**, 300

Limpert, J., Liem, A., Zellmer, H., Tunnerman, A., 2003, Electronics Let. **39**, 135

Lloyd-Hart, M., Georges, J., Angel, R., Brusa, G., Young, P., 2003, SPIE **4494**, 645

Lloyd-Hart, M., Jefferies, S. M., Hege, E. K., Angel, J. R. P., 2001, Opt. Lett. **26**, 402

Max, C.E., Avicola, K., Brase, J.M., Friedman, H.W., Bissinger, H.D., Duff, J., D.T., Gavel, D.T., Horton, J.A., Kiefer, R., Morris, J.R., Olivier, S.S., Presta, R.W., Rapp, D.A., Salmon, J.T., Waltjen, K.E., 1994, JOSA A **11**, 813

Max, C.E., Olivier, S.S., Friedman, H.W., Avicola, J. A. K., Beeman, B.V., Bissenger, H.D., Brase, J.M., Erbert, G.V. , Gavel, D.T., Kanz, K., Liu, M.C., Macintosh, B., Neeb, K.P., Patience, J., Waltjen, K.E., 1997, Science **277**, 1649

McKinnie, I., 2003, *Gemini guide star laser system*, CfAO Fall Retreat, Yosemite, CA

Megie, G., Bos, F., Blamon, J.E., and Chanin, M.L., 1978, Planet Space Sci. **26**, 27

Michaille, L., Clifford, J.B., Dainty, J.C., Gregory, T., Quartel, J.C., Reavell, F.C., Wilson, R.W., Wooder, N.J., 2001, Mon. Not. R. Astron. Soc **328**, 993

Milonni, P.W., Fern, Telle, J.M., Fugate, R.Q., 1999, JOSA A **16** 2555

Milonni, P.W., Fugate, R.Q., Telle, J.M., 1998, JOSA A **15**, 217

Morris, J.R., 1994, JOSA A **11**, 832

d'Orgeville, C., Rigaut, F., Ellerbroek, B., 2000, SPIE **4007**, 131

O'Sullivan, C.M.M., Redfern, R.M., Ageorges, N., Holstenberg, H.C., Hackenberg, W., Rablen, S., Ott, T., Davies, R. , Eckart, A., 1999, in *Astronomy with Adaptive Optics -Present Results and Future Programs*, ESO and OSA, 333

Papen, G.C., Gardner, C.S., Uy, J., 1996, *Adaptive Optics*, OSA Technical Digest Series (Optical Society of America, Washington DC), **13**, 96

Parenti, R.R., Sasiela, R.J., 1994, JOSA A **11**, 288

Payne, S.A., Page, R.H., Ebbers, C.A., Beach, R. J., 2001, Patent application *Synthetic Guide Star Generation*, IL- 10737

Pennington, D.M., Brown, C. , Danforth, P. , Jones, H., Max, C., Chin, J., Lewis, H., Medeiros, D., Nance, C., Stomski, P., Wizinowich, P., 2002, SPIE **4839**

Pique, J.P., Farinotti, S., 2003, JOSA B **20**, 1

Primmerman, C.A., Murphy, D., Page, D., Zollars, B., Barclay, H., 1991, Nature **353**, 141

Quirrenbach, A., Hackenberg, W. , Holstenberg, H., Wilnhammer, N., 1997, SPIE **3126**, 35

Sarrazin, M., *Sodium beacon monitoring- A survey of related activity worldwide*, http:// www.eso.org/genfac/pubs/astclim/lgs/

Simonich, D.M., Clemesha, B.R., Kirchhoff, V.W.J.H., 1979, J. Geophys. Res. **84**, 1543

Summers, D., Gregory, B., Stomski, P.J., Brighton, A., Wainscoat, R.J., Wizinowich, P.L., Gaessler, W., Sebag, J., Boyer, C., Vermeulen, T., Denault, T.J., Simons, D.A., Takami, H., Veillet, C., 2002, SPIE **4839**, 440

Tallon, M., Foy, R., 1990, Astron. Astrophys. Suppl. Ser. **235**, 549

Teehan, R.F., Bienfang, J.C., Denman, C.A., 2000, Appl. Opt. **39**, 3076

Thompson, L.A., Castle, R.M., 1992, Opt. Lett **17**, 1485

Thompson, L.A., Gardner,C.S., 1987, Nature **328**, 229

Tyler, G.A., 1992, in *Laser Guide Star Adaptive Optics Workshop* Proceedings, Albuquerque, NM, 405

Tyson, R.K., 1997, *Principles of Adaptive Optics*, 2nd Edition, Academic Press

U.S. Standard Atmosphere, National Oceanic and Atmospheric Administration, N.A.S.A. and U.S.A.F., Washington, 91, 1976

Vance, J.D., She, C.Y., Moosmller, H., 1998, Appl. Optics **37**, 4891

Welsh, B.M., Gardner, C.S., 1989, JOSA A **6**, 1913

Wizinowich, P., Simons, D., Takami, H., Veillet, C., Wainscoat, R., 1998, SPIE **3353**, 290

Zollars, B.G., 1992, The Lincoln Laboratory Journal **5**, 67

# LASER GUIDE STARS: PRINCIPLE, CONE EFFECT AND TILT MEASUREMENT

Renaud Foy

*Observatoire de Lyon / CRAL*
*9 avenue Charles André, 69561 Saint Genis Laval (France)*
foy@obs.univ-lyon1.fr

**Abstract**     We explain the laser guide star concept and why we do need laser guide stars (LGS) to greatly enhance adaptive optics performances at large telescopes. Then we discuss the two physical limitations of LGS, which are the cone effect and the indetermination of the tilt of the incoming wavefront. Solutions to overcome them are explained.

**Keywords:**     laser guide star, adaptive optics, lasers, cone effect, tilt

## Introduction : Why do we need laser guide stars?

Adaptive optics (AO) aims at providing diffraction limited long exposure images at large telescopes (Ch. 13). Let us briefly recall that it works thanks to deformable mirrors (DM) which correct on line the incoming wavefront corrugated by phase disturbances during its crossing of the turbulent atmosphere (Ch. 1). This correction requires to measure the phase lags onto the telescope pupil, which is achieved by a wavefront sensor. Whatever the kind of wavefront sensor (Ch. 13), it needs enough light to reach the required measurement accuracy, which is typically $\lambda/2\pi$, where $\lambda$ is the wavelength, or much less in case of the so-called extreme AO (Ch. 13).

When the source of interest is too faint to provide the AO wavefront sensor with enough light, on relies on a field star close to the source. It has to lie within the isoplanatism patch of the source of interest, i.e. a few arcsecs to a few tens of arcsecs from the visible to the near infrared (Ch. 13). This constraint put severe limits to the surface of the sky adaptive optics observations can be used. The so-called sky coverage is significantly smaller than 1% at short wavelengths, which is not acceptable for most of astrophysical programmes at high angular resolution. The concept of Laser Guide Star (LGS)

*R. Foy and F. C. Foy (eds.), Optics in Astrophysics, 249–273.*

has been proposed (Foy and Labeyrie, 1985, see also Ch. 14). It is to overcome these drastic limitations, by creating an artificial spot high enough in the
atmosphere to provide the AO device with a proper sampling of atmospheric
turbulent layers. This spot is produced by the scattering of a laser beam shot
within the isoplanatic patch of the source of interest. Possible backscattering
processes, namely Rayleigh scattering up to 20-30km high and fluorescence
by sodium atoms in the mesosphere at $\approx$ 95km , are discussed in detail by D.
Pennington in Ch. 14. The most widely use backscattering process is sodium
fluorescence, relying on the absorption/emission between the two first energy
levels $3S_{1/2}$ and $3P_{3/2}$, corresponding to the so called $D_2$ sodium line (that of
orange lights in the streets). In this chapter, we will consider in most cases the
LGS in the mesosphere.

For all that, the laser guide star does not solve the sky coverage problem,
unfortunately. Why? Mainly for the three following reasons.

THE SATURATION OF THE ABSORPTION. Roughly speaking, saturation
of the absorption occurs when each of the sodium atoms in a column of
section equal to the cross section of the $D_2$ transition have absorbed a
photon per time interval equal to the lifetime of the upper energy level.
There is not so much sodium in the mesosphere ($\approx$ 1 metric ton all
around the Earth). Thus, in spite of the relatively high cross section of
the $3S_{1/2} \rightarrow 3P_{3/2}$ transition, saturation occurs at a quite low level, $\approx$
180 W.m$^{-2}$ (see Ch.14). To get enough return flux at the minimum laser
power, one needs to optimize the laser specifications (continuous wave
or pulsed, pulse width, pulse repetition rate, (average) power, spectral
profile) taking into both saturation, technological, budget and operation
constraints. This is the challenge described in detail in the above mentioned chapter.

THE CONE EFFECT. LGSs are formed at finite distance whereas programme
sources are at infinity. The area of turbulent layers lit by a LGS is included inside a cone of summit the LGS and base the telescope mirror.
The same area lit by a natural star is inside a cylinder: at the telescope
level, the curvature of the wavefront from the LGS is not negligible,
whereas it is for a star. The larger the telescope diameter and the shorter
the wavelength, the stronger the cone effect. Also the higher the turbulent layer and the smaller the coherence length of the wavefront (the
Fried parameter $r_0$) the stronger the cone effect. We discuss in section
15.2 the effect of the poor sampling of turbulent layers, and the way to
overcome it, using arrays of LGS, 3 dimensional mapping of phase lags
(also called tomography) and multiconjugate adaptive optics (MCAO).

THE TILT INDETERMINATION The round trip time of light to the mesosphere is $\approx 0.6$ ms (at zenith). It is significantly shorter than the coher-

ence time of the tilt component of the incoming wavefront, so that the tilt is mostly the same on the ways to and back from the mesosphere. Thus if one launches the laser beam through the main telescope full aperture, the deviation of the upward beam is canceled by the deviation of the return one. From the telescope the LGS appears always in the direction of the upward beam, whatever the tilt, preventing us from measuring it. If the beam is launched from an auxiliary telescope or refractor aside the main telescope, because of the unknown tilt affecting the upgoing beam, one does not know the LGS location, preventing us from disentangling between the upward and downward beams and thus again from measuring the tilt. In section 15.3 we discuss how important it is that AO devices are able to correct for the tilt.

# 1. Sky coverages

Adaptive optics requires a reference source to measure the phase error distribution over the whole telescope pupil, in order to properly control DMs. The sampling of phase measurements depends on the coherence length $r_0$ of the wavefront and of its coherence time $\tau_0$. Both vary with the wavelength $\lambda$ as $\lambda^{6/5}$ (see Ch. 1). Of course the residual error $\sigma_\phi$ in the correction of the incoming wavefront depends on the signal to noise ratio of the phase measurements, and in particular of the photon noise, i.e. of the flux from the reference. This residual error in the phase results in the Strehl ratio following: $S = exp(-\sigma_\phi)$.

Consequently for a given telescope diameter and a given wavefront detector the Strehl ratio depends on the seeing parameters, the wavelength of the observation and the magnitude of the phase reference.

Following a millennial tradition, most astronomers measure optical fluxes in magnitude units. A magnitude is defined as :

$$m = -2.5 \, log(\text{flux}) + \text{constant} \qquad (1)$$

At $\lambda = 500$nm, the flux corresponding to $m_V = 0$ and 10 are respectively $3.6 \; 10^{-8}$ and $3.6 \; 10^{-12}$erg.s$^{-1}$.cm$^{-2}$.nm$^{-1}$. Figure 1 shows the number of photons as a function of $\lambda$ for a star with a temperature of $10^4$K.

Figure 2 describes the theoretical Strehl ratio expected at an 8m telescope with good and average seeing conditions at Cerro Paranal (Le Louarn et al., 1998). The Strehl ratio drops from $V \approx 12$ mag., because of photon noise which prevails in the case of a faint reference source. When there is no star brighter than the threshold corresponding to the required Strehl within the isoplanatic field around the programme object, adaptive optics could not work.

This leads to the concept of sky coverage. It is the probability to find a suitable reference source being the seeing conditions, the wavelength of the observation, the required Strehl, and the number of stars per square degree in the direction of the programme source.

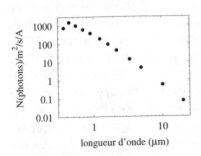

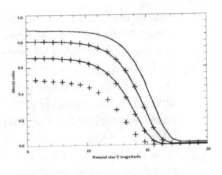

*Figure 1.* Number of photons/m²/s/Å for a star with $m = 0$ and $T_{eff} \approx 10^4$ K

*Figure 2.* Predicted Strehl ratio at an 8m telescope versus magnitude of the reference source at 550 nm. Solid lines $r_0 = 0.2$ m and $\tau_0 = 6.6$ ms, crosses : $r_0 = 0.15$ m and $\tau_0 = 3$ ms. Top to bottom : 2.2, 1.65 and 1.25 $\mu$m.

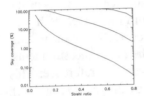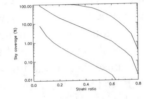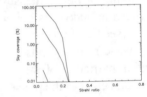

*Figure 3.* Sky coverage of an adaptive optics at an 8m telescope. Pupil sampling by the actuators of the deformable mirror : 0.5m. 0 = 0.17m. From top to bottom: in a direction of galactic latitude $b = 0$ř (the Milky Way), $b = 20$ř and $b = 90$ř (the galactic pole). From left to right : K, J and V bands.

From Fig.3, in the 2.2$\mu$m range sky coverage is very satisfactory in the galactic plane and still acceptable in the direction of the galactic pole. But at shorter wavelengths, is is no longer true. In particular, the sky coverage is ridiculously small at in the visible. This is why artificial phase reference sources have been proposed (Foy and Labeyrie, 1985; Fugate et al., 1991; Primmerman et al., 1991). They are called *Laser Guide Stars* (improperly in my view), or *LGSs*.

The basic principle of a LGS is to launch a laser beam toward the sky very close to the direction of the programme source. The beam is backscattered by some species in the high atmosphere, producing a spot viewed as a "star" from the telescope. Backscattering processes are Mie scattering, due large particles, Rayleigh scattering due to small particles and molecules, mostly $N_2$, and fluorescence, mostly within the sodium $D_2$ line. Mie scattering is not suitable, because above the inversion layer, where are located the best astronomical sites, the density of large particles is very low. Because of the exponential decrease of atmospheric density, Rayleigh scattering is not efficient above $\approx$20-25km, just above the highest turbulent layers. Fluorescence is active in the so called *sodium layer* in the mesosphere, at $\approx$ 92km above sea level. This high altitude allows us to sample all the turbulent layers. Its main limitation is the critical requirement to avoid saturation of the absorption with pulsed lasers, whereas the return flux is barely enough with continuous wave lasers. These processes are explained and discussed in detailed in the D. Pennington lecture in this volume in Ch. 14.

## 2. The cone effect

### 2.1 What is it?

The laser spot in the backscattering layer is located at finite distance from the telescope pupil, whereas the programme source is at infinity. Thus the volume lit by the backscattered flux inscribed between the spot and the telescope pupil is a cone (see Fig. 4). The same volume lit by a natural star is a cylinder. In another wording, the curvature radius of wavefront from the LGS is not negligible, whereas the wavefront from the natural star is plane (not taking into account phase disturbances by turbulence). At the altitude of each of the turbulent layers, the pupil maps homothetically. Consequently, the sampling across the pupil is not preserved at the level of each turbulent layer. The outer ring of the sampled area at each turbulent layer is smaller than the telescope pupil. For instance, as view from a point at the edge of the telescope pupil, the phase measured for a given layer on the line of sight of the LGS is different from the one which one should measure on the line of sight of the programme source at infinity. Optical paths to the pupil from a natural guide star (NGS) and from a LGS are different except on axis. As long as this difference at the levels of turbulent layers is negligible with respect to the coherence length, there is no significant effect on the restored wavefront. It is intuitive that the errors due to the cone effect:

1  increase with the telescope diameter $D$: edges are farther away from the optical axis at large telescopes
2  decrease with the altitude of the backscattering layer $H$
3  increase with the altitude of the turbulent layers
4  increase with the zenith distance $z$ (projected distances to the turbulent layers increase with $z$)
5  decrease with the wavelength (since the coherence length increases)

Let $d_0$ be the diameter of the telescope for which the cone effect causes a phase error $\sigma_\phi^2 = 1\,\mathrm{rad}^2$. Thus for a telescope $D$:the phase error is $\sigma_\phi =$

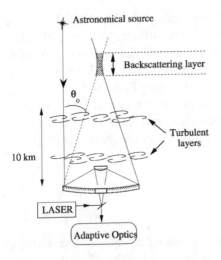

*Figure 4.* Propagation from the laser spot at finite distance and from the astrophysical source at infinity. Beams from the LGS and from the source to the edge of the pupil cross a turbulent layer at distance each other larger than the coherence length (Court. M. Tallon).

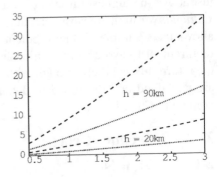

*Figure 5.* Variation of $d_0$ (m) versus wavelength ($\mu$m). Solid lines: optimistic Hufnagel-Valley model for $C_n^2(h)$; dotted lines: pessimistic model; thick lines: $90\,km$ high LGS; thin lines: $20\,km$ high LGS ( Sasiela, 1994).

$(D/d_0)^{5/6}$, and its variance :

$$\sigma_\phi^2 = \frac{2\pi^2}{\lambda^2} \frac{1}{H \cos z} \int_0^H C_n^2(h)\, h^2 (h^{1/3} - 1)\, dh \qquad (2)$$

where $C_n^2(h)$ is the vertical profile of the refractive index structure constant. The behavior of $d_0$ is shown in Fig. 5. At 2.2$\mu$m, $d_0$ is larger than the present time largest telescopes (8 - 10 meters) if the LGS is formed in the mesosphere *and* if the vertical profile of the turbulence strength is favorable (most of the seeing is caused by low altitude layers). Whatever these parameters, a Rayleigh LGS makes an error $\sigma_\phi^2 > 1\,rad^2$, which is too large. At visible wavelengths, the cone effect severely degrades the image quality. This is shown in Fig. 6. It describes modeled Strehl ratios due to the only cone effect (i.e.: all other aberrations being perfectly corrected). Even at infrared wavelengths the degradation of the Strehl ratio is barely acceptable. Thus a full sampling of the uppermost layer is mandatory. It requires several LGSs.

## 2.2 How to overcome the cone effect?

The cone effect has been identified at the beginning of the LGS story (Foy and Labeyrie, 1985). Several methods have been proposed to overcome it.

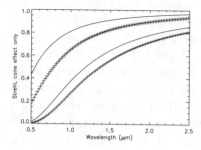

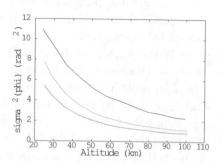

*Figure 6.* Strehl ratio due to the only cone effect, as a function of wavelength. Full and dashed lines: 8 m and 3.6 m telescopes respectively. Top to bottom: optimistic ($r_0 = 0.25$m) and standard models ($r_0 = 0.15$m) for Paranal. Court. of M. Le Louarn

*Figure 7.* Piston and tilt removed phase variance at a 4 m telescope versus altitude. Wavelength: $0.5\mu$m. Solid line: single LGS. Dashed line: at zenith. Dotted line: array of 4 LGSs 45°. Hufnagel-Valley[21] turbulence model.

All of them rely on arrays of LGGs. Most of them belonging to two families : either stitching of the telescope pupil, or mapping of the volume of phase disturbances around the line of sight.

### 2.2.1 Stitching methods.

I will briefly describe this approach, since it seems more and more not to be the right one. The principle is to divide the pupil in subpupils of diameter $\lesssim d_0$, and launch an LGS in front of each subpupil, so that the cone effect is negligible for each beam. It cannot provide us with high Strehl ratios since tilt indetermination (Sect. 15.3) causes discontinuities between restored local wavefronts. Indeed, with the Hufnagel-Valley$_{5/7}$ turbulence model, $d_0(1\mu$m$) \approx 6$ m with a single laser spot and $d_0(1\mu$m$) \approx 15$ m with an infinite spot array. Thus the maximum gain when increasing the array size is given by:

$$\sigma_\infty^2/\sigma_1^2 = (6/15)^{5/3} = 0.217 \qquad (3)$$

(Fried, 1995). The maximum gain with an infinite number of spots is a factor of $\lesssim 5$. But since one increases the volume of the turbulent atmosphere properly sampled due to the number of LGSs, one decreases $\sigma_\phi^2$. There is a trade-off between these two errors. This is shown in Fig. 7 (Sasiela, 1994). The phase variance due to the cone effect is compared to that of the stitching method for a 4m telescope at $0.5\mu$m. At such a short wavelength, the cone effect is so large that the tilt stitching phase error is smaller than the cone effect one, even at $z = 45°$ and for the highest altitude of the LGSs. Of course, it is less favorable for an 8-10 m class telescope.

**2.2.2    3D-mapping methods.**    The 3-dimensional mapping of the phase disturbances in the volume around the line of sight (Tallon and Foy, 1990) is also called "tomography". Consider an array of $m$ LGSs at the altitude $H$ and at locations $(\theta_x^l, \theta_y^l)$, with $l$ the LGS index. Ideally, there are at least as many LGSs as turbulent layers, but in fact it is enough to restrict to dominant ones. The laser spots have to be distributed in such a way that each coherent cell in each layer is lit by at least one LGS as view from the pupil. Thus the whole turbulent volume in between the array and the pupil is sampled. Consequently, the wider the array and the volume, the larger is $m$. Observe the array with a Shack-Hartmann WFS, each of the subpupils observing the whole array (or equivalently with $m$ Shack-Hartmann WFS, each of them observing one of the LGSs). It provides at least $2 \times N \times m$ slope measurements of the $N$ 2-D phase delays through the $m$ turbulent layers. Therefore one should be able to derive the phase delay map and to reconstruct the wavefront at the level of the turbulent layers, over a surface at least as large as the pupil, i.e. with no longer any cone effect. Finally when these delays are used to command $n_{dm}$ DMs conjugated with the turbulent layers, the resulting corrected wavefront leads to a corrected image free from cone effect, over a field possibly larger than the isoplanatic field (Foy and Labeyrie, 1985), depending on $m$. This is *similar to tomography with the additional problem that the equation array is degenerated because of the tilt indetermination* (see 15.3). The use of a cascade of DMs as been called *MultiConjugate Adaptive Optics* (Beckers, 1988).

With the assumption that the amplitude of the influence functions of the DMs is proportional to the command, measurements $b$ and phase delays (i.e.: commands $X$ to the DMs) are related by a matrix linear equation $b = AX$, as in classical AO; it has to be inverted (see Ch. 13).

The interaction matrix can be built following Le Louarn and Tallon, 2002. We assume narrow fields approximation for the propagation from the turbulent layers, i.e.: phase perturbations by the turbulent layers simply add at the level of the pupil. It also means that scintillation is negligible. A point $(x, y)$ in the turbulent layer at altitude $h$ is projected onto the pupil at $(x_{\text{pupil}}, y_{\text{pupil}})$:

$$x = (1 - h/H)x_{\text{pupil}} - h\theta_x^l \qquad (4)$$

$$y = (1 - h/H)y_{\text{pupil}} - h\theta_y^l \qquad (5)$$

A difference with single DM AO is that there is a propagation between the DMs. But since Fresnel diffraction is distributive, influence functions remain proportional to the command after propagation.

Le Louarn and Tallon, 2002 consider triangular influence functions after propagation:

$$f(x, y) = p(x/d_s)p(y/d_s) \text{ with } p(x) = \left\{ \begin{array}{ll} 1 - |x| & \text{if } |x| \leq 1 \\ 0 & \text{otherwise} \end{array} \right. \qquad (6)$$

where $d_s$ is the actuator spacing. They are maximum at the center of the actuators, and at the center of the neighboring ones. The matrix $A$ is built from the influence functions of a DM generated for each DM, shifted and translated according to Eq. 4, resampled on the pupil subaperture grid. Each column of $A$ is filled with the average per subaperture of these functions. This has to be done for each DM and LGS.

Piston and tilts are not sensed from a monochromatic LGS, which causes a degeneracy in the inversion of $A$. Therefore one has to consider no longer measurements from any subaperture $i$ with coordinates $(x_i, y_i)$, but piston and tilts removed measurements in $b$:

$$b'_{il} = b_{il} - \alpha_l x_i - \beta_l y_i \tag{7}$$

where:

$$\alpha_l = \left(\sum_i x_i b_{il}\right) / \left(\sum_i x_i^2\right) \text{ and } \beta_l = \left(\sum_i y_i b_{il}\right) / \left(\sum_i y_i^2\right) \tag{8}$$

are determined by a least square fit of the slope measurements for the LGS $l$. It yields:

$$b'_{il} = \sum_j \left(\delta_{ij} - \left(x_i x_j / \sum_k x_k^2\right) - \left(y_i y_j / \sum_k y_k^2\right)\right) g_{jl}. \tag{9}$$

where $\delta_{ij}$ is the Kronecker's symbol. For the piston removed measurement vector, one has:

$$b''_{il} = \sum_j \left(\delta_{ij} - 1/N_{sub}^2\right) g_{jl}. \tag{10}$$

To have more measurements than unknowns, for a square geometry of the pupil and subpupils, one has to fulfill:

$$m \geq n_{dm} + \sum_{i-1}^{n_{dm}} ((2h_i\theta - (Dh_i/H)) /D)^2 \tag{11}$$

assuming $\theta_x = \theta_y = \theta$. For a zero field of view (correction including that of the cone effect perfect only on axis), Eq. 11 becomes $m \geq n_{dm}$. For a given number of LGSs, the field is limited by the constraint that the center of the pupil projected onto the highest turbulent layer $h_{max}$ be sampled, yielding:

$$\theta \leq (1 - h_{max}/H)(D/2h_{max}). \tag{12}$$

Le Louarn and Tallon (2002) have given an in-depth analysis of the eigenmodes of a 3D mapping system, in the case of a square geometry and 2 deformable mirrors conjugated to the ground layer (5×5 actuators) and to 10km

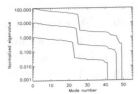 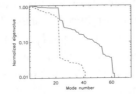 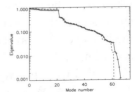

*Figure 8.* Eigenvalues of a 3D mapping system with zero field. Top to bottom: piston and tilts measured by the LGSs, piston not measured and neither piston nor tilts measured. Curves are normalized to 1, 10 and 100 from bottom to top.

*Figure 9.* Eigenvalues of a 3D mapping system with zero field (dashed curve) and non zero field of view (solid curve). Neither piston nor tilts are measured. from the LGSs.

*Figure 10.* Eigenvalues of a 4 LGSs 3D mapping system with a $\neq 0$ field. Neither piston nor tilts are measured from the LGSs. Dashed line: without natural guide star. Solid line: with a NGS to measure tilts, defocus and astigmatisms (Le Louarn and Tallon, 2002).

($7 \times 7$ actuators). Their is then maximum field of view is 66 arcsec, far above the isoplanatic domain. For a given mode, there are obviously two phase corrections, one per deformable mirror. Either they add (even modes) or they subtract each other (odd modes). The last mode is singular.

For zero field of view (Fig. 8), as expected, there are respectively 2, 4 and 8 singular modes when the piston and the tilts are measured from the LGSs, when the piston is not but the tilt is (case of the polychromatic LGS, see 15.3), and when neither the piston nor the tilts are measured (monochromatic LGS case). Even and odd modes correspond respectively to the high and low eigenvalues (lowest and highest modes).

When the piston is assumed to be measured from the LGSs (top curve), the corresponding mode is singular because one cannot measure the contribution of each layer to the total piston. This case is not realistic, since no wavefront sensor measures the piston. When the tilts are measured from the LGSs (case of the polychromatic LGS), the odd piston is not measured again. The even piston is no longer available. And the two odd tilt modes are not also, because whereas the tilt is measured, the differential tilt between the two DMs is not: one does not know where the tilt forms. Thus there are 4 zero eigenvalues.

When neither the piston nor the tilts are measured from the LGSs, there are 5 more 0 eigenvalues: the two even tilt modes, and the odd defocus and astigmatism modes. These last modes result from combinations of not measured tilts in front of the LGSs.

When the field of view is increased(Fig. 9), there are 24 more actuators: 21 more odd ones and 3 more singular ones. They correspond to degenerate odd piston modes: there are more combinations of phases to make waffle modes. The number of even modes remains unchanged.

The degeneracy has to be removed, because light from the programme object does not follow the same path through the turbulent layers as that from a LGS. If the LGSs do not provide the required measurement, a natural guide star (NGS) has to be observed in the field of view to get measurements of these modes. Figure 10 shows that adding the NGS drops the number of the singular from 12 to 7: 2 odd tilts, 1 even piston and 4 odd pistons, which do not need to be corrected. But three of the eigenvalues are very weak. Thus much attention should be paid for noise propagation. In particular it may raises severe constraints about the magnitude of the NGS. Another way is to measure only tilt modes from 3 NGSs in the field (Ellerbroek and Rigaut, 2001).

Without considering photon noise, Le Louarn and Tallon (2002) have found that the 4 LGSs 3D mapping system delivers a Strehl ratio of 80% on axis when a single LGS would have been limited to $S \approx 10\%$. These numbers have to be compared with $S \approx 85\%$ obtained with the same MCAO device fed with a NGS. With a field of view of 100 arcsec, they get $S \approx 30\%$ with little anisoplanatism. Performances are weakly dependent on errors in the altitude of the turbulent layers (which could be measured from the 3D mapping system, at the expenses of the linearity of the equations, since the interaction matrix depends on layer altitudes). Unsensed layers do not produce a significant anisoplanatism, as the central obscuration within which no measurement is available.

Many useful references about MCAO and cone effect are given in Le Louarn and Tallon (2002).

The only 3D mapping system with LGSs under construction today is that for the Gemini South 8m telescope at Cerro Pachon, Chile (Ellerbroek et al., 2002). It consists of 5 LGSs, 3 deformable mirrors with a total of 769 actuators and one 1020 subapertures wavefront sensor per LGS (Fig. 11). It is expected to be the most performing MCAO device.

### 2.2.3    Rayleigh LGSs.

Other methods have been proposed to solve the cone effect problem. Let us briefly mention the following (Lloyd-Hart et al., 2001). A Rayleigh beam at 360nm is focused at 25km with an 8cm projector. The beam is narrower than 1arcsec over $\pm$ 7.5km for a 0.6 arcsec seeing. A fast camera at the telescope records images during $100\mu s$, which are defocused away from the altitude 25km. Then a phase diversity algorithm allows us to recover phases errors. In fact it is a bit more complex, because of the variation of scale with altitude, which requires tomographic techniques to be taken into account. Figure 12 shows the propagation of light backscattered from different regions.

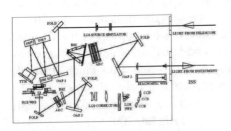

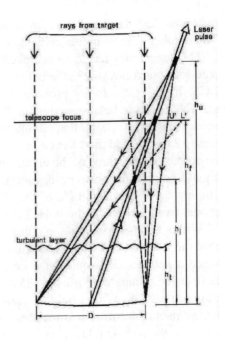

*Figure 11.* Layout of the multiconjugate adaptive optics with 5 LGSs at the Gemini South 8m telescope (F. Rigaut, private communication).

*Figure 12.* Rays from different scattering regions

Spatial resolution is a function of altitude of the backscattering region $H_i$, since the size of the defocus image writes: $D(H_f - H_i)/(H_f H_i)$. With $H_f = 25$km and $H_i = 17$ km and if the spot is observed with a resolution of 1 arcsec, then when projected to the pupil the resolution is $5 \times 10^{-6}(H_f H_i)/(H_f - H_i) \approx 0.26m$. At a 32 m, the defocus image is 2 arcmin wide.

Several LGSs are required. If more than 19 for a 32m, thus one would have to use two slightly different colors to prevent confusion. In spite of the required high number of spots, the authors claim that such a device should be more efficient than devices based on 589nm lasers.

In conclusion of this section, I think that first it is mandatory to remove the cone effect at very large telescopes, and that second this requires:

- Either to measure the first 5 orders from a single NGS,
- Or to measure the tilt for at least 3 NGS,
- Or use polychromatic LGSs.

# 3. The tilt determination

## 3.1 What is the problem?

Let us first consider the case where the laser beam is emitted through the telescope. The outgoing beam undergoes a tilt deflection through the turbulent atmosphere as does the backscattered light. The round trip time to the mesosphere is $\approx 0.6$ ms at zenith. It is much shorter than the tilt coherence time:

$$\tau_{0,\text{tilt}} = 12.33 \, (D/r_0)^{1/6} \, (r_0/v) \tag{13}$$

At an 8m telescope in a good site, $r_0 \approx 0.15$m and a wind velocity of $v \approx 30$m/s, $\tau_{0,\text{tilt}} \approx 0.12$ s at $\lambda = 0.5\mu$m to 1.2 s at $\lambda = 5\mu$m. It scales as $\lambda$. $\tau_{0,\text{tilt}}$ is $\approx 10$ times larger than the wavefront coherence time. Thus both deflections to and from the mesosphere approximately cancel each other and the LGS apparent location is fixed with respect to the optical axis. The principle of inverse return of light applies: one does not know the LGS location in the sky, and the tilt cannot be measured.

Launching the laser beam from the telescope raises technical problems. In addition there is a major concern for the science channel with scattered light in the telescope and in its immediate surrounding. All the current or planned LGSs devices use an auxiliary projector mated to the telescope structure or behind the secondary mirror. In this case, again the location of the LGS is unknown, now because the tilt affecting the upgoing beam is not known. Thus again the tilt cannot be sensed.

This indetermination is a severe problem. Indeed, frames shorter than $\tau_{0,\text{tilt}}$ can be corrected up to the diffraction limit by the AO + LGS device, but because the tilt remains uncorrected images wander from frame to frame, with an rms excursion:

$$\sigma_\theta = 0.062 D^{-1/6} r_0^{-5/6} \cos z^{-2}, \tag{14}$$

The $D^{-1/6}$ term is due to the variation of the phase gradient with the baseline according to the structure function of the air refraction index. At an 8-m telescope, $\sigma_\theta = 0.21''$ and $0.43''$ respectively at zenith and at 45° for $r_0 = 0.15$ m. For comparison the telescope diffraction limit is $0.055''$ and $0.014''$ in K and V. Therefore, the tilt must be corrected.

## 3.2 Use of a natural guide star

One can measure the tip and tilt with a natural guide star for these only modes. Figure 13 shows that the magnitude of the tilt reference star has to be brighter than typically 15 in the near infrared. The sky coverage is limited by the probability to find that star within the isoplanatic patch of the programme source.

These probabilities are shown in Fig. 14 for the case of an AO device with several LGSs and 2 deformable mirrors. The corrected field is 6". 3 NGSs

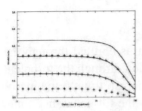

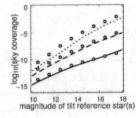

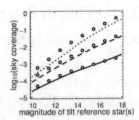

*Figure 13.* Strehl ratio versus V magnitude of the natural reference source for tilt sensing. Solid line: good seeing with $r_0 = 0.2$m. Crosses: standard seeing, with $r_0 = 0.15$m. From top to bottom: K, H and J bands.

*Figure 14.* Sky coverage for a LGSs + 2 deformable mirrors AO. Field: 6". 3 NGS sense the tilt. Dotted, dashed and solid curves: toward $b^{II} = 5°, 20°$ and $90°$, respectively, V magnitude. Open circles : R magnitude.

*Figure 15.* Sky coverage for a LGSs + 1 deformable mirrors AO device. Field is 6" wide. 1 NGS is required to sense the tilt. Same symbols as Fig. 14.

are required to sense the tilt. They are plotted for the V magnitudes (science channel in the R band), and for the R magnitudes (science channel in the V band or in the near IR). For observations in the visible, the limiting magnitude is $\approx$ 12-14, and the sky coverage is $\approx 10^{-12}$ toward the galactic pole, and $\approx 10^{-7} - 10^{-6}$ toward the galactic plane. It is slightly less desperate for IR observations, the highest probability reaching $\approx 1\%$ in the galactic plane with a reasonable limiting magnitude of 17.5 (but $\approx 1.4\,10^{-9}$ toward the pole!) (Fig. 15).

With a single deformable mirror AO, and if a single NGS is used (see Le Louarn and Hubin, 2004), the point spread function is quite peaked in the corrected field, but the probability to find it is much higher: from $\approx 10^{-4}$ toward the galactic pole for observations in the visible to 50% toward the galactic plane for observations in the near infrared.

Here we have considered that the tilt reference stars have to be inside the corrected field, to insure the maximum possible Strehl ratio. Relaxing this requirement would lead to less lower sky coverages, at the expense of the Strehl ratio. I do not discuss here the case of LGSs+AO systems aiming not at providing diffraction limited images, but at increasing the encircled energy, i.e. to decrease the width of the incoherent image ; the global performances of such devices should be compared with those of postprocessing methods, as blind deconvolution (see Ch. 25), which are by far much less expensive.

Thus in most cases, correction of the tilt from the LGSs is mandatory.

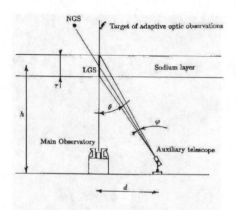

*Figure 17.* A frame a the Calar Alto 2.2m telescope: two NGSs are closely aligned with the laser plume backscattered in the mesosphere from the laser beam launched by the 3.5m telescope ≈ 300m away. The integration time is 10s. Scale: 0.3" per pixel ( Esposito et al., 2000).

*Figure 16.* Tilt measurement from a movable auxiliary telescope looking at a foreground NGS tracked within the isoplanatic patch of the laser plume in the mesosphere( Ragazzoni et al., 1995).

## 3.3    Perspective method

The tilt can be measured from the deviation of the mesospheric plume with respect to a foreground NGS observed at the focus of an auxiliary telescope located in such a way that the NGS lies within the isoplanatic patch $\epsilon$ of any portion of the plume (Ragazzoni et al., 1995).

Mesospheric sodium atoms excited at the $3P_{3/2}$ level scatter light in every direction. The backscattered beam observed at an auxiliary telescope $B$ meters away from the main one looks like a plume strip with an angular length $\phi \approx B\,\delta h\,/\,H^2$ where $\delta h$ stands for the thickness of the sodium layer. The tilt of the wavefront at the auxiliary telescope and vibrations equally affects the plume and the NGS. Thus departures of the plume from the average NGS location is due to the only tilt on the upward laser beam. Therefore measuring this departure allows us to know the actual location of the LGS, and to derive the tilt. Because of Earth rotation and of perspective effects, the auxiliary telescope has to track the diurnal rotation, and simultaneously to move on the ground to keep aligned the NGS and the LGS plume. Two mobile auxiliary telescopes are necessary for the two components of the tilt.

In addition to the LGS brightness, a tilt isoplanatic angle $\phi$ causes another flux limitation: only the fraction $\phi/\epsilon$ of the plume can be used to sense one component of the tilt. From preliminary studies, the telescope diameter of the auxiliary telescopes should range around ≈ 25cm. This concept should provide us with a sky coverage ≳ 50%, provided that the auxiliary telescopes can move across several hundred meters baselines.

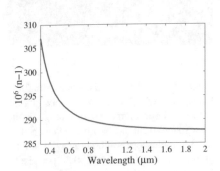

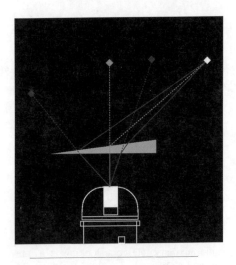

*Figure 18.* Air refraction index versus wavelength.

*Figure 19.* The polychromatic LGS. The laser beam is tilted on its way to the meso-sphere, where it produces a spectrum spanning from the UV to the infrared. The optical path of the backscattered light depends on the wavelength.

An experiment has been done between the two 3.5 and 2.2m telescopes at Calar Alto (Esposito et al., 2000), using the ALFA LGS device (Davies et al., 2000). A promising reduction by a factor of 2.5 was obtained for the tilt variance. As far as I know, this approach is no longer carried on.

### 3.4  The polychromatic LGS

The goal of the polychromatic LGS (PLGS) is to provide us with a 100% sky coverage, down to visible wavelengths. Its basic principle relies on the chromatic property of the air refraction index $n$ (Fig. 18).

If a laser beam produces in the outer atmosphere a spectrum spanning from the ultraviolet to at least the red, then the return light will follow different optical paths depending on the wavelength (Fig. 19). The air refraction index is a function of air temperature $T$ and pressure $P$:

$$n(\lambda, P, T) - 1 = f(\lambda) \times g(P, T) \tag{15}$$

By derivating:

$$\Delta n/(n-1) = \Delta f(\lambda)/\lambda \tag{16}$$

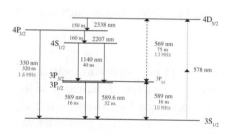

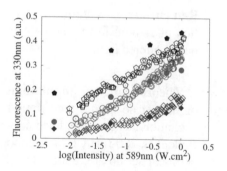

*Figure 20.* Energy levels of neutral sodium atom mostly involved in the resonant incoherent 2-photon excitation for the polychromatic LGS. Wavelengths (nm), lifetimes (ns) and homogeneous widths (MHz).

*Figure 21.* Theoretical (filled symbols) versus observed fluxes at SILVA CEA lab (open symbols). Without spectral modulation (bottom), with modulation at 589nm (middle) and with modulation at 589 and 569nm (top) Bellenger, 2002.

and applying to the tip tilt $\theta$:

$$\theta_{\lambda_3} = \Delta\theta_{\lambda_1,\lambda_2} \times (n_{\lambda_3} - 1)/\Delta n_{\lambda_1,\lambda_2} \qquad (17)$$

Therefore the tip-tilt at $\lambda_i$ expresses in terms of the tip-tilt difference between any two wavelengths (Foy et al., 1995).

### 3.4.1 What excitation process?.

The physical process which creates the PLGS has to span a large $\lambda$ interval, including the deepest possible ultraviolet. It is again the excitation of mesospheric sodium! Figure 20 shows the process in the case of a resonant incoherent 2-photon excitation. The valence electron of sodium atom is raised from the $3S_{1/2}$ ground level to the $3P_{3/2}$ level by a laser beam locked on the 589nm $D_2$ transition. Almost simultaneously, the electric field of a second beam, locked on 569nm, raises it up to the high energy level $4D_{5/2}$ ($\approx 0.5$eV from ionization). Then there is radiative cascade back to the ground state through the $4P_{1/2}$ level. The line spectrum spans from 330 nm, close to the absorption cutoff of atmosphere ozone, to the near infrared at 2.3$\mu$m. This spectrum is well suitable for the PLGS.

### 3.4.2 Feasibility study.

A feasibility study of the PLGS has been carried out in France. It is the *Étoile Laser Polychromatique pour l'Optique Adaptative* (ELP-OA). It has addressed the following main topics: i/ return flux at 330nm, ii/ accuracy of tilt measurements, iii/ telescope vibrations, iv/ wide spectral profile of lasers and vi/ budget link. I will shortly review them in the following

**Return flux @ 330nm.** The question was : does the non linear excitation of sodium in the mesosphere works, and then does it generates enough backscattered light?

**Theoretical level populations.** Since there are population variations on time scale shorter than some level lifetimes, a complete description of the excitation has been modeled solving optical Bloch equations (*Beacon* model, Bellenger, 2002) at CEA. The model has been compared with a laboratory experiment set up at CEA/Saclay (Fig. 21). The reasonable discrepancy when both beams at 589 and 569 nm are phase modulated is very likely to spectral jitter, which is not modeled: velocity classes of Na atoms excited at the intermediate level cannot be excited to the uppermost level because the spectral profile of the 569 nm beam does not match the peaks of that of the 589 nm beam.

**Experiments on the sky.** Two experiments have been carried out at the sky, using two laser installations built for the American and French programmes for Uranium isotope separation, respectively AVLIS at the Lawrence Livermore Nat'l Lab (California) in 1996 and SILVA at CEA/Pierrelatte (Southern France) in 1999. The average power was high : $\approx 2 \times 175$ W, with a pulse repetition rate of 12.9 and 4.3 kHz, a pulse width of 40 ns and a spectral width of 1 and 3 GHz. Polarization was linear. The return flux was $\lesssim 5\ 10^5$ photons/m$^2$/s (Foy et al., 2000). Thus incoherent two-photon resonant absorption works, with a behavior consistent with models. But we do need lower powers at observatories!

At CEA/Pierrelatte, the laser average power has been varied, from 5 to 50 W, which are values closer to what we can reasonably install at astronomical sites. The spectral width was 1 GHz, the pulse width $\approx 40$ ns and the pulse repetition rate 5 kHz. Polarization was linear again. Figure 23 shows the measured 330 nm return fluxes compared with the prediction from the Beacon model. Beacon overestimates measurements by a factor of $\approx 3$. This can be due to the Na column density, unknown at the time of the experiment, to a slight misalignment of the two beams resulting in non fully overlapping spots at the mesosphere, in laser instabilities (variable power and modulation, time lag between 569 and 589nm and spectral jitter) and in the assumption of a hypergaussian profile of the model beam. Being these uncertainties, the agreement between observations and model is satisfactory.

**Accuracy in differential tilt measurements.** Since one measures the differential tilt to get the tilt itself, one needs an accuracy $(n-1)/\Delta n$ higher than it would be required if the tilt could be measured directly. If the wave-

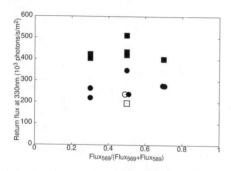

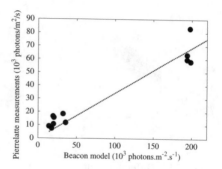

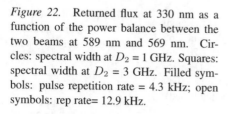

*Figure 22.* Returned flux at 330 nm as a function of the power balance between the two beams at 589 nm and 569 nm. Circles: spectral width at $D_2$ = 1 GHz. Squares: spectral width at $D_2$ = 3 GHz. Filled symbols: pulse repetition rate = 4.3 kHz; open symbols: rep rate= 12.9 kHz.

*Figure 23.* Return flux 330nm: Measurements at Pierrelatte versus Beacon model ( Bellenger, 2002).

length base is 330 - 589.6nm, $(n-1)/\Delta n \approx 25$. Can we measure the tilt with such an high accuracy ? An experiment has been made at the 1.52 m telescope at Observatoire de Haute-Provence (OHP). Natural stars were used to mimic polychromatic LGSs. A set of dichroic beam splitters and filters divided the image into 4 images at 4 wavelengths from 365 to 700 nm. Figure 24 shows the correlation between the differential tilts 700 nm - 390 nm and 570 nm - 365 nm. This correlation is independent from telescope vibrations. Its is due to the chromatic variation of $n - 1$ . The standard deviation of the differential tilt versus tilt relation is $\approx$ 1 Airy disk at 650 nm (Fig. 25). It could be improved with more accurate flat fielding, and by using a phase restoration algorithm ( Vaillant et al., 2000).

Thus the tilt can be measured with enough accuracy from the chromatic differential tilt.

**Measurements of telescope vibrations.** The polychromatic LGS is not sensitive to mechanical motions, since for each frame the whole set of chromatic images is fixed with respect to the optical axis. A pendular seismometer has been specially developed (Tokovinin, 2000) to measure low frequency vibrations (Koehler, 1998; Altarac et al., 2001), which are inaccurately sensed by accelerometers. It consists in an aluminum cylinder with an 1 kg mass at each edge, and supported at its center by a pair of Bendix strings. It is servoed at equilibrium. The signal from the capacitive sensors of the servo system provides us with the telescope vibrations of motions. The frequency ranges is $\approx 0.5 - 50$ Hz. The standard deviation of vibration measurements is $\approx 3$ marcsec, or $\approx 1/4$ of the Airy disc of an 8m telescope at 550nm. Thus we are

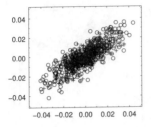

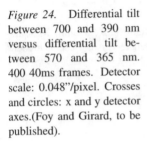

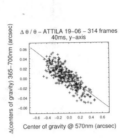

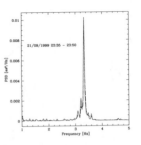

*Figure 24.* Differential tilt between 700 and 390 nm versus differential tilt between 570 and 365 nm. 400 40ms frames. Detector scale: 0.048"/pixel. Crosses and circles: x and y detector axes.(Foy and Girard, to be published).

*Figure 25.* Differential tilt between 700 and 365 nm versus tilt at 570 nm. 31540ms frames. Detector scale: 0.048"/pixel. (Foy and Girard, to be published).

*Figure 26.* Vibrations of the OHP 1.52m telescope (tracking, dome open). The records at 23h35 (solid line)and at 23h50 (dotted line) show the signal stability. Amplitude is 66 mas rms (Tokovinin, 2000).

now able to measure telescope vibrations/motions to the accuracy required at an 8m class telescope.

**Which lasers?.** The above mentioned accuracy of the tilt measurements can be achieve if there are enough return photons. The average laser power required to get them is $\approx 2 \times 20$ W. Up to now, there is no cw laser available that powerful (see Ch. 14). In addition it raised the problem of saturation of the absorption by Na atoms in the $D_2$ transition. These two problems have justified the development of the *modeless laser* (LSM) at LSP (Pique and Farinotti, 2003).

In the LSM oscillator (Fig. 27) an acousto-optic crystal shifts the frequency of the beam each times the beam crosses it. It breaks interferences in the cavity, resulting in a broad continuous spectral profile. The FWHM is $\approx 3$ GHz, matching very well the hyperfine + Doppler width of the $D_2$ line formed in the mesosphere. The LSM for the 569 nm transition is under development. It has to match a FWHM of 1 GHz, since there is the hyperfine structure of the $3P_{3/2}$ energy level is negligible with respect to the Doppler width. The large profile makes much easier the optical stabilization: an electronic servoloop is no longer necessary. The return flux measured at laboratory is $\approx 6$ times higher than that for a monomode oscillator, and $\approx 3$ times that of a phase modulated one (e.g.: the LSM would result in an LGS brighter by 1.2 magnitude of the Keck LGS). These are the gains expected from theory and measured at laboratory (Fig. 28). The expected gain for ELP-OA with respect to 2 phase modulated oscillators is $\approx 10$, since there are two transitions.

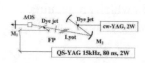

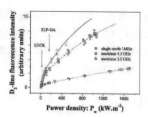

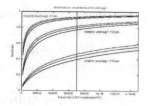

*Figure 27.* Layout of the cavity of the modeless oscillator. The acousto-optics frequency shifter prevents interferences in the cavity, resulting in a broad continuous spectral profile (Pique and Farinotti, 2003).

*Figure 28.* Return flux from the modeless laser in the $D_2$ line versus the input power. Top to bottom: spectral FWHM = 3 GHz, 4.5 GHz and 1 MHz single mode (Pique and Farinotti, 2003).

*Figure 29.* Maximum Strehl ratio versus return flux above atmosphere at 330 nm at an 8m telescope. Top to bottom: tilt correction in K, J and V bands. Solid, dashed and dotted lines: $r_0$: 15, 20 cm and 25 cm. (Schöck et al., 2002).

**Budget link.** We have developed an end-to-end model to estimate the performances of a PLGS at an 8m telescope. Measured fluxes with the LLNL and CEA experiments have been used as input. Figure 29 describes the expected Strehl ratio due to the only tilt in the K, J and V bands (Schöck et al., 2002). A modeless laser with an output average power of 25 W is assumed. The maximum Strehl modeled in this configuration is $\approx 0.5$ at 550 nm, and $\gtrsim 0.9$ in K at 2.2$\mu$m, whatever the direction in the sky. These performances rely on the most powerful LSM one can reasonably build, and on a center of gravity algorithm to measure $\delta\theta$s. Requirements would be significantly less severe if $\delta\theta$s are measured with the much more efficient phase restoration algorithm.

All these results assert the feasibility of the PLGS concept. A demonstrator to evaluate performances on the sky is now being studied. It will be first run at the coudé focus of the 1.5m telescope at OHP.

### 3.4.3 Non resonant coherent excitation.

Another excitation process of the $4P_{5/2}$ energy level relies on 2-photon non resonant coherent processes (Foy et al., 1995). Fast adiabatic passages or $\pi$ pulses can be used to excite the uppermost level directly, without populating the intermediate $3P_{3/2}$ level. Theoretically, the whole population of sodium atoms would be raised up to the $4P_{5/2}$ level, resulting in an increase by a factor of 6 in the return flux with respect to the resonant incoherent excitation. A theoretical study (Froc et al., 2000) concluded that it should work. An in depth study of the process have been carried on, both theoretically (Biegert et al., 2000) and experimentally ( Biegert and Diels, 2003) at UNM/Albuquerque. An experiment has also been carried out at the CAR/Toulouse laboratory (de Beauvoir and Girard, private communication). All these works conclude in the feasibility of this excitation process in terms of return photons. Of course the lasers which are required are more critical than in the case of the incoherent resonant excitation, but Biegert (private communication) is optimistic.

## 3.5    Other concepts

Other concepts have been proposed to measure the tilt without any NGS limiting the sky coverage. Let us briefly mention the following ones:

the dual adaptive concept: a second AO+LGS shoots at the sky within the isoplanatic patch of the programme source close to a comparatively faint NGS. It is used to measure the tilt with a higher accuracy, because it is shrunk by the AO (Rigaut and Gendron, 1992). This concept is not so far from the use of a NGS and an array of LGSs feeding a MCAO device.

the Rayleigh and the fluorescent $D_2$ plumes: measuring from the same beam the differential location of Rayleigh scattered light at moderate altitude and of the $D_2$ one at the mesosphere level provides a measurement of the tilt for the outgoing beam, thus allowing us to derive the tilt (Angel et al., 2001).

radiowave interferences can be produced at an altitude of $\approx 250$km with radiotelescopes 30-100 km apart. They break species in the ionosphere, in particular N and O (Ribak et al., 2003). The fringe pattern is not affected by any tilt on the upward radio beams. The moderate altitude increases the cone effect.

## 4.    Conclusion

A long path has been followed to overcome these two physical problems raised by the laser guide star, namely the cone effect and the tilt determination. And there is still a lot of work to be carried out before everything is solved. But I would say that the mathematics and the physics involved in the solutions we have today in hands are mastered. it is now clear that:

**MCAO without LGSs** is limited to the infrared domain, with a better probability close to the Galactic plane

**LGS without MCAO** is of limited interest, except for "small" telescopes in the IR

**MCAO with LGSs** definitely requires to determine the tilt. We can hardly rely on natural guide stars, because they are too few, except again in the infrared and relatively close to the galactic plane, and except if the goal is to increase the only encircled energy, not to reach the diffraction limit. The highest interest of the combination MCAO+LGS lies in observations at visible/red wavelengths.

Next steps toward experimental works will be particularly exciting: the MCAO MAD bench at ESO, the MCAO+LGSs device at Gemini South, experiments at the MMT and at Mt Wilson, the polychromatic LGS demonstrator in France. Also we still have a lot to learn with the feedback from the devices running now or very shortly, at Lick, Keck and ESO. But the biggest challenge of the coming years will be to extrapolate this know how to the case of decametric or even hectometric optical telescopes.

# References

Altarac, S., Berlioz-Arthaud, P., Thiébaut, É., Foy, R., Balega, Y. Y., Dainty, J. C., and Fuensalida, J. J. , 2001 *Effect of telescope vibrations upon high angular resolution imaging*, Mon. Not. R. Astron. Soc **322**, 141

Angel, R., Lloyd-Hart, M., Hege, K., Sarlot, R., and Peng, C., 2001, *The 20/20 telescope: MCAO imaging at the individual and combined foci Beyond Conventional Adaptive Optics*, pages 1

Beckers, J. M., 1988, *Increasing the size of the isoplanatic patch with multiconjugate adaptive optics*, in ESO conferences, M.-H., Ulrich, ed. **30**, 693

Bellenger, V., 2002, *Étude de l'interaction laser sodium dans le cadre du projet de l'Étoile Laser Polychromatique pour l'Optique Adaptative*, PhD thesis, Univ. Paris VI, Paris, France

Biegert, J. and Diels, J.-C., 2003, *Feasibility study to create a polychromatic guidestar in atomic sodium*, Phys.Rev A **67**, 043403.

Biegert, J., Diels, J.-C., and Milonni, P. W., 2000, *Bi-chromatic 2-photon coherent excitation of sodium to provide a dual wavelength guide star*, Optics L. **25** 683

Davies, R., Hippler, S., Hackenberg, W., Ott, T., Butler, D., Kasper, M., and Quirrenbach, A. 2000, *The alfa laser guide star: operation and results* Exp.A. **10**, 103

Ellerbroek, B. and Rigaut, F., 2001, *Methods for correcting tilt anisoplanatism in laser-guide-star-based multi-conjugate adaptive optics* Optics Com. **18** 2539

Ellerbroek, B. L., Rigaut, F., Bauman, B., Boyer, C., Browne, S., Buchroeder, R., Catone, J., Clark, P., d'Orgeville, C., Gavel, D., Hunten, G. Herriot M. R., James, E., Kibblewhite, E., McKinnie, I., Murray, J., Rabaut, D., Saddlemeyer, L., Sebag, J., Stillburn, J., Telle, J., and Véran, J.-P. ,2002, *MCAO for Gemini-South*, SPIE **4839**, 55

Esposito, S., Ragazzoni, R., Riccardi, A., O'Sullivan, C., Ageorges, N., Redfern, M., and Davies, R., 2000, *Absolute tilt from a laser guide star: a first experiment*, Exp. A., **10**, 135

Foy, R. and Labeyrie, A., 1985, *Feasibility of adaptive telescope with laser probe*, A.A, **152**, L29

Foy, R., Migus, A., Biraben, F., Grynberg, G., McCullough, P. R., and Tallon, M., 1995, *The polychromatic artificial sodium star: a new concept for correcting the atmospheric tilt*, A. A. Suppl. Ser., **111**, 569

Foy, R., Tallon, M., Tallon-Bosc, I., Thiébaut, E., Vaillant, J., Foy, F.-C., Robert, D., Friedman, H., Biraben, F., Grynberg, G., Gex, J.-P., Mens, A., Migus, A., Weulersse, J.-M., and Butler, D. J., 2000, *Photometric observations of a polychromatic laser guide star*, J. Opt. Soc. Am. A, **17**, 2236

Fried, D.L., 1995, *Focus anisoplanatism in the limit of infinitely many artificial-guide-star reference spots*, J. Opt. Soc. Am. A, **12**, 939

Froc, G., Rosencher, E., Atal-Trétout, B., and Michau, V., 2000, *Photon return analysis of a polychromatic laser guide star*, Optics Com., **178**, 405

Fugate, R. Q., Fried, D. L., Amer, G. A., Boek, B. R., Browne, S. L., Roberts, P.H., Ruane, R. E., Tyler, G. A., and Wopat, L. M., 1991, *Measurement of atmospheric wavefront distortion using scattered light from a laser-guide-star*, Nature, **353**, 144

Koehler, B., 1998, *VLT unit telescope suitability for interferometry: first results from acceptance tests on subsystems*, SPIE **3350**, 403

Le Louarn, M., Foy, R., Hubin, N., and Tallon, M., 1998, *Laser guide star for 3.6m and 8m telescopes: Performances and astrophysical implications*, Mon. Not. R. Astron. Soc, **295**, 756

Le Louarn, M. and Hubin, N., 2004, *Wide-field adaptive optics for deep-field spectroscopy in the visible*, Mon. Not. R. Astron. Soc, **349**, 1009

Le Louarn, M. and Tallon, M., 2002, *Analysis of the modes and behavior of a Multiconjugate Adaptive Optics system*, JOSA.A **19**, 912

Lloyd-Hart, M., Jefferies, S. M., Roger, J., Angel, P., and Hege, E. K., 2001, *Wave-front sensing with time-of-flight phase diversity*, Optics Let. **26**, 402

Pique, J.-P. and Farinotti, S., 2003, *An efficient modeless laser for a meso-spheric sodium laser guide star*, JOSA. B, **20**, 2093

Primmerman, C. A., Murphy, D. V., Page, D. A., Zollars, B. G., and Barclay, H. T., 1991, *Compensation of atmospheric optical distortions using a syn-thetic beacon*, Nature, **353**, 141

Ragazzoni, R., Esposito, S., and Marchetti, E., 1995, *Auxiliary telescopes for the absolute tip-tilt determination of a laser guide star*, Mon. Not. R. Astron. Soc, **276**, L76

Ribak, E. N., Ragazzoni, R., and Parfenov, V. A., 2003, *Radio plasma fringes as guide stars: Tracking the global tilt*, A.A **410**, 365

Rigaut, F. and Gendron, É., 1992, *Laser guide star in adaptive optics: the tilt determination problem*, Astron. Astrophys. **261**, 677

Sasiela, R. J., 1994, *Wave-front correction by one or more synthetic beacons*, JOSA. A **11**, 379

Schöck, M., Foy, R., Tallon, M., Noethe, L., and Pique, J.-P., 2002, *Performance analysis of polychromatic laser guide stars used for wavefront tilt sensing*, MNRAS **337**, 910

Tallon, M. and Foy, R., 1990, *Adaptive optics with laser probe: isoplanatism and cone effect*, Astron. Astrophys. **235**, 549

Tokovinin, A., 2000, *Pendular seismometer for correcting telescope vibrations*, MNRAS **316**, 637

Vaillant, J., Thiébaut, É., and Tallon, M., 2000, *ELPOA: Data processing of chromatic differences of the tilt measured with a polychromatic laser guide star*, SPIE **4007**, 308

# INTERFEROMETRY

## An Introduction to Multiple Telescope Array Interferometry at Optical Wavelengths

Oskar von der Lühe

*Kiepenheuer–Institut für Sonnenphysik*
*Schöneckstrasse 6-7, 79104 Freiburg im Breisgau, Germany*
ovdluhe@kis.uni-freiburg.de
http://www.kis.uni-freiburg.de/~ovdluhe/

**Abstract**     Fundamentals of amplitude interferometry are given, complementing animated text and figures available on the web. Concepts as the degree of coherence of a source are introduced, and the theorem of van Cittert – Zernike is explained. Responses of an interferometer to a spatially extended source and to a spectrally extended one are described. Then the main methods to combine the beams from the telescopes are discussed, as well as the observable parameters – vibilities and phase closures.

**Keywords:**     amplitude interferometry, stellar interferometry, coherence, interferometers, beam recombinaison, phase closure, visibility

## Introduction

This article complements the three lectures on "Interferometry" at the CNRS-/NATO/EC Summer school "Optics in Astrophysics" in Cargèse, Corse, in September 2002. The original lectures make extensive use of the features of Powerpoint, including animated text and figures. Therefore the original Powerpoint files, as well as PDF versions thereof, are available for download from the web sites of the Summer school and of the author. Several animations were prepared using MathCAD to produce animations in "avi" format, each of which is several MB in size. These animations are separately available for download at the same web sites.

The organisation of this article is as follows:

- each of the three lectures comprises a section of this article.

*R. Foy and F. C. Foy (eds.), Optics in Astrophysics, 275–287.*

- Subsections and subsubsections are organised thematically, with references to one or more viewgraphs of a lecture.
- (I.20) refers to lecture I, viewgraph 20, (I.6-9) to lecture I, viewgraphs 6 to 9.

The reader of this article is therefore encouraged to avail herself (himself) of the lecture material to be viewed along with reading this article.

## 1.    Lecture I: Fundamentals of Interferometry

### 1.1    Why Interferometry?

**What Is Interferometry (I.3)**   Interferometry deals with the physical phenomena which result from the superposition of electromagnetic (e.m.) waves. Practically, interferometry is used throughout the electromagnetic spectrum; astronomers use predominantly the spectral regime from radio to the near UV. Essential to interferometry is that the radiation emerges from a single source and travels along different paths to the point where it is detected. The spatio-temporal coherence characteristics of the radiation is studied with the interferometer to obtain information about the physical nature of the source.

Interferometry in astronomy is used to surpass the limitations on angular resolution set by the Earth's atmosphere (i. e., speckle interferometry), or by the diffraction of the aperture of a single telescope. We will focus in this lecture on interferometry with multiple telescope arrays with which it is possible to obtain information on spatial scales of the source beyond the diffraction limit of its member telescopes.

**Diffraction–limited Imaging (I.4)**   The resolution of a single telescope is limited by diffraction at its aperture stop, which in many cases is formed by the primary mirror. The image of a point source (e. g., a very distant star) at a wavelength $\lambda$ is ideally an extended patch of light in a focal plane, surrounded by a regular sequence of rings. The radius of the innermost dark ring sets a limit for the resolution of the telescope in the sense of a lower limit for the angle between two point sources at the celestial sphere which appear separated. This limit is approximately $\lambda/D$, where $D$ is the diameter of the *entrance pupil* of the telescope – i. e., the image of the aperture stop in the space towards the source. Its inverse, $\mathcal{R}_\alpha = \lambda/D$, is called the *diffraction limit* of the telescope. The diffraction limit can be overcome only by making the telescope larger or by decreasing the wavelength. If the latter is undesireable and the former impossible, two or more telescopes with their optical beams combined by interferometry may improve the resolution.

## 1.2 Interference Basics

**Young's Interference Experiment (I.5)** A Young's interferometer basically consists of a stop with two holes ("array elements"), illuminated by a distant point source, and a screen which picks up the light at a sufficiently large distance behind the holes. The light patch produced by the stop is extended and shows a set of dark fringes which are oriented perpendicularly to the direction which connects the centers of the two openings.

The diameter $D$ of the openings (we assume that they are equal), the *baseline* $B$ (distance between the centers of the holes), the separation $F$ between the stop and the screen, and the wavelength $\lambda$ fully characterise the intensity distribution which is observed on the screen. While the characteristic dimension of the light patch is approximately $\lambda F/D$, the fringe period is $\lambda F/B$.

In order to observe fringes, the screen should be placed in the regime of *Fraunhofer diffraction* where $F/B >> B/\lambda$. In practice, such an interferometer can be realized by placing the stop immediately in front of a collecting optics, e. g., a lens or a telescope, and by observing the fringes in its focal plane ($F = f_{\text{eff}}$).

**Basic Interferometer Properties (I.6-9)** Although the relationship between element aperture diameter, baseline, and wavelength is quite simple, it is instructive to visualise the influence of each of these characteristics. To this end, we consider a Young's interferometer with element diameters $D = 1$m, a baseline $B = 10$m at a wavelength $\lambda = 1\mu$m in the animations. The intensity profile across the fringe pattern on the detector (screen) is shown with linear and logarithmic intensity scales in the lower two panels. The blue line represents the intensity pattern produced without interference by a single element.

*Variation of baseline between 5 and 15 m:* note that the fringe pattern contracts while the baseline increases, but the blue envelope remains constant.

*Variation of element diameter between 0.5 and 1.5 m:* the overall patch size decreases with increasing diameter while the fringe period is unaffected. The first animation shows the effect with normalized peak intensity. The second animation shows the effect of increasing collecting power; the peak intensity actually increases proportional to $D^4$.

*Variation of wavelength between 0.5 and 1.5 μm:* both fringe pattern and envelope expand as the wavelength increases, the ratio between envelope width and fringe period remains constant at $B/D$.

**Geometric Relation between Source and Interferometer (I.10-13)** If the position of the source changes with respect to the orientation of the interferometer, the position of the illuminated area on the detector changes proportionally.

The fringe also moves at the same rate, in e., the intensity pattern moves as an entity.

A relative motion between envelope and fringe pattern can be produced through a change of the *differential delay* between the two interferometer arms by, e. g., inserting a piece of glass into the beam behind one of the holes. This time the fringe pattern is displaced with respect to the position of the overall envelope. If $d$ is the thickness of the glass and $n$ is its index of refraction, the differential delay produced is $\delta = d(n-1)$. The animation on page 13 shows how the fringe pattern moves as the delay is changed by $\pm 2\mu$m while the position of the envelope does not change.

## 1.3 Diffraction Theory and Coherence

If we wish to include spectral coverage and extended sources into our considerations, we need to review the theory of diffraction and to introduce some concepts of coherence.

**Monochromatic Waves (I.14)** A *monochromatic* e.m. wave $V_\omega(\mathbf{r}, t)$ can be decomposed into the product of a time–independent, complex-valued term $U_\omega(\mathbf{r})$ and a purely time–dependent complex factor $\exp j\omega t$ with unity magnitude. The time–independent term $U_\omega$ is a solution of the Helmholtz equation. Sets of base functions which are solutions of the Helmholtz equation are plane waves (constant wave vector $\mathbf{k}$ and spherical waves whose amplitude varies with the inverse of the distance of their centers.

**Propagation of Field (I.15)** A popular method to solve the Helmholtz equation in the presence of a diffracting aperture is the *Kirchhoff–Fresnel integral*. It is based on the Huyghens' principle, where the field is represented by a superposition of spherical waves. The field amplitude $U_\omega$ at a position $\mathbf{r}_0$ in an extended volume behind the aperture is computed from the field amplitude $U_\omega$ at a position $\mathbf{r}_1$ inside the aperture. This is done by integration of the field inside the aperture $\Sigma$, weighted by the amplitude and phase term of the spherical wave, $1/r \exp(2\pi j\, r/\lambda)$ with $j = \sqrt{-1}$. The term $\chi(\vartheta)$ is an inclination term which is 1 for small values of the angle $\vartheta$ that the direction of the line connecting $\mathbf{r}_0$ and $\mathbf{r}_1$ makes with the general direction of light propagation, and which vanishes for increasing $\vartheta$.

**Nonmonochromatic Waves (I.16)** Diffraction theory is readily expandable to non-monochromatic light. A formulation of the Kirchhoff–Fresnel integral which applies to quasi–monochromatic conditions involves the superposition of retarded field amplitudes.

**Intensity at a Point of Superposition (I.17)** The measurable physical parameter of an optical wave is its energy density or intensity. If two fields are superimposed, the measured intensity is given by the sum of the individual intensities plus aterm which describes the long term *correlation* of the field amplitudes. Long term means time scales which are large compared to the inverse of the mean frequency $\bar{\omega}/2\pi$ (about $10^{-12}$ s); the time scale is set by the time resolution of the detector. This is why the field product term is expressed in the form of an *ergodic mean* $\langle \cdots \rangle$. An interferometer produces superimposed fields, the correlation of which carries the desired information about the astronomical source. We will discuss exactly how this happens in the following sections.

**Concepts of Coherence (I.18-21)** The correlation term in the expression for the intensity of superimposed fields, and in general the ergodic mean of the product of field amplitudes sampled at different points in space and time, is called *mutual intensity*. If the conditions are temporally stationary – which they would be if the observed sources are stable with time – it is the time difference $\tau$ which matters.

Usually we are only interested in mutual intensity suitably normalised to account for the magnitude of the fields, which is called the *complex degree of coherence* $\gamma_{12}(\tau)$. This quantity is complex valued with a magnitude between 0 and 1, and describes the degree of "likeness" of two e. m. waves at positions $\mathbf{r}_1$ and $\mathbf{r}_2$ in space separated by a time difference $\tau$. A value of 0 represents complete decorrelation ("incoherence") and a value of 1 represents complete correlation ("perfect coherence") while the complex argument represents a difference in optical phase of the fields. Special cases are the *complex degree of self coherence* $\gamma_{11}(\tau)$ where a field is compared with itself at the same position but different times, and the *complex coherence factor* $\mu_{12} = \gamma_{12}(0)$ which refers to the case where a field is correlated at two positions at the same time.

In the first case, the degree of self coherence depends on the spectral characteristics of the source. The *coherence time* $\tau_c$ represents the time scale over which a field remains correlated; this time is inversely proportional to the spectral bandwidth $\Delta\omega$ of the detected light. A more quantitative definition of quasi–monochromatic conditions is based on the coherence time; all relevant delays within the interferometer should be much shorter than the *coherence length* $c\tau_c$. A practical way to measure temporal coherence is to use a Michelson interferometer. As we shall see, in the second case the *spatial coherence* depends on the apparent extent of a source.

One should keep in mind that e. m. waves are transverse vector fields and are polarized. Strictly speaking, one has to consider the polarisation characteristics of the light throughout this treatment. Polarisation plays an important practical role as instrumental depolarisation can change the coherence characteristics of the light and therefore advesely influence measurements. However, it is not really important in the understanding of most celestial sources which is why we will not consider polarisation in this lecture.

**Extended Spectral Distribution (I.22-23)**   Let us now turn back to our interferometer experiment. Only the central fringe of a Young's interferometer represents the position of zero delay in time between the superimposed beams, for each maximum to the side there is a corresponding number of oscillation periods $(2\pi/\omega)$ of delay. It should therefore be no surprise that the *contrast* of the fringe pattern is reduced if the spectral distribution of the source is changed from monochromatic to extended. Note how the fringe pattern contrast remains high near the central fringe while it is reduced in the wings as the source's spectral distribution becomes broader and the *spectral resolution* $R_\lambda = \bar{\lambda}/\Delta\lambda = \bar{\omega}/\Delta\omega$ becomes smaller. The reduction of contrast is caused by the superposition of interference patterns with different fringe spacings which remain centered on the position of zero delay.

A delay error shifts the position of zero delay with respect to the overall intensity envelope, resulting in a substantial reduction of overall contrast. The contrast may vanish entirely if the zero delay position coincides with a minimum. Therefore, there is a relation between the allowable delay error $\delta_{\max}$ and the spectral bandwidth $\Delta\omega$ of the detected radiation if the amplitude error of the fringe modulation is to remain small, i. e., $\delta_{\max} \leq \bar{\lambda} \cdot R_\lambda = \bar{\lambda}^2/\Delta\lambda$.

## 1.4    Interferometry of Extended Sources

We now turn our attention to extended sources which we will assume monochromatic for the time being.

**Extended Incoherent Sources (I.24-26)**   First, we investigate how the coherence properties of light propagates by inserting the Kirchhoff–Fresnel formula into the expression for mutual intensity. We note that the propagation of coherence between two planes $\Sigma_1$ and $\Sigma_2$ involves a fourfold integration over two coordinates within $\Sigma_1$. Next, we note that *extended* sources can be decomposed into a distribution of point sources whose brightness is given by the intensity distribution $I(\mathbf{r}_1)$ of the extended source.

We still need to consider the coherence properties of astronomical sources. The vast majority of sources in the optical spectral regime are thermal radiators. Here, the emission processes are uncorrelated at the atomic level, and the source can be assumed incoherent, i. e., $J_{12} = \lambda^2/\pi \, I(\mathbf{r}_1) \cdot \delta(\mathbf{r}_2 - \mathbf{r}_1)$, where $\delta(\mathbf{r})$ denotes the Dirac distribution. In short, the general source can be decomposed into a set of *incoherent* point sources, each of which produces a fringe pattern in the Young's interferometer, weighted by its intensity, and shifted to a position according to its position in the sky. Since the sources are incoherent, the intensity patterns add directly without further interference, and the fringe contrast decreases.

**Theorem of van Cittert – Zernike (I.27-28)** In order to quantify the reduction in fringe contrast one first transports mutual intensity from the source to the position of the interferometer. All celestial sources of interest have large distances $S$ compared to their dimensions, the differences in the propagation distances $S_1$ and $S_2$ can be expanded in first order to simplify the propagation equation. Inclination terms $\chi(\vartheta)$ are set to unity. Linear distances $\mathbf{r}$ in the source surface are replaced by apparent angles $\vartheta = \mathbf{r}/S$. The coherence transport equation can then be simplified to involve an integral over the source of the source intensity, weighted by a term $\exp -2\pi j \vartheta (\mathbf{x}_1 - \mathbf{x}_2)/\lambda$, where $\mathbf{x}_1$ and $\mathbf{x}_2$ are the positions of the two apertures of the interferometer.

We are now ready to derive an expression for the intensity pattern observed with the Young's interferometer. The correlation term is replaced by the complex coherence factor transported to the interferometer from the source, and which contains the *baseline* $\mathbf{B} = \mathbf{x}_1 - \mathbf{x}_2$. Exactly this term quantifies the contrast of the interference fringes. Upon closer inspection it becomes apparent that the complex coherence factor contains the two-dimensional Fourier transform of the apparent source distribution $I(\vartheta)$ taken at a spatial frequency $\mathbf{s} = \mathbf{B}/\bar{\lambda}$ (with units "line pairs per radian"). The notion that the fringe contrast in an interferometer is determined by the Fourier transform of the source intensity distribution is the essence of the **theorem of van Cittert – Zernike**.

**Response of the Interferometer to Extended Sources (I.29-32)** We will now examine the response of the interferometer to the simplest source, a double point source of equal brightness with apparent separation $\delta\alpha$ varying between 0 and 0.1 arcsec. The Fourier transform of the source intensity is a cosine function with a period proportional to $\delta\alpha$ and is sampled by the interferometer at a constant frequency. Note how the contrast of the fringe varies periodically as the source separation increases and the interferometer scans the cosine visibility function. Each time the cosine term becomees negative, the "contrast" of the fringe pattern changes sign and central fringe, which is bright for a zero separation, becomes dark. Eventually the separation of the double source becomes so large that it is resolved by the individual elements of t he interferometer and the two Airy functions separate.

The next source we investigate is the surface of an extended, limb–darkened star whose apparent diameter increases from 1 to 25 milli-arcseconds. The visibility function of such a source resembles the Airy function, varying periodically between zero and a maximum value which decreases with increasing frequency. Note how the fringe contrast vanishes repeatedly to rise again without reaching its previous maximum value as the source's apparent diameter increases. A detailed investigation of the visibility function permits deriving the stellar apparent diameter and the limb darkening coefficients.

However, the response of a two-way interferometer to an extended source for a single baseline is not unique. The decrease in contrast indicates that the source is resolved by the baseline but does not give any information about the source's structure. The full information can only be obtained by measuring the two-dimensional visibility function as completely as possible, which implies measurements with many baselines.

## 1.5    Summary and Conclusion (I.33-34)

The fundamental quantity for interferometry is the source's visibility function. The spatial coherence properties of the source is connected with the two-dimensional Fourier transform of the spatial intensity distribution on the cesetial sphere by virtue of the van Cittert – Zernike theorem. The measured fringe contrast is given by the source's visibility at a spatial frequency $B/\lambda$, measured in units *line pairs per radian*. The temporal coherence properties is determined by the spectral distribution of the detected radiation. The measured fringe contrast therefore also depends on the spectral properties of the source and the instrument.

## 2.    Lecture II: Concepts of Interferometry

## 2.1    Elements of an Interferometer (II.3)

The diagram presents the essential elements which constitute an interferometer – here in particular the VLT Interferometer. There are two telescopes which collect the light to be coherently combined. Each light path needs optics to direct the beams into a stationary laboratory where combinaiton takes place. The optics include delay lines which guarantee equal optical path length from the source through the two telescopes up to the point of detection. This configuration can be extended to more than two telescopes, each of which is equipped with its own delay line.

## 2.2    Baselines and Projected Baselines (II.4-8)

The diameter of a telescope entrance pupil or the distance between two telescopes determine the baseline, which determines the resolution of the interferometer in combination with the detected wavelength. The table compares the resolution of single telescopes and interferometers at optical and radio wavelengths. Note that the resolution of optical interferometers is comparable to that of radio very long baseline interferometry (*VLBI*).

The spatial frequency measured by the interferometer is a two-dimensional quantity. It depends on the three-dimensional baseline $\mathbf{B}_{ik}$ between telescopes

$i, k$ and the direction **S** of the source (directional cosines). Since the direction of the source changes with time with respect to the baseline of a multiple telescope interferometer, the visibility function is sampled at the *projected baseline* $\mathbf{B}'_{ik}$. Two of the three components of $\mathbf{B}'_{ik}$ are the sampled frequencies projected onto the celestial sphere, traditionally denoted $u$ and $v$ ("UV plane"). The third component, $w$ is the *geometric delay* which needs to be compensated by the delay lines. The relationship between telescope positions and projected baseline can be formulated as a matrix equation where the matrix depends on site latitude, and source declination and hour angle. Note that the geometric delay is generated in vacuum above the atmosphere, and therefore delay lines should operate in vacuum as well for the interferometer to be fully achromatic.

## 2.3 Array Configuration and Earth-rotational Synthesis (II.9-16)

This section presents the layout of several interferometers, and the range of the visibility function with are sampled through the course of a night while the source transverses the sky (*uv coverage*). The map of frequencies covered by a reasonable range of source hour angles on the UV plane depends on the array configuration and is called a "sausage diagram". There are $N(N-1)/2$ pairs "sausages" for an interferometer with $N$ telescopes; each pair belongs to a projected baseline ($\mathbf{B}'_{ik}$ and $-\mathbf{B}'_{ik}$). The goal of the array configuration is to obtain an optimum sampling of the UV plane up to a limit set by the longest baselines.

The projected baseline of the LBT only rotates while its length remains constant, independent of source declination, since the two telescopes share the same mount. The VLT interferometer produces six sausages when all four main telescopes are combined; the prominent diagonal orientation reflects the asymmetric array layout. The pattern can be made substantially denser when auxiliary telescopes are included into the array (thinner sausages). The coverage is also strongly dependent on source declination, note how the pattern becomes more circular for sources with southern declinations (the latitude of the VLTI is 24.5° south). The Keck interferometer features one baseline between its 10m telescopes (thick sausages), eight baselines between the 10m and the four 1.5m outrigger telescopes (medium sausages), and six baselines between the outriggers themselves (thin sausages). The thickness of the sausage is an indication for sensitivity. Generally, arrays with independent telescopes have a deficit in coverage for short baselines because telescope domes mutually obstruct the view.

## 2.4    Beam Combination (II.17-18)

There are many ways to combine the beans which are collected by several telescopes in order to produce a fringe. A fundamental distinction for interferometry at optical wavelengths is the size of combined field of view. One refers to "multi–mode" beam combination if the combined field exceeds the resolution element of a single telescope, and to "single mode" beam combination if the combined field does not exceed the resolution element. The term "mode" is related to optical throughput ("etendue"), the product of field area and the solid angle covered by a focal beam. For a single resolution element of a telescope with diameter $D$ and focal length $f$ this product becomes $(\lambda f/D)^2 (D/f)^2 = \lambda^2$. In a multi–mode beamcombiner the throughput is just many times $\lambda^2$.

**Single Mode Beam Combination (II.19-21)**    The realisation of single beam combiners based on single mode optical fibers is quite popular because light is easily transported once it has been transferred into the fiber, and micro-optical elements can be used for beam combination. Furthermore, the light level in each fiber is easily monitored, making high precision measurements of the fringe contrast possible. Measurements of visibility with the highest accuracy are done using fiber beamcombiners. Since the measured field is constrained to the size of the Airy disk ("antenna pattern"), this mode bears the most similarity to radio interferometry. The IOTA interferometer with FLUOR and the VLTI VINCI and AMBER beamcombiners are examples for fiber based single mode beamcombiners.

**Multi–Mode Beam Combination (II.22-27)**    It is possible to distinguish between "pupil plane" and "image plane" beam combination in a multi–mode beamcombiner depending on whether the beams intersect in overlapping images of the aperture stops or the sources; such a distinction is irrelevant for single mode beam combination. In an image plane, the light originating from each (Airy-disk sized) point in the source may carry its own interference pattern which can be detected separately and simultaneously with a high resolution camera. The Fourier transform of such a detection is resolved at each baseline; the number of resolution elements is given by the area of the detected field divided by the area of the Airy disk. The GI2T interferometer uses image plane beam combination.

The beams which originate from different locations intersect the pupil plane with different angles, where they all produce the same intensity pattern, although with a phase which depends on their position in the field. The simultaneous detection of a pathlength modulated fringe results in the visibility due to all sources whose light is allowed to reach the detector, representing a single point in the Fourier plane. Examples for multi–mode pupil plane beamcombiners include NPOI, COAST and VLTI MIDI.

**Fizeau- and Michelson Beam Combination (II.28-30)**   A special case of an image plane multi–mode beamcombiner is a *Fizeau interferometer*. The optical configuration is equivalent to a large telescope whose entrance aperture is covered with a mask with holes which represent the interferometer array and where the fringes are observed in the focal plane. Such a configuration preserves the *Helmholtz-Lagrange invariant* within each array element and the array as a whole. An image plane beam combiner where this is not the case is called a *Michelson (stellar) interferometer*. Another way to describe these two types of beam combination is that for the Fizeau interferometer, the exit pupil is an exactly scaled and oriented replica of the entrance pupil of the interferometer *as seen by the source* ("homothetic mapping"). This is easy to realize with a masked telescope, but difficult for a distributed array.

The external geometric differential delay (see below) of an off axis source is exactly balanced within a Fizeau interferometer, resulting in fringes with the same phase on top of each source in the field. The position of a source may differ from the "position of zero OPD" in a Michelson interferometer depending on how dissimilar entrance and exit pupils are. The fringe contrast of off-axis sources also depend on the temporal degree of coherence of the detected light.

**Differential Delay Tracking (III.3-4)**   Consider two nearby sources (stars) which are observed by a two-element interferometer. Both elements point at the same star, resulting in a projected baseline $B'$ and a geometric delay $w$ which is canceled by a delay line to detect the zeroth order (white light) fringe at the beamcombiner. The second star appears at an angle $\alpha$ from the first star, in direction of the projected baseline. There is a differential geometric delay $\delta w = \alpha B'$ which, when left uncompensated, results in a phase shift $2\pi \delta w / \lambda$ at the beamcombiner. A second, differential delay line can be used to cancel the differential delay and to produce a white light fringe also for the second star. The differential delay can be measured with high precision, permitting high precision measurements of the angle between both stars. The position of one of the two stars which is unresolved to the interfeometer may serve as a *phase reference*, making it possible to measure directly the object phase of the other source. Monitoring of secular variations of relative positions of two stars caused by, e. g., the reflex motion due to low mass companions, to the precision of a few micro-arcsec is the most compelling use for off source phase referencing.

## 2.5    Observables (III.5)

The physical quantity directly detected by the interferometer is the correlated flux and the complex visibility for a given wavelength, corresponding to an angle frequency on the celestial sphere which depends on the projected baseline. Let us summarise which kind of quantities an interferometer can deliver and which kind of information on the observed source can be extracted.

**Group delay (III.6)**   Group delay encodes information on the source position. This observable can be calibrated using reference stars or suitable metrology. It is used for high precision astrometry.

**Visibility Amplitude (III.7)**   This is the most readily measured observable, the maximum contrast of the interferogram. It contains essential photometric information for images of the source and all information for circularly (or elliptically) symmetric sources. A high precision measurment requires calibrator sources with known visibilities and / or monitoring of system parameters for calibration.

**Referenced Phase (III.8)**   The directly detected phase is not a good observable because its errors cannot be controlled. There are several possibilities to create references for the phase which are measured simultaneously, including the phase of a second source or the phase of the same source at a different wavelength. Phases are essential for generating images (maps) of extended sources of any degree of complexity. Phases can be substantially transformed by shifting the origin of the source coordinate system.

**Closure Phases (III.9-10)**   Closure phases are obtained by triple products of the complex visibilities from the baselines of any subset of three apertures of a multi–element interferometer. Element-dependent phase errors cancel in these products, leaving baseline dependent errors which can be minimised by careful designs. Although there are many closure relations in a multi–element array, there are always fewer independent closure phases than baselines. Closure phases are essential for imaging if no referenced phases are available.

## 2.6    Sensitivity (III.11-12)

A simple expression for the signal-to-noise ratio (SNR) of a measurement of visibility amplitude involves several parameters relating to interferometer and source properties. The formula presented here provides the fundamental sensitivity limit. Contrast loss arising from instrumental jitter and seeing are summarised in a common factor "system Strehl", which is the ratio of the number of photons which can be used for a coherent measurement to the total number of photons. Other sources of error include photon noise, background photons, and detector noise which is expressed here in the form of an equivalent photon count. The redundancy of the considered baseline (number of baselines with the same baseline vector) and the number of array elements influence the measured signal. One can distinguish between *coherent* and *incoherent* measurements which can be combined to improve the SNR. It suffices to increase the number of detected photons for combining coherent measurements (e.g., increasing the exposure time), while the SNR only increases with the square root of the number of incoherent measurements (e.g., adding the moduli squared). The graphs show the SNR as function of source magnitude in the H band ($\lambda = 1.6\mu m$) for arrays with 1.8m and 8m diameter elements and for various instrumental conditions and source visibilities.

## 3. Lecture III: Practical Interferometry (III.13-35)

This lecture provides an overview of exisiting interferometers at the time of the course, Summer 2002. Although just recently becoming a significant tool for observational astrophysics, stellar interferometry at optical wavelengths looks back to a history of almost 140 years! Since there is substantial progress, any information provided here would be superseded quickly. Also, links on the world-wide web have the tendency to become obsolete. A good and well maintained, up-to-date resource are the "Optical Long Baseline Interferometry News", edited by Peter Lawson (http://olbin.jpl.nasa.gov/).

# GUIDED OPTICS FOR STELLAR INTERFEROMETRY

François Reynaud and Laurent Delage
*Equipe optique IRCOM*
*123 rue A. Thomas*
*87060 Limoges CEDEX France*
reynaud@ircom.unilim.fr

**Abstract**     We propose a general presentation of waveguides use in the frame of high resolution imaging by stellar interferometry. Brief recalls on guiding principle are followed by the required functions to be implemented. This paper is concluded by a present status on the experimental developments.

**Keywords:**     High resolution imaging; stellar interferometry; optical fibers; guided and integrated optics.

## 1.     GENERAL PROPERTIES OF OPTICAL WAVEGUIDES

The purpose of this chapter is to propose a general overview of waveguides properties. This short introduction cannot include detailed presentations. For additional information you can refer to a large number of books. We suggest you Jeunhomme, 1990.

### 1.1     Guiding light

Figure 1 gives a scheme of an optical fibre. The guiding properties result from the refractive index difference between the core (high refractive index $n_1$) and the cladding (low refractive index $n_2$). By means of total reflection, the light is trapped over the fibre core area $S$ if its propagation direction is included in an angular cone defined by $\beta$ angle.

$$S = \pi/4w^2 \tag{1}$$

289

*R. Foy and F. C. Foy (eds.), Optics in Astrophysics, 289–306.*
© 2006 *Springer. Printed in the Netherlands.*

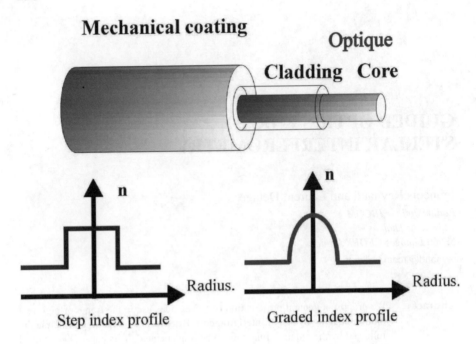

**Mechanical coating**

**Optique**

**Cladding   Core**

Step index profile                          Graded index profile

*Figure 1.*   Optical fibre general scheme and refractive index profile.

$$sin(\beta) = \sqrt{n_1^2 - n_2^2} \tag{2}$$

where $w$ is the core diameter. It results in a limitation of the position over a surface $S$ and a restriction on the angular domain over a solid angle $\Delta\Omega$. Consequently such waveguide can be used as beam spatial filter in order to reduce the number of spatial degree of freedom.

The number of spatial degree of freedom N, so called *étendue*, is given by

$$N = \frac{S \times \Delta\Omega}{\lambda^2} \tag{3}$$

with

$$\Delta\Omega = 2\pi(1 - cos\beta) \tag{4}$$

It corresponds to the number of optical path to be used for light beams trapping and it depends on the wavelength $\lambda$. By using a small diameter fibre with a small core/cladding refractive index difference, it is possible to design a monomode waveguide with $N = 1$. The main interest of such waveguides is to have a perfect control of the spatial geometry of the beam propagating in the core. By this way, the optical path is fully controlled and the disturbances over the propagation are minimized. These disturbances are called *dispersion*

**Large variety of direction**

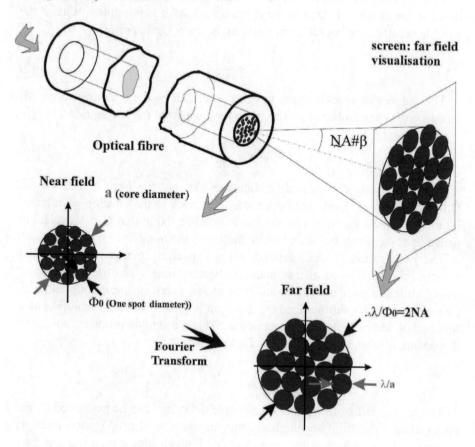

*Figure 2.* Near field and far field of a multimode fibre. The number of speckle structures allows to roughly determine the number of modes.

and can result from chromatic effects, polarization properties and multimode behaviors.

## 1.2 Characterization of the light propagation

By deriving or computing the Maxwell equation in the frame of a cylindrical geometry, it is possible to determine the modal structure for any refractive index shape. In this paragraph we are going to give a more intuitive model to determine the number of modes to be propagated. The refractive index profile allows to determine $w$ and the numerical aperture $NA = sin(\beta)$ , as defined in equation 2. The near field (fiber output) and far field (diffracted beam) are related by a Fourier transform relationship: **Far field = TF(Near field)**.

Consequently the outlines of the far field are characteristics of the smallest detail in the near field. Due to the diffraction laws it is possible to link $\Phi$ the speckle substructure typical dimension to the numerical aperture:

$$\Phi \approx \frac{\lambda}{2\beta} \tag{5}$$

The number of speckle spots to be squeezed in $S$, the core area, approximately gives the number of modes to be propagated in the waveguide.

$$N \approx \frac{S}{\frac{\pi \Phi^2}{4}} \tag{6}$$

If the core only corresponds to one speckle spot the fibre is monomode. Otherwise the waveguide is multimode. This back to the envelop calculation intuitively shows the origin of the mode number. Note that N is wavelength dependent: the larger the wavelength, the lower the number of mode.

The propagation of light depends on the spectral domain because of the wavelength dependence of the material refractive index and the evolution of modal structure over the spectrum. The global result of these coupled complex effects can be summarized by the $\beta$ propagation constant evolution as a function of wavelength $\lambda$ or frequency $\omega$. This waveguide property, so-called dispersion, is often expressed using Taylor's series:

$$\beta(\nu) = \beta_0 + \frac{d\beta}{d\omega}(\omega - \omega_0) + \frac{d^2\beta}{d\omega^2} \times \frac{(\omega - \omega_0)^2}{2} + \dots \tag{7}$$

The second term corresponds to a simple delay and can be generated by air propagation. The third and higher terms are specific of propagation through material. They induce pulse spreading for optical telecommunications and degradation of the interferences contrast in the frame of interferometry.

The last point to consider is the birefringent properties of waveguides due to mechanical or thermal stresses. The propagation of each spatial mode is degenerated and the two neutral axis are characterized by their specific propagation constant. These properties are wavelength dependent leading to a very complex situation when broadband sources are launched in the waveguide.

The effects of dispersion and birefringence on stellar interferometry will be discussed in Sections 17.2.3 and 17.2.4. New kind of fibres has been design to manage the dispersion properties using a silica / air structure. These fibres, so called *Photonic Crystal Fibres*, are very promising for many applications (Peyrilloux et al., 2002).

## 1.3    Fibre material

Silica is the more conventional material used for fibre manufacturing. The optical fibre telecommunications concentrated the main efforts for waveguide

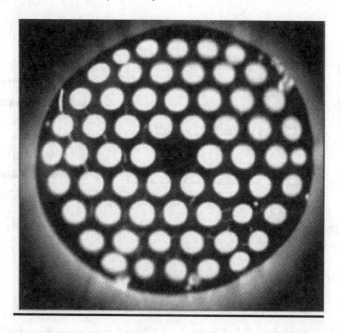

*Figure 3.* Photonic Crystal Fibre (PCF). An air hole / silica arrangement allows to manage the dispersion properties.

*Table 1.* Polarization preserving fibres available

| Material | Operating wavelength | attenuation |
|---|---|---|
| Chalcogenide | 3-10$\mu$ m | 0.5 dB/m @ 6$\mu$ m |
| Fluoride | 0.5- 4.3 $\mu$ m | 0.02 dB/m @ 2.6$\mu$ m |
| Sapphire | 0.2-4 $\mu$ m | 20 dB/m @ 3$\mu$ m |
| AgBr/Cl | 3.3-15$\mu$ m | 0.7 dB/m @ 10.6$\mu$ m |
| Silica* | 0.2- 3$\mu$ m | 0.2 dB/km @ 1.5$\mu$ m |
| PPMA | 0.4-0.8 $\mu$ m | 1 dB/m @ 0.8$\mu$ m |

research on silica fibres leading to very high quality productions. Nevertheless, for more specific applications, a large set of waveguide type have been developed as shown in the following table (Tapanes, 1991).

## 2. Context of Waveguides Use in Stellar Interferometry

## 2.1 Basic information

The stellar interferometry is based on the spatial coherence analysis of the source by mixing the optical fields E1, E2 collected by two or more separated telescopes (see Ch. 16). In a two telescopes configuration, the corresponding interferometric signal is given by:

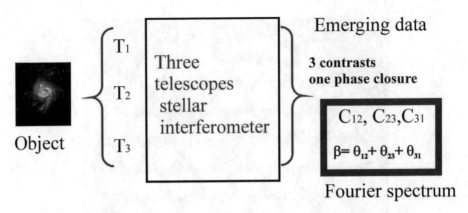

*Figure 4.* The fringe contrasts measure the resemblance (cross-correlation) of the two mixed fields. (right).

$$I(\delta) = I_1 + I_2 + 2 \int E_1(t)E_2^*(t + \delta/c)dt \qquad (8)$$

with $I_1$ and $I_2$ the intensity of beam 1 and 2 and $\delta$ the optical path difference. The last term gives the coherence of the optical fields and can be expressed as the cross-correlation of $E_1$ and $E_2$. Consequently this interferometric measurement is based on the analysis of the resemblance between the fields to be mixed. The complex visibility of fringes is given by the Zernike and Van Citert theorem and is equal to the Fourier transform of the intensity distribution of the object.

Obviously, the analysis of the correlation between the two fields emerging from the telescope and related devices makes necessary to avoid dissymmetry between the interferometric arms. Otherwise, it may result in confusion between a low correlation due to a low spatial coherence of the source and a degradation of the fringe contrast due to defects of the interferometer. The following paragraphs summarize the parameters to be controlled in order to get calibrated data.

## 2.2    Optical paths stabilization

The fringes can be spatially or temporally displayed but during the integration time, in the first case, or over the full scan, in the second case, phase shifts may induce contrast degradation or chirp on fringes. Consequently a servo control system is necessary. In order to generate an error signal it is necessary to measure the optical path of each arm using a metrology laser. The simultaneous propagation of the science and metrology signals is possible by means of spectral multiplexing using dichroic plates. An example is given in Sect. 17.3.3. If the error signal is available, it is possible to compensate the optical path fluctuation by using actuators such as optical fibre modulator. The servo loop includes:

- Error signal generation.
- Data processing with a PID filter.
- Power stage with high voltage amplifiers.
- PZT actuator.

An example for the first step of this process is developed in Sect. 17.3.3.

## 2.3 Differential dispersion compensation

The fringes contrasts are subject to degradation resulting from dissymmetry in the interferometer. The optical fields to be mixed are characterized by a broadband spectrum so that differential dispersion may induce a variation of the differential phase over the spectrum. Detectors are sensitive to the superposition of the different spectral contributions. If differential dispersion shifts the fringes patterns for the different frequency, the global interferogramme is blurred and the contrast decreases. Fig. 5 shows corresponding experimental results.

This behavior can be analyzed using the channeled spectrum method Shang, 1981 as shown in Fig. 6. A spectral analysis of the interferometric output signal allows to visualize the different interferometric status over the spectrum. If the phase shift is independent of the frequency (top), all the spectral components give synchronized fringes leading to a good contrast. Conversely, second order dispersion term leads to a degradation of the fringes visibility because of unsynchronized fringes (middle and bottom). The compensation of dispersion effects is easy for the first order term that can be canceled by an air delay line. Conversely, for higher order terms, the only solution is to cut the waveguides in order to cancel the differential effect. This technique is efficient for homogeneous fibres (Reynaud and Lagorceix, 1996) but is more difficult for long propagation lengths (Perrin et al., 2000).

## 2.4 Birefringence control and compensation

The evolution of polarization status of optical fields propagating in a standard waveguide is complex and sensitive to local stresses. The compensation of these fluctuations is possible using fibre loops (LeFevre, 1980) but this technique is restricted to narrow spectral bandwidth. When a broadband spectrum propagates in a fibre the only solution consists in the use of polarization preserving fibres based on high internal birefringence able to mask small perturbations. By this way, two polarization modes with a very low cross talk allow to preserve the polarization coherence. Figure 7 shows experimental results on the degradation of the fringes contrast with standard fibres (right) and the

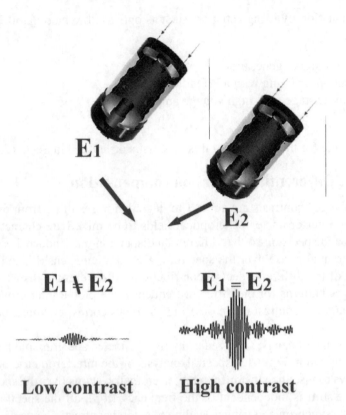

*Figure 5.* Evolution of the fringes contrast C as a function of the differential dispersion (a.u.)
The maximum of this function corresponds to the cancellation of differential dispersion between
the fibre arms of the interferometer.

contrast improvement using polarization preserving fibres (left). Due to group
velocity dispersion between the two polarization modes a birefringent compensator is necessary. A fibre version, so called optical fibre compensator, has
been developed using a fibre coil squeezed in order to enhance the internal
stresses (Lagorceix and Reynaud, 1995).

## 2.5    Spatial filtering

One of the attractive features of single mode waveguides is their ability to
filter the spatial field distribution. All the wavefront aberrations only result in
a photometric fluctuation easy to monitor. It results in a very good calibration
of the interferometric data as firstly demonstrated by FLUOR, as seen in Fig.
8 (Coudé du Foresto et al., 1998). Nevertheless, care has to be taken to keep
in mind that turbulence may have a spectral selectivity in the launching pro-

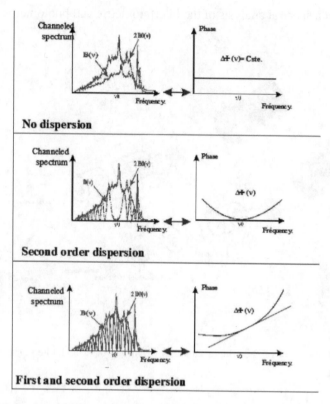

*Figure 6.* Channel spectrum and related spectral phase shifts. Top: compensated dispersion; the phase is constant over the spectrum. Middle and bottom: Second order effect with or without first order. The spectral phase variation induces channel in the spectrum.

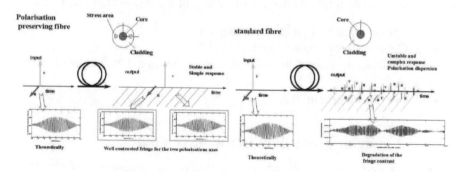

*Figure 7.* Influence of the polarization preservation in a fibre interferometer. Right: Standard fibre use leads to a complex pulse response associated with fringe degradations with polarization crosstalk. Left: Using polarization preserving fibre, the two polarization modes give rise to contrasted fringes and can be separated using a polarizer.

cess so that a spectral analysis of the interferometric and photometric output is required.

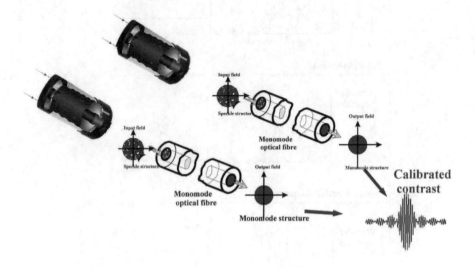

*Figure 8.*   Calibration of the fringes contrasts by means of spatial filtering in monomode waveguides.

## 3.    BASIC FUNCTIONS

Figure 9 gives a global scheme of an all guided stellar interferometer. The four functions to be implemented are discussed in the following paragraphs.

- 1. Launching star light in the waveguides.
- 2. Coherent propagation in the optical fibres.
- 3. Synchronization of the optical fields to be mixed by means of delay lines.
- 4. Interferometric mixing.

## 3.1    Launching star light in the wave-guides

This first step makes necessary a correction of the atmosphere aberrations by means of an adaptive optics or at the minimum a tip tilt device. If the turbulence induces high aberrations the coupling efficiency is decreased by a factor $1/N$ where $N$ is the number of spatial modes of the input beam. Note that tilt correction is also mandatory in a space mission as long as instabilities of the mission platform may induce pointing errors. Figure 10 (left) illustrates the spatial filtering operation. This function allows a very good calibration of

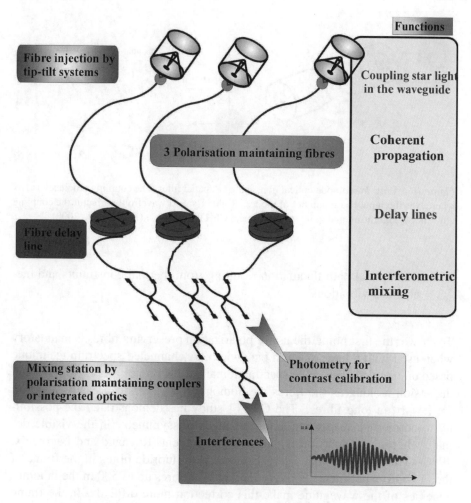

**Functions**

**Fibre injection by tip-tilt systems**

**3 Polarisation maintaining fibres**

**Coupling star light in the waveguide**

**Coherent propagation**

**Delay lines**

**Fibre delay line**

**Interferometric mixing**

**Mixing station by polarisation maintaining couplers or integrated optics**

**Photometry for contrast calibration**

**Interferences**

*Figure 9.* General scheme of a guided stellar interferometer.

the spatial parameters but may induce high losses for an aberrated input beam. Figure 10 (right) gives an example of star light launching on CFH Telescope with the help of the PUEO adaptive optics in the frame of the OHANA project.

## 3.2 Light coherent propagation

The coherent propagation of star light through optical fibres over long distance has been mainly investigated at IRCOM and kilometric linkages are now possible. The two main points to be managed are:

- The differential dispersion and polarization properties of optical fibres.

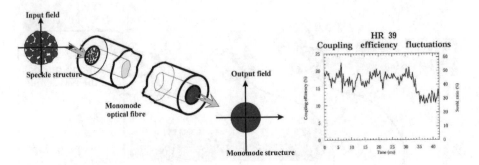

*Figure 10.* Left: Monomode optical fibre acts as a spatial filter. The coupling efficiency is 1/N where N is the input beam number of modes. Right: Using a wavefront corrector the coupling efficiency is significant and quite stable (K band CFHT/ OHANA) (Perrin et al., 2000).

■ The optical path fluctuation resulting from thermal constraints and mechanical vibrations.

To answer the first point, the use of polarization preserving fibres is mandatory when used with a large spectral bandwidth. A channeled spectrum technique based on interferences in the spectral domain allows correcting the differential dispersion as long as the fibres are homogeneous. This can easily achieved for few 10 m long fibres. The OAST1 study has demonstrated the possibility to correct the dispersion and birefringent dissymmetry in the visible do-
main where the dispersion is particularly stringent (Reynaud and Lagorceix, 1996). This technique has been also used with fluoride fibres in the frame of the FLUOR experiment. For very long optical fibres up to 500 m the inhomo-
geneities of the waveguide make this correction more difficult. In the frame of the OHANA project, IRCOM has compensated two 300 m fibres leading to fringe contrasts in the range of 80% over the J band (Vergnole et al., 2004). For the K band, the same job is achieved by LESIA team from Meudon and Verre Fluoré.

The second point deals with stability of the instrument. The optical fibre lengths have to be servo in order to cancel the disturbing effects of thermal and mechanical perturbations. This servo control system need a specific design as shown on Fig. 11. As long as the fibre ends at the telescopes foci are far from each other, the optical path of each waveguide is controlled by launching the metrological light to do a round trip from the mixing station to the telescope foci. This optical path is compare to a reference fibre and, after processing the interferometric signal, a set of PZT actuator enable to correct the fluctuations. Typical nm stability has been obtained in the different experiments (OAST1, OAST2...).

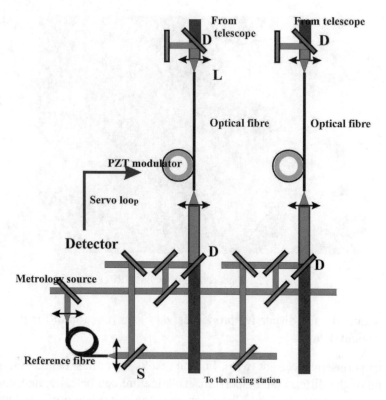

*Figure 11.* Principle of servo control system for optical paths stabilization in a fibre linked interferometer.

## 3.3    Optical fibre delay lines

The optical fibre delay lines have been developed at IRCOM (Simohamed and Reynaud, 1999). The principle of these devices takes advantage of the good elasticity of silica fibres. A mechanical stretcher induces a geometrical elongation of the waveguide up to 2%. Figure 12 shows a picture of the update version. The waveguide is glued on a rubber cylinder and a radial expansion enables to induce the optical delay. The resulting properties have been investigated in the frame work of coherent propagation for interferometry:

- The attenuations resulting from this mechanical elongation are not significant and difficult to be accurately measured. They are lower than 1% (Fig. 13).
- The variation of the delay is proportional to the applied stress. It results in a mechanical command easy to manage.
- The optical delay results form the augmentation of the length of a dispersive material and the deformation of the optical waveguide. Con-

*Figure 12.* Optical fibre delay line (IRCOM).

sequently, the dispersive properties vary as a function of the generated optical delay.

The main consequences are twice. First, it results in contrast degradations as a function of the differential dispersion. This feature can be calibrated in order to correct this bias. The only limit concerns the degradation of the signal to noise ratio associated with the fringe modulation decay. The second drawback is an error on the phase closure acquisition. It results from the superposition of the phasor corresponding to the spectral channels. The wrapping and the nonlinearity of this process lead to a phase shift that is not compensated in the phase closure process. This effect depends on the three differential dispersions and on the spectral distribution. These effects have been demonstrated for the first time in the ISTROG experiment (Huss et al., 2001) at IRCOM as shown in Fig. 14.

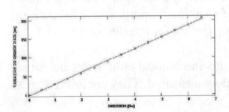 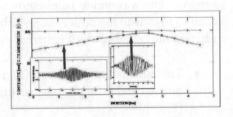

*Figure 13.* Left: Optical delay as a function of the stretching parameter. Right: Transmission close to 100 % upper curve and contrasts as function of the generated delay. Low contrast corresponds to a fringes packet spreading.

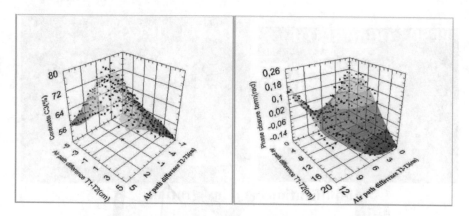

*Figure 14.* One of the three contrasts (left) and phase closure (right)evolution as functions of the optical delay to be generated between the telescopes of a three aperture stellar interferometer(OPD between T1 - T2 and T2 - T3 reported in the horizontal plane).

## 3.4 Beams combination

The different studies are mainly reported on the OBLIN site (NASA). The main goal is to simplify and integrate all the functions to be achieved in the beam combiner by using integrated optics and optical fibre components. Consequently, the compactness, the stability and the reliability are significantly improved. Figure 15 illustrates the simplification of the experimental setup.

Such devices have to provide interferometric mixing and photometry of the incoming beam in order to remove the photometric visibility factor from the acquired data.

This work has been developed by the FLUOR team (Kervella et al., 2003) (Observatoire de Meudon) in the K-band with a different experiment on the IOTA interferometer. This experiment is now implemented on CHARA. Figure 16 gives a schematic representation and a photo of this combiner.

The component are $2 \times 2$ fluoride fibre couplers connected together in order to provide the photometric and interferometric outputs. A replica of this combiner has been implemented of the VLTI and first results have been obtained recently on the sky (Berger et al., 1999). For the J and H-band the use of integrated optics components have been successfully tested on the sky by IMEP and LAOG teams from Grenoble. This project, so called IONIC, has been implemented on IOTA (Traub et al., 2003; Rousselet-Perraut et al., 2002) and has provided experimental results including the contrasts and phase closure acquisition. Figure 17 gives the idea of the outline dimensions of this combiner.

Other laboratory studies conducted at IRCOM have compared the coupler and integrated optics in the visible (670 nm) in the frame of the ISTROG instrument (Huss et al., 2001; Fig. 18). The high dispersion effect in this spectral

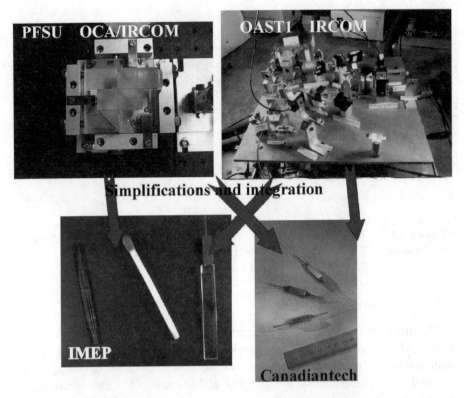

*Figure 15.* Classical recombining devices (up) and guided optics alternatives (down).

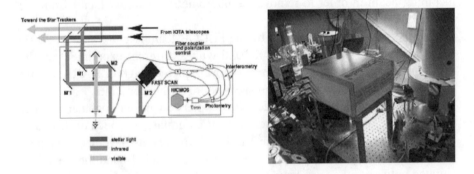

*Figure 16.* The FLUOR scheme and the corresponding experimental assembly.

domain leads to phase closure bias to be corrected. Similar results have been obtained with the two kinds of components but the potential of integrated optics components seems higher even if technological efforts are necessary for the short wavelengths.

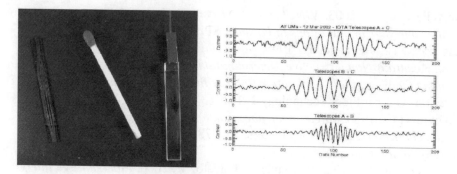

*Figure 17.* The IONIC combiner and the first three telescopes results on IOTA.

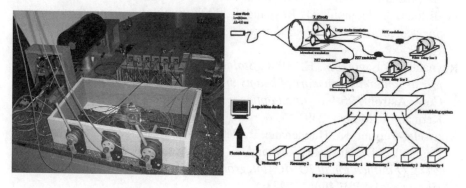

*Figure 18.* ISTROG Interferometer. The high improvement on the accuracy of contrast measurements with such devices mainly results from the efficient spatial filtering and the spatial stability of the recombining assembly.

The high improvement on the accuracy of contrast measurements with such devices mainly results from the efficient spatial filtering and the spatial stability of the recombining assembly.

## References

Berger J.-P., Rousselet-Perraut K., Kern P., Malbet F., Schanen-Duport I., Reynaud F., Haguenauer P., Benech P., 1999, *Integrated optics for astronomical interferometry. II- First laboratory white-light interferograms*, Astron. Astrophys. Sup.**139**, 173

Coudé Du Foresto, V., Perrin, G., Ruilier, C., Mennesson, B. P., Traub, W. A., Lacasse, M., 1998, *FLUOR fibered instrument at the IOTA interferometer*, SPIE **3350**, 856

Huss, G., Reynaud, F., Delage, L., 2001, *An all guided three arms interferometer for stellar interferometry*, Optics Com. **196**, 55

Jeunhomme L., 1990, *Single-mode fiber optics*, Marcel Dekker Inc.

Kervella P. et al., 2003, *VINCI, the VLTI commissioning instrument: status after one year of operations at Paranal*, SPIE **4838**, 858

Lagorceix, H., Reynaud, F., 1995, *Squeezing highly Birefringent Fiber for an Accurate Nondestructive Determination of Principal-Axis Orientation along the Waveguide: application to Fiber Babinet Compensator Implementations* Optical Fiber Technology **1**, 171

LeFevre, H.C., 1980, *Single-mode fibre fractional wave devices and polarization controllers*, Electronics Letters, **16**, 778

*http://olbin.jpl.nasa.gov*

Perrin, G., Lai, O., Léna, P., Coudé du Foresto, V., 2000, *Fibered large interferometer on top of Mauna Kea: OHANA, the optical Hawaiian array for nanoradian astronomy*, SPIE **4006**, 708

Peyrilloux, A., Pagnoux, D., Reynaud, F., 2002, *Evaluation of photonic crystal fiber potential for fibre linked version of stellar interferometers*; SPIE **4838**, 1334

Reynaud, F., Lagorceix, H., 1996, *Stabilization and control of a fiber array for the coherent transport of beams in a stellar interferometer*, Proc. of the Conf. Astrofib'96 *Integrated optics for astronomical interferometry*, Ed. P. Kern and F. Malbet, 249

Rousselet-Perraut, K., Haguenaue, P., Petmezakis, P., Berger, J., Mourard, D., Ragland, S.D., Huss, G., Reynaud, F., LeCoarer, E., Kern, P.Y., Malbet, F., 2002, *Qualification of IONIC (integrated optics near-infrared interferometric camera)*, SPIE **4006**, 1042

Shang, H.-T., 1981, *Chromatic dispersion measurement by white light interferometry on meter-length single-mode optical fiber*, J. Opt. Soc. Am., **71**, 1587

Simohamed, L.M., Reynaud, F., 1999, *Characterization of the dispersion evolution in a large stroke optical fiber delay line*, Optics Comm. **159**, 118

Tapanes, E., 1991, SPIE **1588**, 356

Traub, W. A., Ahearn, A., Carleton, N. P., Berger, J-P., Brewer, M. K., Hofmann, K-H., Kern, P. Y., Lacasse, M. G., Malbet, F., Millan-Gabet, R. Monnier, J. D., Ohnaka, K., Pedretti, E., Ragland, S., Schloerb, F. P., Souccar, K., Weigelt, G., 2003, *New Beam-Combination Techniques at IOTA*, SPIE **4838** , 45

Vergnole, S., Delage, L., Reynaud, F., 2004, *Accurate measurement of differential chromatic dispersion and contrast in an hectometric fibre interferometer in the frame of OHANA project.* Optics Comm. submitted

# AN INTRODUCTION TO
# GRAVITATIONAL-WAVE ASTRONOMY

Vincent Loriette

*ESPCI, Laboratoire d'optique, CNRS UPR5*
*10, rue Vauquelin, 75005 Paris, France*
loriette@optique.espci.fr

**Abstract**     This is an introduction to gravitational wave astronomy. The physical bases of gravitation and the generation and detection of gravitational waves are recalled and then kilometric detectors under construction are described.

**Keywords:**     gravitational waves, PSR1913+16 pulsar, relativistic objects, gravitational wave detectors

## 1.     Introduction

This course is an introduction to gravitational wave astronomy for students not involved in the field. It aims at covering two different and somehow disconnected aspects of this branch of experimental gravitation: first it starts by a very brief approach of the physical bases of gravitation and the generation and detection of gravitational waves (gw), then it covers a technical description of the giant detectors (observatories) that are being built on earth, focusing on optics. Just because those two aspects of gravitational wave astronomy can easily be disconnected (one could easily work on a interferometric gravitational wave detector without knowing nor understanding anything about gw), it is very important they are not. The concept of gravitational waves cannot be fully seized if one separates the theoretical picture of the Einstein theory of gravitation from the technical achievements of Virgo LIGO, GEO, TAMA, and soon LISA. But do not think it is simply a question of intellectual satisfaction, the reason lies in gravitation itself. In general relativity the concept of measurement is crucial, often misleading. The description of a measurement procedure, the choice of a particular reference frame used by the observer influences, not the measurement by itself, but the interpretation of the physical

*R. Foy and F. C. Foy (eds.), Optics in Astrophysics, 307–325.*

effect. This is why a question as seemingly simple as "what happens when a gravitational wave crosses an interferometer ?" is, by far, not a trivial one.

As a short course intended to give a first insight into gravitational wave astronomy, this lecture will try to answer the following questions:

- What are gravitational waves?
- How are they generated?
- What can we learn from them?
- How can they be detected?
- How does a gravitational wave observatory work?

## 2. Gravitational Waves

## 2.1 What is a gravitational wave ?

Before we enter into a mathematical representation of gravitational waves, it is useful to summarize their main characteristics:

- gw are periodical perturbations of the gravitational field;
- they are generated by fast displacements of large amounts of matter;
- they propagate at the speed of light;
- they are nearly not absorbed by matter;
- gravitational waves carry energy, so **they carry information**.

What is very interesting for us is that the physical phenomenons that generate gw usually do not generate electromagnetic waves, or are localized in regions inside the sources from where electromagnetic radiation cannot escape. They are our only messenger from those regions. In order to formalize those characteristics we must look at the relativistic theory of gravitation, the General Relativity (GR). The GR is founded on the Einstein Equivalence Principle (EEP), which endows the Newton equivalence principle and states that the Special theory of Relativity (SR) is locally valid. This means that an observer can always build, in a small region of spacetime (in a small 3-space volume and for a short duration), a minkowskian coordinate system fixed in the laboratory (comoving with the laboratory). In this coordinate system the SR holds, which means that the form of the fundamental element (proper time) is the one of SR:

$$d\tau^2 = c^2 dt^2 - dx^2 - dy^2 - dz^2 = \eta_{\mu\nu} dx^\mu dx^\nu \qquad (1)$$

where $dx = (cdt, dx, dy, dz)$ is the separation between two events and $\eta = diag\{1, -1, -1, -1\}$ is the Minkowski tensor. The fundamental property of light is that $d\tau_\gamma^2 = 0$. This coordinate system is in free-fall and the world-line of the observer is a geodesic. The problem is that EEP is strictly valid only in a infinitely small region of spacetime. In practice laboratories on earth or inside satellites have a finite spatial and temporal extension, so if an observer builds a coordinate system in its laboratory, it will not be minkowskian, and its world-line will not be a geodesic. The immediate effect is that the fundamental element is not in the one of SR, it has a more general form:

$$d\tau^2 = g_{\mu\nu} dx^\mu dx^\nu \tag{2}$$

where $g$ is a tensor whose non-diagonal elements may not be null. If the observer carries an experiment inside the laboratory, he will find discrepancies between the experimental results and the SR. Nevertheless it can be shown that the fundamental property of light : $d\tau_\gamma^2 = 0$ still holds. Now we know that on earth, and in the solar system, SR works pretty well, so $g$ in a laboratory cannot be very different from $\eta$. We write the elements of $g$ as

$$g_{\mu\nu} = \eta_{\mu\nu} + k_{\mu\nu} \tag{3}$$

where the elements of $k$ are small so that $k$ can be seen as a perturbation of $\eta$. The last thing we need to know is the content of $k$. The difference between the coordinate system in which SR holds ($k_{\mu\nu} = 0$) and the operational coordinate system ($k_{\mu\nu} \neq 0$) is that the first one is free falling, while in the second the observer can measure the acceleration in the gravitational field. That means that what is contained in $k$ must be the gravitational field. For example the gravitational field of earth give a correction to $\eta_{00}$ and

$$k_{00} = -\frac{2GM_\oplus}{r_\oplus c^2} \tag{4}$$

The relation between the value of $g$ and the energy and mass content of the Universe is given by the field equations of GR:

$$G^{\mu\nu} = \frac{8\pi G}{c^4} T^{\nu\mu} \tag{5}$$

where $T$ is the energy-momentum tensor, and $G$ is a tensor whose elements depend only on $g$ and its partial derivative up to second order. Equation 5 is thus a set of 16 non-linear differential equations that link the elements of $g$ to the sources of gravitation in the Universe (In fact there are only four independent elements in $G$, because $T$ is a symmetric tensor, and verifies $\nabla_\mu T^{\mu\nu} = 0$ ). In the case where $g$ is close to $\eta$ (weak field approximation) these equations can be written

$$\Box k_{\mu\nu} = -\frac{16\pi G}{c^4} \left( T_{\mu\nu} - \frac{1}{2} \eta_{\mu\nu} T \right) \tag{6}$$

where $\Box$ is the d'Alembertian operator. Just like in the case of electromagnetism, one is free to use a particular gauge transformation in order to simplify the Einstein equations. In the preceding equation we used a particular gauge *equivalent* to the Lorentz gauge in electromagnetism. Compare these equations with the electromagnetic field equations written in their covariant form:

$$\Box A_\mu = j_\mu \tag{7}$$

The analogy with electromagnetic waves is evident. In empty space Eq.(6) becomes:

$$\Box k_{\mu\nu} = 0 \tag{8}$$

which admits solutions in the form of **plane waves**, hence the name gravitational waves, which **propagate at the speed of light**. Before we end with this part we want to be able to compute the values of the elements of **k** knowing the elements of **T**. In the case where the source moves slowly ($v^k \ll c$) then $T^{00} \approx \rho c^2$, the elements of **k** contain static parts (the earth gravitational field) and time dependent parts that we note $h_{\mu\nu}$. In a system of cartesian coordinates centered on the observer oriented such as the $z$-axis points along the line of sight, these elements are:

$$h_{xx} = -h_{yy} = h_+ \left( t - \frac{R}{c} \right), \ h_{xy} = h_{yx} = h_\times \left( t - \frac{R}{c} \right), \tag{9}$$

$$h_{xz} = h_{yz} = h_{zz} = 0$$

where

$$h_+ = \frac{-G}{c^4 R} \left( \ddot{I}^{xx} - \ddot{I}^{yy} \right), \ h_\times = \frac{-2G}{c^4 R} \ddot{I}^{xy} \tag{10}$$

The $I$ are the reduced quadrupoles :

$$I^{jk} = \iiint \rho(t, \mathbf{x}) \left( x^j x^k - \frac{1}{3} \delta^{jk} \mathbf{x}^2 \right) d^3 x \tag{11}$$

that must be evaluated at the retarded time $t - R/c$, where $R/c$ is the delay between the emission and detection of the wave. In order to find this last result it was necessary to perform multipolar expansion of the impulsion-energy tensor. We see here a fundamental difference between electromagnetic and gravitational waves. While the first source term for electromagnetic waves is the dipole, it is the quadrupole for gravitational radiation. The fact that the perturbation **k** appeared in the metric because the observer and its coordinate system does not follow a geodesic, could lead one to conclude that after all gravitational waves are only a mathematical artefact that appears when one performs a change of coordinates. This is not true, although it may not evident to get convinced, the fact is that gw can induce effects on an apparatus whatever the coordinate system which is used to describe the experiment. The sources loose energy while generating gravitational waves. The power radiated out is:

$$L_{gw} = \frac{G}{5c^5} \sum_{j,k} \left( \dddot{I}^{jk} \right)^2 \tag{12}$$

## 2.2 A proof of the existence of gravitational waves

Take a system of two gravitationally bounded stars. This system being non-spherically symmetric and non-static, looses potential energy via emission of gw. This in turn results in a decrease of the distance $r$ between the two stars and and increase of the angular velocity $\omega$. With the following definitions :

$$M = M_1 + M_2 \qquad\qquad m = M_1 M_2 / M$$
$$\boldsymbol{M} = M^{2/5} m^{3/5} \qquad\qquad \tau = G\boldsymbol{M}/c^3$$

one may express the angular velocity as

$$\omega(t) = \frac{1}{8\tau} \left( \frac{t}{5\tau} \right)^{-3/8} \tag{13}$$

where $t$ is here the time remaining before the two stars merge. Of course, for the variations of the orbital parameters to be detectable, the amount of energy emitted in the form of gw has to be a non-negligible fraction of the internal power flow of the system. A gravitationally bound system of mass $M$ and size $R$ has a kinetic energy $E \approx GM^2/R$, a characteristic speed $v \approx (GM/R)^{1/2}$ and the characteristic time for the masses to move is $T \approx R/v \approx R^{3/2}/(GM)^{1/2}$ then has an internal power flow

$$L_{\text{int}} \approx G^{3/2} \left( \frac{M}{R} \right)^{5/2} \tag{14}$$

the third time derivative of the reduced quadrupole is of the order of $MR^2/T^3$ so

$$L_{gw} \approx \frac{G}{5c^5} L_{\text{int}}^2 \tag{15}$$

A binary system will be efficient if

$$\frac{G}{c^5} L_{\text{int}} = \left( \frac{GM}{Rc^2} \right)^{5/2} \rightarrow 1 \tag{16}$$

so it must be a compact system of massive stars. The change of orbital parameters because of the energy loss via gravitational wave emission is a purely relativistic effect, of course absent from the newtonian theory of gravitation. Moreover the rate of change of the orbital parameters is dictated by the field equation of GR, thus if one could observe such variations and at the same time be able to measure precisely the mass parameters of the binary system, one could compare those variations with the ones predicted by the theory. If both agree, one could claim to have observed, although indirectly, gravitational waves. But there is a problem: in order to evaluate the mass of the stars the measurement of orbital frequency is not enough, all other orbital parameters (masses, eccentricity) must be known. The quest seems hopeless...

**2.2.1    The binary pulsar PSR1913+16.**    In 1974 Russell Hulse and
Joseph Taylor discovered a very special pulsar: PSR1916+13. A striking fea-
ture that was revealed just after its discovery was a large shift of the pulses
rate with a period of a few hours. This was explained by Doppler shift, so
PSR1916+13 is in fact a binary system in which a pulsar orbits about the cen-
ter of mass of the system. Thanks to the presence of this very stable local clock,
Hulse and Taylor were able to obtain all the orbital parameters of this system.
By studying the variations of those parameters over nearly twenty years, they

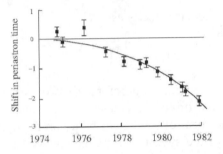

*Figure 1.*    Shift of the periastron of PSR1913+16 recorded during 8 years. The horizontal line
(no shift) is the newtonian prediction, while the curve is the prediction of GR.

saw a decrease of the orbital phase shift with a quadratic behavior (Fig. 1).
This result shows that the orbital period is not constant, because if it was, then
the phase shift would just be a consequence of an inaccurate determination of
the period, and it would have a linear time dependence. When comparing the
predicted phase shift to the observed one, the two results agree to within one
percent.

## 2.3    Sources of gravitational waves

First of all, let us see if we can build a home-made source of gravitational
waves. For this we build an apparatus in which a mass $m$ moves periodically
at the angular frequency $\omega$ along a characteristic length $D$. The quadrupole
moment has a magnitude of the order of $I \approx mD^2$ and a time derivative of the
order of $mD^2\omega$. Eqs. 10 and 12 give us orders of magnitude for the amplitude of the gravitational waves and the luminosity $h \approx (GmD^2\omega^2)/(rc^4)$
and $L_{gw} \approx (Gm^2D^4\omega^6)/(5c^5)$. Taking $D = 10$ m, $m = 10^4$ kg, and
$\omega = 628$ rad/s we end with $h \approx 3 \times 10^{-34}$ and $L \approx 3 \times 10^{-25}$ J. The
detectors built today aim at detecting amplitudes not smaller that $10^{-21}$, so a
few years will pass until we will be able to detect gravitational waves emitted by terrestrial sources. The same kind of order of magnitude calculations
lead us to conclude that gravitational events in the solar system will be out of
reach during our near future. Let us point out the required features of potential
sources:

- The sources are to be found outside the solar system.
- The sources must be as massive as possible (star-like objects).
- The source must be in a non-steady state (because of the second time derivative of **I**).
- The motion of the source must be fast (possibly relativistic).
- A spherical object cannot radiate gravitational waves (quadrupolar source term).

Points two and four seem incompatible, usually the massive objects are slower than light ones for a given density. Thus good quantity to look at is not mass but compactness C (see Bonazzola in Barone et al., Eds., 2000). Compactness can be defined as:

$$C = R_S/R \qquad (17)$$

with $R$ being the radius of the source and $R_S$ its Swarzschild radius. The best candidates as gravitational wave emitters are : supernovae, neutron stars and black holes. The amplitude of gw emitted by real astrophysical objects, and the rate of detectable events are very difficult to estimate accurately, because we have so few information about the sources, and also because reliable estimations of gravitational waveform require to perform calculation in full GR, and not with the linearized equations just presented.

### 2.3.1 Neutron stars.

Neutron stars are compact ($C_{NS} \approx 0.3$) relativistic objects. They can participate to the emission of gravitational waves following different mechanisms, either alone or in binary systems.

**Rotation of a neutron star.** A rotating neutron star can emit periodic gw if some phenomenon breaks its axial symmetry. We can think of some anisotropy of the solid outer layer of the star, like mountains, or a deformation of the star induced by its own magnetic field (Crab pulsar), or by accretion (X-ray binary). Although the estimated amplitude of gw is small for this kind of isolated object, we can use the fact that it is periodical with a remarkable stability over years, so the gw signal could be integrated over large periods. Four years of integration should allow to detect the gw signal of the crab pulsar if its ellipticity is larger than $5 \times 10^{-6}$. The detection of such gw will allow us to probe the interior of neutron stars, in particular to know wether it is superconductor. This information will in turn help to refine the models of pulsar magnetospheres.

**Coalescing neutron star binaries.** Coalescing of neutron stars (or black holes) is foreseen to be the the most powerful source of detectable gw. The frequency of such events is estimated to be $\sim 1y^{-1}(D/200 \text{ Mpc})^{-3}$ and their amplitude will allow detection of sources as far as 50 Mpc. We are thus waiting for about one event every 60 years with the current sensitivity of detectors.

This number depends a lot on the knowledge of the gravitational waveform, especially during the last seconds before the two stars merge into one single object. Therefore with accurate models at our disposal the number of detected events could increase significantly. The successful detection of neutron star coalescence will give precious informations about the equation of state of those objects the density of which is 4 to 10 times the nuclear density. The stiffness of the equation of state fixes the instant at which the two bodies lose their coherence, which should appear as a sharp decrease of the gravitational signal. Also the number of detected events should help to refine the stellar evolution models.

**2.3.2    Black holes.**    One great achievement of gravitational wave astronomy would be the first detection of a signal coming directly from a black hole. Just like with neutron stars, black holes can emit gw either alone or in binary systems.

**Oscillations of black holes.**    Non-radial oscillations of black holes can be excited when a mass is captured by the black hole. The so called quasi-normal modes have eigenfrequencies and damping times which are characteristic of black holes, and very different of eigenfrequencies and damping times of quasi normal modes of stars having the same mass. Also the eigenmodes being different for a star and a black hole, the associated gw will also exhibit characteristic features.

**Coalescence of black holes.**    The coalescence of two black holes will generate even more gravitational waves than neutron stars coalescence, and coalescence of two $10 M_{\odot}$ black holes will be detectable up to 500 Mpc

## 2.4    Idealized optical detection of gravitational waves

One important point to keep in mind is that as long as masses are free falling, ie as long as no force acts on them, they keep on moving in spacetime following geodesics, so keeping their coordinates, whether or not local spacetime curvature is time-dependent (in presence of gravitational waves). Although the speed of light in vacuum is $c$ even in the presence of gw, the time needed for light to go back and forth between two free falling masses depends on the gravitational field (curvature of spacetime) between them. Let us suppose that in absence of gw an observer at the origin of its system of coordinates measures the distance between one mirror having spatial coordinates $(0,0,0)$ and another mirror having spatial coordinates $(L,0,0)$ by measuring its proper time $\tau_{obs}$ (its ageing) while the light performs a round trip. The observer being at rest during the measurement ($dx_{obs} = 0$) we have

$$d\tau_{obs} = cdt \tag{18}$$

Using the fact that

$$d\tau_\gamma^2 = c^2 dt^2 - dx^2 = 0 \tag{19}$$

he will measure

$$T_{obs} = \frac{\tau_{obs}}{c} = \frac{2}{c} \int_0^L dx = \frac{2L}{c} \tag{20}$$

and conclude that the value of the *spatial distance* $D_0$ between the two mirrors equals: $D_0 \equiv cT_{obs}/2 = L$. Now if a gravitational wave propagating along the $z$ axis of the local coordinate system is present, the $d\tau_\gamma$ is still equal to zero but we have

$$d\tau_\gamma^2 = c^2 dt^2 - (1 + h_{xx}(t))dx^2 = 0 \tag{21}$$

The observer measures:

$$T_{obs} = \frac{2}{c} \sqrt{1 + h_+(t)} \int_0^L dx \approx \left(1 + \frac{h_{xx}(t)}{2}\right) \frac{2L}{c} \tag{22}$$

(In order to remove the perturbation $h(t)$ from the integral we supposed that the period of the gravitational wave was much longer than the round trip time of light). The observer now concludes that the *spatial distance* $L_{gw}$ between the two mirrors is:

$$L_{x,gw} \equiv \frac{cT_{obs}}{2} = \left(1 + \frac{h_{xx}(t)}{2}\right) L_0 \tag{23}$$

Now if there is a third mirror with spatial coordinates (0,L,0), the *spatial distance* between this mirror and the mirror at (0,0,0) will be

$$L_{y,gw} \equiv \frac{cT_{obs}}{2} = \left(1 + \frac{h_{yy}(t)}{2}\right) L_0 \tag{24}$$

The observer concludes that the presence of the gw modifies the *distance* between free falling masses. The last thing to do in order to detect gw is thus to measure with a great accuracy the *distance* between two masses. But do not forget that the amplitudes of gw have been estimated in the range $h \approx 10^{-21}$ to $h \approx 10^{-27}$ so even if the two masses are 1 km apart, the change in length will be at most $\delta L = 5 \times 10^{-19}m$ ! At first sight we may be doubtful at the possibility to detect such a small variation of distance. Before the development of laser the task seemed out of reach, but since then it seems we have everything at hand to succeed.

**2.4.1    A simple Michelson interferometer.**    If we place two mirrors at the end of two orthogonal arms of length $L$ oriented along the $x$ and $y$ directions, a beamsplitter plate at the origin of our coordinate system and send photons in both arms trough the beamsplitter. Photons that were sent simultaneously will return on the beamsplitter with a time delay which will depend on which arm they propagated in. The round trip time difference, measured at the beamsplitter location, between photons that went in the $x$-arm ($x$-beam) and photons that went in the $y$ arm ($y$-beam) is

$$\delta t = \frac{L_{x,gw} - L_{y,gw}}{c} = \frac{2L}{c}\left(\frac{h_{xx}(t)}{2} - \frac{h_{yy}(t)}{2}\right) = \frac{2L}{c}h_+(t) \qquad (25)$$

In the case were the arms are oriented at $45°$ from the $x, y$ axes, the same equation holds but $h_+$ has to be replaced by $h_\times$. If the incident light is a monochromatic plane wave of frequency $\nu_{opt}$, this time delay will appear as a phase shift $\delta\varphi$ between the $x$-beam and $y$-beam :

$$\delta\varphi = 2\pi\nu_{opt}\frac{2L}{c}h_+(t) = 2\pi\frac{2L}{\lambda_{opt}}h_+(t) \qquad (26)$$

We see here the gain brought by the use of light: the small value of $h$ is partially counterweighted by $L/\lambda_{opt}$. If $L \approx 1$ km and $\lambda_{opt} \approx 10^{-6}$m we gain a factor $10^9$ and the challenge is now to measure phase shifts of the order of $10^{-11}$ rad. We may also note at this point that we perform a differential measurement as we measure the phase difference between waves that followed different paths.

We now end the theoretical description of gravitational waves, and start the practical description of gravitational wave detectors.

# 3.    Interferometric gravitational wave detectors

## 3.1    Properties of real optical detectors

From what we have learn in the preceding sections we can already outline some features we will expect of a real interferometric detector :

- The detector must be able to measure optical phases of about $10^{-11}$ rad.
- The effect of gw being a modification of phase of light after a round trip in the arms, any disturbance that mimics this signal must be cancelled.
- The measurement being differential, the noise which in uncorrelated between both arms is much more dangerous than correlated noise.

In fact we could find many other requirements, but those three are directly related to the optical properties of the detector. At this point we could simply describe the various optical components of the detector, but it is more interesting to understand what led to the choice of some particular properties of those components. In order to do this we need to investigate the main limitations of real gw detectors. The number of potential noise sources is impressive : seismic noise, thermal noise, shot noise, etc, contribute to limit the sensitivity of gw detectors. We will only focus on the ones most closely related to the performances of optical components and laser source. By doing this we certainly dismiss many of the characteristics of passive and active components, but in the other hand we will have a better understanding of why real gw detectors are built the way they are, and why they are so similar.

### 3.1.1     Laser properties and optics quality requirements.    It seems

that any kind of light source could be used to perform the phase shift measurement. But looking at Eq.(26) we see that the longer the arms, the larger the phase shift to detect, so in practice the need is for kilometer long arms. It would be impossible to realize a beam of non-laser light well collimated enough to propagate back and forth over kilometers, and non laser light having a small coherence length, a few microns up to a centimeter say, it would be impossible to build two kilometer-long arms having the same length to within a fraction of the coherence length of the source. On the other hand lasers can have coherence lengths of tens of meters and their divergence is limited so that they can propagate without difficulty over kilometers. Now it is well known that the field amplitude $A(x, y, z)$ of a real laser beam can be represented as a superposition of functions (modes) of the form:

$$
A_{mn}(x, y, z) = H_m\left(\frac{x\sqrt{2}}{w(z)}\right) H_n\left(\frac{y\sqrt{2}}{w(z)}\right) \frac{w(0)}{w(z)} \exp\left(\frac{-(x^2 + y^2)}{w(z)^2}\right)
$$
$$
\times \exp\left(-i\left(kz + \frac{kr^2}{2R(z)} - (1 + m + n)\tan^{-1}\left(\frac{z}{z_0}\right)\right)\right)
$$

$$(27)$$

where $w(0)$ is the beam waist, $w(z)$ is the beam size at a distance $z$ from the beam waist position, $k$ is the wavenumber, $z_0$ is the Rayleigh range, $R(z)$ is the beam radius of curvature and the functions $H_{m, n}$ are Hermite polynomials of order $m$ and $n$. $w(z)$, $R(z)$, and $z_0$ are functions of $w(0)$, $\lambda$, $z$ only ($z_0$ does not depend on $z$). The $A_{mn}$ form a complete orthonormal basis in the $(x, y)$ plane. We can remark that the phase term in these functions depends on the values of the two integers $m$ and $n$. Thus in order to obtain the required $10^{-11}$ rad sensitivity, only one single mode must be present. The transverse section of a mode is roughly equal to $S = \pi w(z)\sqrt{1 + m}\sqrt{1 + n}$. The $m = n = 0$, or $TEM_{00}$ mode is chosen in practice because it is the simplest to generate in a laser, but chiefly because it is the smallest one. The obligation to work with a single-mode beam puts severe requirements on the surface quality of mirrors. When a laser beam interacts with an optical component, either being transmitted, or reflected, or both, its shape is usually modified. A beam incident on a

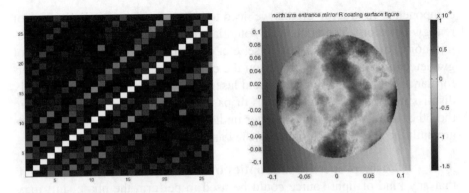

*Figure 2.*   Right figure: amplitude of the $M_{ijkl}$ coefficients of a real mirror. If the mirror were perfect, then only diagonal elements would be non-null. Left figure: surface figure map of a Virgo mirror.

component, represented by a complex amplitude $A_0 = \sum_{i,j} a_{ij}^{(0)} A_{ij}$ with $a_{ij}^{(0)}$ being complex numbers, will be converted in a reflected or transmitted beam $A_1 = \sum_{i,j} a_{ij}^{(1)} A_{ij}$. The effect of an optical component is thus to change the incident amplitude vector $a_{ij}^{(0)}$ into a new amplitude vector $a_{ij}^{(1)}$, so the optical component acts as an operator $M$ on the laser beam:

$$a_{ij}^{(1)} = \sum_{kl} M_{ijkl} a_{kl}^{(0)} \tag{28}$$

For example, if a mirror is characterized by a reflectance $r(x, y)$ and a surface figure $h(x, y)$, the reflection operator $R$ is:

$$R_{ijkl} = \iint_{x,y} A_{ij}^{*}(x, y, z)\, A_{kl}(x, y, z)\, r(x, y) \exp(\frac{4i\pi h(x, y)}{\lambda}) dx dy \tag{29}$$

The requirement of single-mode operation is that for each component an each type of interaction (reflection, beam splitting, transmission), one has

$$M_{00kl} \ll 1,\ \forall (k, l) \neq (0, 0) \tag{30}$$

If this happens, then the components are said to be matched to the beam (see Fig. 2). The requirements of Eq. (30) can be converted into surface quality requirements, concerning chiefly the surface figure and the homogeneity of the reflection factor. In terms familiar to astronomers, the surface figure requirements for gw detectors are in the $(\lambda/100, \lambda/200)$ range on about $1000\,\text{cm}^2$ surfaces (see Fig. 2).This requirement is in practice very difficult to fulfill, and new mirror manufacturing methods had to be developed (see Ch. 19).

**3.1.2    Scattered light noise.**    It is common in optics to separate the effects of large scale aberrations and the ones of small scale defects. Although this distinction is mathematically meaningless, it is in fact sensitive in practice for two reasons:

- When performing optical simulations of laser beam propagation, using either the modal representation presented before, or fast Fourier transform algorithms, the available number of modes, or complex exponentials, is not infinite, and this imposes a frequency cutoff in the simulations. All defects with frequencies larger than this cutoff frequency are not represented in the simulations, and their effects must be represented by scalar parameters.
- The polishing of a mirror is divided into major steps : shaping, polishing, superpolishing, each using different techniques to correct surface defects at different scales (see Ch. 8). The so-called surface figure is fixed by the shaping process, while roughness is fixed during the superpolishing process, so there is a natural distinction between the two kinds of defects, but the spatial frequency frontier between the two is not clearly defined.

**3.1.3    Shot noise.**    Although a gw modifies the phase of light, what is detected at the output port of the interferometer is a change of power. In a perfect interferometer the output power is:

$$P_{out} = P_{in} \cos^2\left(2\pi \frac{\delta L}{\lambda}\right) \tag{31}$$

where $\delta L$ is the arm length difference. $P_{out}$ can change either when $\delta L$ changes, for example when a gw passes, or when $P_{in}$ changes. The rate of arrival of photons on the detector is $\bar{n}$ ($\bar{n} = \frac{\lambda P_{out}}{2\pi \hbar c}$), the number of photons collected in a time $\tau$ is $\bar{N} = \bar{n}\tau$. Because this photon counting is characterized by a Poisson distribution, the fluctuations of this measurement are given by:

$$\frac{\sigma_{\bar{N}}}{\bar{N}} = \frac{1}{\sqrt{\bar{n}\tau}} \tag{32}$$

In order to convert this fluctuation into an equivalent length fluctuation, we have to fix the working point of the interferometer. The simplest choice is to work at the point where $dP_{out}/d\delta L$ is maximum. At that point $P_{out} = P_{in}/2$ and

$$\frac{dP_{out}}{d\delta L} = \frac{2\pi}{\lambda} P_{in} \tag{33}$$

The fluctuation of the photon number $\sigma_{\bar{N}}$ is interpreted as a length fluctuation $\sigma_{\delta L}$, and

$$\sigma_{\delta L} = \frac{\sigma_{\bar{N}}/\bar{N}}{\frac{1}{P_{out}} \frac{dP_{out}}{dL}} \tag{34}$$

that we can write using Eq. (32):

$$\sigma_{\delta L} = \sqrt{\frac{\hbar c \lambda}{4\pi P_{in}\tau}} \tag{35}$$

This can be converted in an equivalent gravitational wave noise using $\sigma_h = \sigma_L/L$

$$\sigma_h = \frac{1}{L}\sqrt{\frac{\hbar c \lambda}{4\pi P_{in}\tau}} \tag{36}$$

or

$$\sigma_h = 1.64 \times 10^{-21} \left(\frac{100\text{km}}{L}\right)\sqrt{\left(\frac{\lambda}{1064\text{nm}}\right)\left(\frac{10\text{W}}{P_{in}}\right)\left(\frac{10\text{ms}}{\tau}\right)} \tag{37}$$

The value of $\sigma_h$ is inversely proportional to the square root of the integration time, so in terms of spectral density (one-sided) it has a white amplitude:

$$\sigma_h(f) = \frac{1}{L}\sqrt{\frac{\hbar c \lambda}{2\pi P_{in}}} \tag{38}$$

or

$$\sigma_h(f) = 2.3 \times 10^{-22} Hz^{-1/2} \left(\frac{100\text{km}}{L}\right)\sqrt{\left(\frac{\lambda}{1064\text{nm}}\right)\left(\frac{10\text{W}}{P_{in}}\right)} \tag{39}$$

Remember that the aim is to detect amplitude in the range $10^{-21}$ to $10^{-27}$, so values given in Eq. (37) are the low limits of what is required. From this last equation we can draw some conclusions. We have only three parameters to play with to lower the photon shot noise: arm length, laser power and integration time.

- For technical reasons the choice of Nd-YAG laser ($\lambda = 1064$ nm) is common to all detectors. Reliable 10 W Nd-YAG lasers are available today but 10 W is not enough.
- Interferometric detectors have arms with physical length of only a few kilometers. In order to use longer arms the laser beam has to be folded. The common choice is to use Fabry-Perot cavities in the arms. This has also the advantage of increasing the amount of stored power, and thus brings a solution to the preceding problem. With Fabry-Perot cavities the stored power can reach a few kilowatts. In Virgo, the arms are 3 kilometer long cavities with a finesse of 50, equivalent to 150 kilometer unfolded arms. The use of kilometer long arms also imposes requirements on the beam diameter. For technical, (and financial) reasons, the

smaller the mirrors the better. The diameter of a laser beam at a distance $z$ from the beam waist position is given by the formula:

$$w\left(z\right) = w\left(0\right)\sqrt{1 + \left(\frac{\lambda z}{\pi w\left(0\right)^2}\right)^2} \tag{40}$$

so, for a given arm length $L$ there is an optimal value of the beam waist that gives the smallest beam spot size $w(L)$. The minimum beam radius (at $e^{-2}$) is of the order of $4 \sim 5\,\mathrm{cm}$, so the diameter of the mirrors must be around $6w(L) \approx 30\,\mathrm{cm}$.

- Even if the length and power problem is solved, the integration time cannot be reduced well below $10\,\mathrm{ms}$ so the detection is limited to signals with frequencies below a few hundreds of hertz (in fact a few kilohertz).

We have seen how the presence of shot noise dictates some key choices : minimum laser power, beam and mirror diameter, necessity to use Fabry-Perot cavities in the arms. Other noise sources will fix other important optical parameters.

### 3.1.4 Radiation pressure noise.

When dealing with shot noise we only assumed that this noise only affected the number of collected photons, but shot noise has another subtle effect. The laser beam exerts a force on each mirror equal to:

$$F = \frac{P}{c} \tag{41}$$

so fluctuations of the laser power induce fluctuations of this force:

$$\sigma_F = \frac{\sigma_P}{c} \tag{42}$$

If the mass of the mirror is $M$ then the displacement will be $x$ with $m\ddot{x} = F$, its spectrum will be:

$$x\left(f\right) = \frac{P\left(f\right)}{mc \times \left(2\pi f\right)^2} \tag{43}$$

or

$$x\left(f\right) = \frac{1}{mf^2}\sqrt{\frac{\hbar P_{in}}{8\pi^3 c\lambda}} \tag{44}$$

Each mirror will be affected by this noise and it can be shown that the effects will be uncorrelated in the two arms. In terms of equivalent gw spectral density we have

$$\sigma_h\left(f\right) = 2\frac{x\left(f\right)}{L} = \frac{1}{Lmf^2}\sqrt{\frac{\hbar P_{in}}{2\pi^3 c\lambda}} \tag{45}$$

or

$$\sigma_h(f) = 2.3 \times 10^{-24} Hz^{-1/2} \left(\frac{100\text{km}}{L}\right) \left(\frac{1\text{kg}}{m}\right) \left(\frac{1\text{Hz}}{f}\right)^2$$
$$\times \sqrt{\left(\frac{P_{in}}{10\text{W}}\right) \left(\frac{1064\text{nm}}{\lambda}\right)} \tag{46}$$

The radiation pressure noise decreases rapidly with frequency, so we focus our attention on the worst case, the low frequency domain. Because of the seismic noise which is very large at low frequency, it will be very difficult to detect gw below a few hertz, so we put this as a limit. Equation (46) exhibits interesting features:

- we see that the radiation pressure noise increases with increasing power, so just as shot noise places a lower limit to laser power, radiation pressure noise puts an upper limit. At 2 Hz with 10 kg mirrors, the radiation pressure noise becomes dominant when $P_{in}$ is higher than 40 kW, so in order to work in an optimal configuration at low frequency, the laser power in the arms should not exceed a few tens of kilowatts. In Virgo the intra-cavity power is around 15 kW.
- The noise is inversely proportional to the mirror mass, so a mass of 1, preferably 10 or more kilograms is required.

**3.1.5    Frequency noise.**    If the two interferometer arms are not exactly symmetric (if they have different lengths $\delta L \neq 0$ or finesses $\delta F \neq 0$), then the laser frequency noise $\sigma_\nu$ can mimic a gravitational wave signal

$$\sigma_h = \frac{\lambda}{c} \left(\frac{\delta L}{L} + \frac{\delta F}{F}\right) \sigma_\nu \tag{47}$$

At 10 Hz in a typical Nd-YAG laser $\sigma_\nu \approx 1000\,\text{Hz}/\sqrt{\text{Hz}}$, and the typical finesse asymmetry is of the order of one percent. In order to detect a gw signal the laser frequency noise has to be lowered by six orders of magnitudes (compared to the noise of a free running laser), and the two arms made as identical as possible. In order to achieve this complex frequency stabilization methods are employed in all interferometric detectors, and in order to insure the perfect symmetry of the interferometer, all pairs of Virgo optical components are coated during the same run (both Fabry-Perot input mirrors then both end mirrors are coated simultaneously).

**3.1.6    Internal thermal noise of mirrors.**    Mirror substrates are transparent material cylinders. They are affected by thermal noise and each one of their modes of vibration can be represented by an harmonic oscillator. The study of thermal noise in solids is a complex task, made difficult because no simple mathematical model can mimic with a sufficient accuracy the behavior of a real mirror. Nevertheless, using a simple harmonic oscillator representation, it is relatively easy to show that the displacement noise of a mirror is given by:

$$\sigma_x^2 = \frac{k_B T}{2\pi^3 Q(f) M(f_0)} \frac{f_0}{f^2 \left( \frac{f_0^2}{Q^2} + \left( \frac{f^2 - f_0^2}{f} \right)^2 \right)} \tag{48}$$

where $T$ is the temperature, $k_B$ is the Boltzmann constant, $f_0$ the resonance frequency, $M(f_0)$ is the *equivalent* mass of the vibration mode, and $Q(f)$ is the mechanical quality factor of the resonance. A simple way to represent internal damping in solids is to add an imaginary part $\phi$, called *loss angle*, to the oscillator spring constant. In this case the mechanical quality factor reads:

$$Q = \frac{f}{\phi f_0} \tag{49}$$

The lowest resonance frequency of 10 kg, 30 cm diameter mirrors is about 6 kHz, and we have seen that we will not be able to detect signals with frequencies higher than a few kHz, so we are always in the case $f \ll f_0$. In this case Eq.(48) simplifies to

$$\sigma_x^2 = \frac{k_B T \phi}{2\pi^3 M f_0^2 f} \tag{50}$$

or

$$\sigma_h = \frac{2}{L} \sqrt{\frac{k_B T \phi}{2\pi^3 M f_0^2}} \frac{1}{\sqrt{f}} \tag{51}$$

taking $f_0 = 6$ kHz we obtain

$$\sigma_h = 1.7 \times 10^{-23} \mathrm{Hz}^{-1/2} \left( \frac{100\mathrm{km}}{L} \right) \sqrt{\left( \frac{T}{300\mathrm{K}} \right) \left( \frac{\phi}{10^{-6}} \right) \left( \frac{5\mathrm{kg}}{M} \right) \left( \frac{1\mathrm{Hz}}{f} \right)} \tag{52}$$

So in order to minimize this noise one possibility is to work at low temperature, this is what is done in the Japanese TAMA project. The other solution is to use optical materials with a low value of the loss angle, or equivalently a high quality factor.

### 3.1.7 Compilation of results.

In order to reduce as much as possible the effects of possible sources of noise we were forced to fix some properties of passive and active optical components. Let us summarize what we have learned. The laser has the following properties: $\lambda = 1064$ nm, $P > 10$ W, single mode operation, $w(0) \approx 3$ cm. The mirrors must have the following properties: $m > 10$ kg, ffl $\approx 30$ cm, surface figure $\lambda/100$, roughness 0.1 Å rms, made with high Q material The interferometer arms: must be Fabry-Perot cavities, $L$ in the kilometer range, giving an equivalent optical path length of about 100 km. By adding (quadratically) all noise sources we can estimate the achievable sensitivity of interferometric detectors. The sensitivity curve of Virgo is shown on Fig. 3.

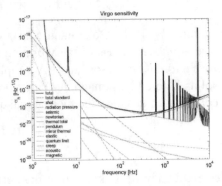

*Figure 3.*   Virgo sensitivity curve with contribution from all identified noise sources

## 3.2     The future of gw detection

### 3.2.1     Obligatory improvements.     All the actual interferometric detectors of gw share more or less the same optical characteristics, except for the length of their arms that differs, from 300 meters for TAMA, up to 4000 m for LIGO, chiefly for geographical and financial reasons. Their designs were frozen during the last decade, and were based on the technology that was available at that time, or that was thought to be available by the end of the century. It was the fate of all interferometric gw detectors to be based on years old technology. From what we have learned in the preceding sections, it is clear that the sensitivity of current detectors is really close to the minimum required one, we have currently no safety margin, so if one single subsystem of those complex detectors fails to reach its performance objectives, then detection will become impossible. That is why, since the starting of all projects, it was foreseen to regularly improve their sensitivity by using up to date technology and replace some parts of the detectors. In parallel to the construction of all the actual large interferometers, a lot of work was made to improve the performances of optical components and materials. The main efforts were aimed at understanding and improving the mechanical quality factor of mirrors (including their holders), improving the coating processes and characterization methods, finding ways to work at low temperature, studying new optical configurations, developing higher power laser sources...

# References

Proc. of the International summer school on *Experimental physics of gravitational waves*, (Barone, M. et al. Eds., World Scientific, London 2000). Contains a valuable chapter on General relativity by P. Tourrenc (contains a precise description of the various coordinates systems and their use, OBLIGATORY), a chapter by S. Bonazzola and E. Gourgoulhon on compact sources, in particular neutron stars, and a chapter by Jean-Yves Vinet on numerical simulations of interferometric gw detectors.

Misner, C.W., Thorne, K.S., Wheller, J.A., 1973, *Gravitation*, (Freeman Ed., New-York), The MTW, a reference book that can be read at two levels (following two pathes), with a very visual approach to gravitation.

Saulson, P.R.,1994, *Fundamentals of interferometric gravitational wave detectors*, (World Scientific, London), provides a very complete technical description of interferometric gravitational wave detectors. No previous knowledge in relativity is required.

Svelto, O., 1989, *Principles of lasers, 3^{rd} edition*, (Plenum, New-York), A simple yet complete introductory book on laser physics and technology.

Weinberg, S., 1972, *Gravitation and Cosmology*, (John Wiley & Sons, New York), covers special and general relativity and applications in the fields of astrophysics and cosmology.

Will, C.M., 1993, *Theory and experiment in gravitational physics*, (Cambridge University Press, Cambridge), contains descriptions of the various experiments in the field of gravitation. Requires some background in general relativity (for example, path one of the MTW or the book by Weinberg).

# COATINGS PRINCIPLES

Jean-Marie Mackowski

*LMA-VIRGO*
*Université Claude Bernard Lyon 1*
*22 Bd Niels Bohr, 69622 Villeurbanne cedex (France)*
mackowksi@lma.in2p3.fr

**Abstract**      The principles of coatings to either enhance reflectivity of mirrors or to enhance transmission of glass optics are described. Then the ion assisted deposition and ion beam sputtering techniques are addressed. Performances of these techniquesand their limitations are illustrated with the characteristics of the VIRGO mirrors coated at LMA. The importance of metrology is emphasized.

**Keywords:**   dielectric coatings, metallic coatings, deposition, ion beam sputtering, ion assisted deposition

## 1.      Introduction

Coatings of optical surfaces to improve either their reflectivity or transmission are critical to make a sound use of the scarce photons collected big telescopes. Poor reflectivity has been in the past of strong limitation to the development of telescopes (see Ch. 3). Most of the telescope mirrors today are aluminum coated, with reflectivity of ≈85%. Being the number of mirrors in trains of mirrors to bring the light from the sky to e.g. the coudé focus, the total throughput is frustrating low. The same concern applies to focal instrumentation, of which the throughput rarely exceeds 30%. Because of the cost of the present very large telescopes, and still more that of the forthcoming decametric or hectometric ones, one has to ensure that each of their components is as much efficient as possible, starting by the first one light hit, the primary mirror. Also one needs to have a robust surface, keeping its optical quality for a rather long time, depending on the application. Astronomical telescope mirrors are exposed to a relatively severe environment: temperature fluctuations, dust, humidity, ... Because of dust in particular, the mirrors have to be cleaned quite frequently (typically once a year). Passivation methods have been little

*R. Foy and F. C. Foy (eds.), Optics in Astrophysics, 327–341.*
© 2006 *Springer. Printed in the Netherlands.*

studied in the case of telescopes, presumably because of a lack of cooperation between opticians in Astrophysics and those in the field of coatings.

What is a coating on a substrate? It is a thin layer, of a stack of thin layers, made of materials with a refraction index different from that of the substrate. Part of the light is reflected at each interface between the layers or between air and the first layer or between the last one and the substrate. Reflected beams interfere so that for a given wavelength and a given thickness, reflected light can be either canceled out or maximized. It corresponds to maximum transmission or reflectivity of the layer respectively. These materials and their indexes are chosen in order to fit as close as possible the specifications for the final coated optics.

In this lecture we recall the basic principles of the two main coating families: dielectric and metallic coatings. In a second part, we describe the coatings deposition techniques and we address their performances and limitation. Most of the examples given in this lecture have been developed for the gravitational waves interferometers VIRGO and LIGO (see Ch. 18).

## 2.    Principles : Dielectric Coatings

Layers with a thickness of $\lambda/4$, where $\lambda$ is the wavelength of interest, or so-called *quarterwave layers* give maximum interference effect (because of the reflexion, the phases of incoming and reflected beams are opposed) over the maximum wavelength interval. Halfwave layers are absentee layers: they have no effect on the phase (but of course they decrease the amplitude). The thicker the layer, the higher is the frequency of the interference pattern and the shorter the wavelength interval of maximum reflectivity per period. The reflectivity of dielectric layers and of metal layers varies slowly and strongly respectively with increasing $\lambda$. The optical thickness of a quarterwave layer is $\lambda_0/4n_f$, where $\lambda_0$ is the working wavelength and $n_f$ the refraction index of the filter layer. It transforms the refraction index following $n_t = n_f^2/n_s$, where $n_s$ is the substrate refractive index. Interference calculations for two waves is straightforward when the Waves are combined exactly in or out of phase. Resultant amplitudes are the sum or the difference of individual amplitudes respectively. All other cases are intermediate (see Ch. 2). The phase lag suffered by a wave traveling across a film of thickness $d$ and index $n$ is $\delta = 2\pi nd/\lambda$. In case of a single film coating, reflection losses are canceled out if $n_0/n_f = n_f/n_s$, or $n_f = (n_0 n_s)^{1/2}$.

Let us now consider a stack consisting of $x$ quarterwave layers of high index $n_H$ and of $x-1$ quarterwave layers of low index $n_L$; it is denoted $n_0|HLHL \ldots LH|n_s$. Its reflectance $R$ is

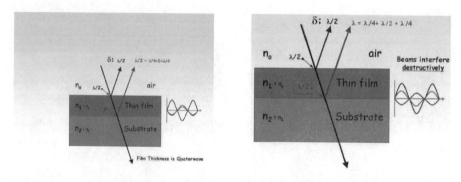

*Figure 1.* Incoming light undergoes a $\lambda/2$ phase delay $\delta$ when reflected at the interface between a low index medium and a high index medium. There is a null phase delay at the interface between a high index medium and a low index medium. In case of a quarterwave film, the phase lag is $\lambda/4$ on the way to the substrate, then 0 (left) or $\lambda/2$ (right) at the substrate surface and again is $\lambda/4$ on the way back, so that $\delta = \lambda/2$ (left) or $\lambda$ (right). Both reflected beams either have the same phase and they interfere *constructively* (left) or have a $\lambda/2$ phase delay and they interfere *destructively*. This is the goal of a coating respectively for a mirror (maximum reflectivity) or for a lens (maximum transmittance). Since the beam reflected at the substrate surface is slightly attenuated, in the film material, the destruction can never be total: the reflectivity is never enhanced up to 100%.

$$R = \frac{n_0 - \dfrac{n_H^{2x}}{n_L^{2(x-1)}} n_s}{n_0 + \dfrac{n_H^{2x}}{n_L^{2(x-1)}} n_s} \tag{1}$$

Thus the phase delay increases by $\lambda/2$ every $L|H$ interface, leading to complex interference patterns (Fig. 2). Ripples on the blue and red edges of the high reflectivity area in Fig. 2 are usually strongly decreased by adding several layers at each end of the coating and refining them in a matching structure, in this case, adding a two pairs of a halfwave low index layer plus a $0.1\lambda/4$ wave high index layer on the entrance side (denoted $(2L.1H)\hat{\ }2$) and symmetrically a $(.1H2L)\hat{\ }2$ coating on the substrate side; they act as impedance adapters. In this notation, the exponent is the number of layer systems in the parentheses, $L$ and $H$ are respectively low and high index layers and the preceding number is the thickness of the layer in quarterwave unit. Further decreases of the ripple pattern can be achieved with more complex adaptation of indexes on both sides of the coating. Figure 2 shows that adding the layers to decrease ripples widens the high reflectivity area, and shifts it toward the red (see also Fig. 10).

**Broadband coatings.** Broader bandwidth reflective coatings can be achieved. Figure 3 shows a 400 nm wide coating consisting of 22 layers arranged as $(HL)\hat{\ }5 \, 1.2L(1.4H1.4L)\hat{\ }5 \, 1.4H$, where layers are made of cryolite $Na_3AlF_6$ (low index, $n = 1.35$) and of $ZnS$ (high index, $n = 2.35$). This is a school

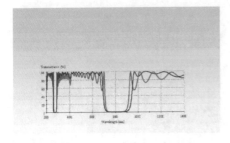

*Figure 2.* Transmittance spectral profile of a coating consisting of a quarterwave stack of 23 layer stack centered on 800 nm. Light gray: without ripple control. Dark gray: with ripple control. It can be used either as a intermediate band filter, or a shortwave dichroic beam splitter or a longwave one.

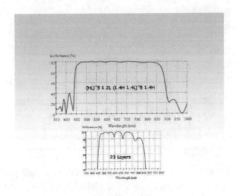

*Figure 3.* A broadband reflecting coating, achieved with 22 layers (top). Simply adding a poorly matched layer lead to a significantly lower reflectance (bottom)

case, since material can interact at their interface, so that the index gradient is not infinite; this is particularly true with $ZnS$, which is a inhomogeneous material. Adding another layer, with the purpose of decreasing the ripples, can degrade the reflectance significantly, as shown in the bottom part of Fig. 3. Thus the trade-off of the arrangement of layers is critical.

**Multicavity filters.** Multicavity Fabry-Perot filters are used to make very narrow transmission filters. A simple Fabry-Perot cavity (see Ch. 2) consists of a halfwave layer surrounded by two reflectors of typically 10 layers each. Figure 4 shows three transmission profiles obtained with one, two or three cavity filters. The three cavity $((HL)^5HH(LH)^5)^3$ filter has a 1.2 nm bandwidth. It has $\approx 60$ layers. Note that the three-peak top of the transmittance. Each cavity has to be well adapted to the following one; if not the resulting transmittance can be very poor. Such cavities are broadly used in telecoms in between arrays of antennas for cell phones.

**Oblique incidence and wide angle.** At oblique incidence, the path difference between the beams is reduced. Their amplitudes is increased for s-polarized light and decreased for p- one (Fig. 5). As a result, for instance for a reflective coating, the filter is blue shifted with s-polarized light at $45°$ incidence with respect to the normal incidence case, and the ripple pattern is strongly enhanced. The shift is a little bit stronger for p-polarized light, with a not so strong enhancement of ripples; but the bandwidth is significantly reduced. Figure 7 shows an example of an antiflection coating for $\lambda = 510$ nm.

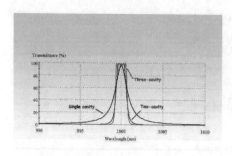

*Figure 4.* Transmittance of a multicavity Fabry-Perot filter.

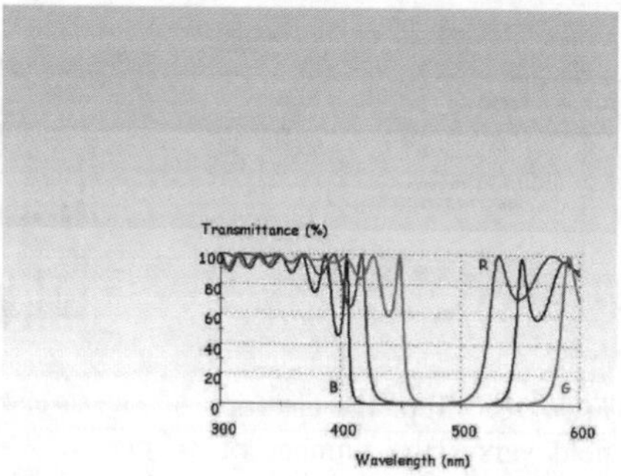

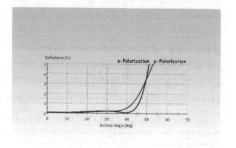

*Figure 5.* Effect of the incidence angle on the spectral profile of a transmission coating. G: normal incidence. R: p-polarization. B: s-polarization.

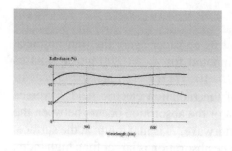

*Figure 6.* Reflectance for s- and p- polarization of a 8 layers 45° beam splitter.

*Figure 7.* Reflectance versus semi-angular field for both s- and p- polarization at 510 nm.

Three materials are used: $MgF_2$ ($n \approx 1.35$), $Al_2O_3$ ($n \approx 1.8$) and $TiO_2$ ($n \approx 2.35$). The reflectance is very good, and almost constant over a field of $\pm 35°$ for both polarizations: this is a huge field for astrophysical instruments!

**Non-polarizing filters.** The design of dielectric coatings in order to have equal p- and s- polarization over a wide spectral region is exceptionally difficult. Figure 6 shows the two polarization components for a 45° beam splitter in the 500-600 nm region. Layers are made of 8 layers of $TiO_2$, $Al_2O_3$ and $SiO_2$, in order of decreasing $n$. Playing with the numbers and the parameters of the layers allows us to translate the curves one with respect to the other, but they can never have the same shape.

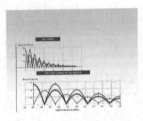

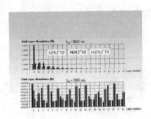

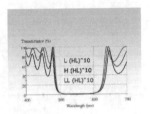

*Figure 8.* Top: the electric field versus the number of the layer. Bottom: magnification of the left part of the top curve; each lobe corresponds to a single layer. Three coatings are shown, with 3 different entrance layers: a low index and a high index quarterwave, and a halfwave.

*Figure 9.* Absorption per layer for the coatings of Fig. 8. Top: at the working wavelength $\lambda_0 = 800$ nm. Bottom: at $\lambda = 550$ nm. Dark: high index quarterwave entrance layer. Gray: high index quarterwave. Light gray: halfwave.

*Figure 10.* Transmittance of the reflection coating of Fig. 8. The coating with a halfwave entrance layer gives the better transmittance curve.

**The role of the entrance layer.**   let us consider again the transmittance of a coating deposited on a mirror, computed for $\lambda_0 = 800$ nm in 3 cases with 3 different entrance layers: $L(HL)^{\wedge}10$, $H(HL)^{\wedge}10$ and $LL(HL)^{\wedge}10$ (Fig. 8). The square of the integral of the radiation electric field $E$ versus the layer number is the absorption. Most of the absorption occurs between the first layers. When the first layer is a quarterwave, $E$ is maximum at the surface. This is bad since dust will be attracted. The absorption is larger for a high index quarterwave than for a low index one, whatever the layer (Fig. 9). At contrary, with halfwave first layer, $E = 0$ at the surface, which minimizes the risk of pollution of the coating by dust. And the absorption is also minimum. The transmittance curves are given in Fig. 10. The bandwidth is slightly narrower with the halfwave entrance layer than with quarterwave ones. The reflectance is slightly better. The main difference are the the weaker ripples. Thus the entrance layer of a reflecting coating should be a halfwave.

The absorptance per layer (Fig. 9) is smaller at 550 nm than at $\lambda_0$. But its sum over the whole stack is significantly higher. Indeed the stack is optimized for $\lambda_0$ in order that $E$ fades when the incoming radiation crosses the coating. It reaches nearly 0 from the $10^{\text{th}}$ layer with a low index halfwave entrance layer, or slightly more with a high index quarterwave one (Fig. 11). But at $\lambda = 550$ nm, $E \approx 20$ V/m whatever the number of the layer and whatever the entrance layer, up to enter the substrate, which results in a low reflectance.

To summarize this section about dielectric coatings, let us recall their advantages and drawbacks. Advantages are:

■ Reflectances can be $> 99.9\%$

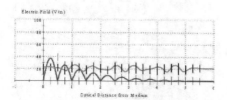

*Figure 11.* Radiation electric field in the coating of Fig. 8 with a low index halfwave entrance layer. At λ = 550 nm the electric field never vanishes.

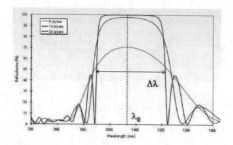

*Figure 12.* Reflectance of a coating centered on $\lambda_0 = 1.064\mu$m, for three numbers of layers.

- Absorption losses are low, $<10^{-5}$ in the visible and the infrared.

But drawbacks are :

- High efficiency coatings have more than 30 layers, for which the deposition time is long. Thus they are expensive.
- High reflectances are obtained over a short domain ($\Delta\lambda \approx 250$ nm).

A typical reflectance response is given in Fig. 12, for 6, 14 and 26 layers. The reflectance is given by

$$R = \frac{\left(1 - n_H/n_L\right)^{2x} n_H^2/n_S\right)^2}{\left(1 + n_H/n_L\right)^{2x} n_H^2/n_S\right)^2} \tag{2}$$

and the FWHM bandwidth by

$$\Delta\lambda = \pi\lambda_0 \arcsin\left((n_H - n_L)/(n_H + N_L)\right) \tag{3}$$

The higher the number of layers, the sharper are the edges of the filter, and the more contrasted the ripple pattern.

## 3.      Coatings including metallic layers

Now we describe metallic layer coatings, which are widely used in astronomy. Gold is slightly better at infrared wavelengths than silver. At short wavelengths aluminum is by far the best of the 3 metals (Fig. 13), but with a marked minimum around 800 nm. The driving parameter is the thermal conductivity.

Silver tarnishes rapidly. Thus a passivation layer is required to keep its reflectance at its value when freshly deposited. It can be done sandwiching the $Ag$ layer between two quarterwave $SiO_2$ layers (Fig. 14): the reflectance is not significantly altered (except in the ultraviolet, in a region where silver is not good), but now the $Ag$ layer is protected against oxidation. One can

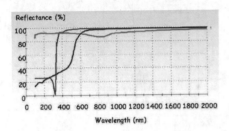

*Figure 13.* Reflectance of a 100 nm thick coating. Dark: gold, gray, silver and light gray: aluminum.

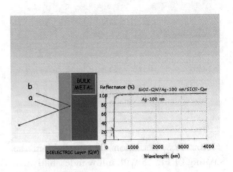

*Figure 14.* Reflectance of a 100 nm thick silver coating enhanced with two quarter-wave $Si_O2$ layers.

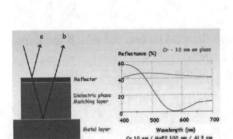

*Figure 15.* Antireflection glass mirror coated with a 10 nm layer of $Cr$, with a 100 nm layer of $MgF_2$ plus a 3 nm layer of $Al$.

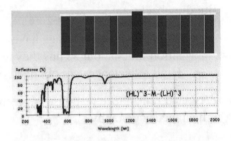

*Figure 16.* Reflectance of a coating on a glass substrate, made of 3 doublets of high/low index quarterwave layers symmetrically deposited on each side of 100 nm $Cr$ layer.

decrease the reflectance as well (Fig. 15): 10 nm thick layer of $Cr$ deposited on a glass substrate has its reflectance enhanced in the blue, but almost zeroed in the 550-600 nm region. This is obtained with a dielectric phase matching 100 nm layer of $MgF_2$ coated with a thin film of $Al$. The wavelength of the low reflectance region can be shifted by tuning the width of the $Al$ coating. It is amazing that a metallic layer can act as an antireflection coating. More complex designs can be achieved, as shown in Fig. 16. Here a 100 nm thick layer of $Cr$ is surrounded on each side with three doublets of high-low index quarterwave layers, deposited on a glass substrate; the reflectance profile is similar to that of a dichroic beam splitter. A higher (odd) number of layers would produce sharper edges of the transmission region.

The popular $Al$ coating can be enhanced significantly with a low index ($SiO_2$ or $MgF_2$) and a high index ($TiO_2$) coating. The higher the number

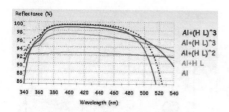

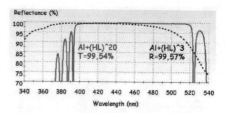

*Figure 17.* Enhanced reflectance of an *Al* coating on a glass substrate, with one or several doublets of high ($TiO_2$) and low ($SiO_2$ or $MgF_2$) index quarterwave layers.

*Figure 18.* An example of an *Al* layer on a glass substrate coated with 20 doublets of $TiO_2 - SiO_2$ layers, leading to a reflectance of 99.54% flat over 120 nm, compared with the 3-doublet case of Fig. 17.

of doublets of $HL$ layers, the higher the maximum reflectance, at the expense of the bandwidth (Fig. 17).

To summarize this section about metallic coatings, let us recall their advantages and drawbacks. High reflectances (>90%) are obtained

- over a large range of incident angles
- over a wide spectral range, from the ultraviolet to the infrared.

But metallic coatings have high absorption losses. But one can remedy them with dielectric coatings, which are necessary also to passivate the metal, avoiding oxidation. Aluminum is the best metal for the ultraviolet; in addition it adheres on most substrates; it needs passivation. Silver is easy to deposit; it has the highest reflectance in the visible and the infrared; since it tarnish rapidly, passivation is mandatory. Gold is the best material beyond $\approx 700$ nm, and it is considered that it does not tarnish, which is not true actually because its surface is not that stable on the long term.

## 4.    Coating deposition techniques

We now describe the current techniques of deposition. A coating process involves several parameters. There is the nature of the substrate: a crystal or an amorphous material, the quality of its polishing and its temperature. There are also the characteristics of the source, as temperature and emission law, and those of the medium in between, as its pressure and composition. In evaporation process the energy of particles is $\approx 0.1$ eV, or 1100 K; their impact velocity is in the range of m.s$^{-1}$. With sputtering techniques, the energy lies in between 10-50 eV and the impact velocity is in the range of km.s$^{-1}$.

The most widely deposition technique is the ion assisted deposition (IAD). A material in a melting-pot is vaporized by heating either with an electron beam, or by Joule effect, or with a laser beam, or with microwaves, or whatever else. The vapor flow condensates on the substrate. In the same time, an ion

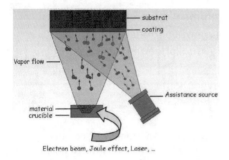

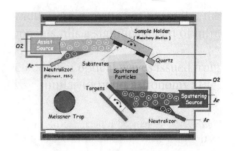

*Figure 19.* Schematic view of the ion assisted deposition technique. As the vapor condenses on the substrate, ions hammer the coating in formation to compress it.

*Figure 20.* Schematic view of the ion beam sputtering (IBS) technique. See text.

source is used to pack the coating as it is deposited (Fig. 19). The ion source is loaded with either a reactive gas -either N if the vapor is a nitrure, or O in case it is an oxide, in order to control the stoichiometry, or with a neutral gas, e.g. argon or krypton with the drawback that they are absorbing materials which consequently change the absorption coefficient. Ions are used as hammers to improve the volume factor of the coating.

The lowest losses are obtained with the ion beam sputtering (IBS) technique. It is used for the coatings of the gravitational waves interferometers as **VIRGO** (see Ch. 18). A source projects ions on a target so that materials for the coating are sputtered (Fig. 20). Ions are generally argon or krypton, depending on the mass of the target atom, which has to be close to that of the projected ions. At LMA, the target area is $80 \times 60$ m$^2$ for VIRGO coatings. Ions are accelerated up to 1.5 - 2 kV, and the intensity is $\approx 1$ A. Thus to prevent the beam from diverging, one neutralizes the charge of the beam with a particular ion source. Then the sputtered particles condense on the substrate. As in the IAD technique, an assist source(with $O_2$, with its own neutralizer (with argon), tunes the stoichiometry by compressing the coating as it forms. Losses as low as of $\approx 10^{-7}$ have been obtained, at 1.064 $\mu$m YAG wavelength with a process time of $\approx 70$ hours. The process is fully automatized. For VIRGO, a tantalum oxide is used; in this case the target is a $Ta$ plate, and $O_2$ is injected in the sputtered particles beam to form $Ta_2O_5$.

Figure 21 shows the large coater for the VIRGO optics at LMA (Lyon). The technology is DIBS. Its size is $2 \times 2.2 \times 2.2$ m$^3$. The inner pressure is $2 \times 10^{-8}$ mBar. The coater is located in a 150 m$^2$ class 1 room whith metrology and control. In a class 1 room, tolerance is <10 particles of $0.1 \mu$m and <1 particle of 0.5 $\mu$m $\times$foot$^{-3} \times$s$^-1$.

The uniformity of the layer thickness is a strong requirement in order to get high quality wavefront surfaces. It is controlled thanks to a mask with

*Figure 21.* The VIRGO large coater at LMA.

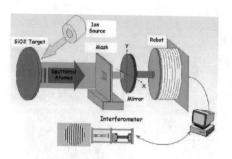

*Figure 22.* The mask device at LMA. A computer controlled robot moves the mirror behind a mask in order to get a uniform thickness coating.

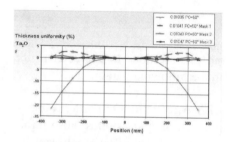

*Figure 23.* Improvement of the relative thickness uniformity of a coating layer with the mask device shown in Fig. 22. Bottom curve: without mask. Top curves: thickness uniformity with several mask configurations.

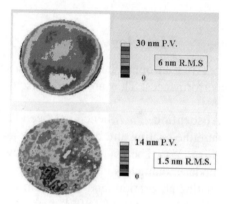

*Figure 24.* Improvement of the thickness uniformity. Bottom: without correction. Top: with correction.

an aperture moving in front of the substrate during the deposition process. After several iterations of the deposition, the thickness profile becomes very smooth. Positions and velocity are computer controlled. Figure 23 shows the thickness of a $Ta_2O_5$ layer and its optimization over a diameter of up to 700 mm. Whereas a relative uniformity of 6 $10^{-3}$ is achieved without masking over a diameter of 350 mm, it can reach 6 $10^{-3}$ over a diameter of 700 mm with a mask in front of the coated optics moved by computer controlled robot. Figure 24 shows a reduction from 30 to 14 nm peak-to-valley or 6 to 1.5 rms.

The VIRGO IBS coater allows us to get at 1064 nm an absorption as low as 0.1 $10^{-6}$ and a process scattering in the range $0.6 - 4 \ 10^{-6}$. The repeatability of the wavelength centering is $\pm 0.4\%$. Nowadays the wavefront error after corrective coating is $\approx 1$ nm rms.

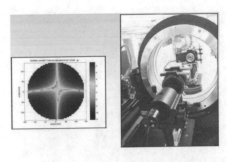

*Figure 26.* Bench for s- and p- polarization measurements of the losses due to birefringence in s- and p- polarization (dex scale). The cross in the left image is due to the mirror support.

*Figure 25.* The absorption bench for VIRGO coatings.

## 5. Metrology

After an optics has been coated, several tests are done to characterize its performances. They are briefly discussed in the following paragraphs.

**Absorption.**   A bench has been built at LMA for VIRGO in order to measure the coating and bulk (or blank as said in astronomical optics) absorption (Fig. 25). The source is a 28 W cw YAG:ND$^{3+}$ laser. The bench accommodates diameters up to 400 mm for which the sensitivity is $2 \ 10^{-8}$ for the coating absorption, and $2 \ 10^{-8} \mathrm{cm}^{-1}$ for the bulk. A typical absorption for a VIRGO mirror is $0.63 \ 10^{-6}$ over a diameter of 150 mm.

**Polarization.**   The amplitude and the principle axis direction both in s- and in p- polarization can be mapped up to a diameter of 400 mm (Fig. 26). The sensitivity is $10^{-5} \ \mathrm{rad.cm}^{-1}$.

**Relative reflexion.**   Measurements of the map of the relative reflexion are done with a sensitivity of $2 \ 10^{-5}$ (Fig. 27).

**Scattering and transmission.**   Figure 28 shows the layout of the scattering measurement bench. Scattering is measured over almost $4\pi$ steradians, with two $\approx 1°$ dead angles, in the direction of the source and of the detector. A typical value of the amount of scattered light is $4 \ 10^{-6}$ over $150^2 \ \mathrm{mm}^2$. On the back side, one measures the transmission. A typical value of the transmission obtained at LMA for VIRGO is $42.9 \pm 0.210^{-6}$ over a diameter of 50 mm.

**Wavefront.**   A critical parameter of coatings, for VIRGO or other instruments, is the wavefront rms error. It is measured with the bench in Fig. 29,

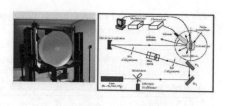

*Figure 27.* The relative reflexion bench for VIRGO coatings.

*Figure 28.* Layout of the bench for scattered light measurements (right), and a VIRGO bulk being measured (left).

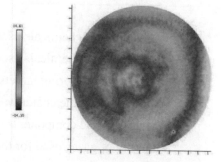

*Figure 29.* The bench for wavefront measurements, located close to the air filters.

*Figure 30.* An example of the map of departures from the mean wavefront over 150 mm in diameter. The peak-to-valley error is 29 nm.

located near the ceiling close to the air filters of the room. Indeed measurements are typically 2-day long, so that dust pollution is a concern in spite of the high cleanness quality of the room. The wavefront error is 3.8 rms over a diameter of 150 mm in Fig. 30, or $\lambda/280$, within the VIRGO specs.

Finally, microroughness and defects are micromapped, down to a threshold of 0.3 $\mu$m. Table 1 summarizes the LMA-VIRGO measurements of optics for the VIRGO end mirror, and compares them with VIRGO requirements. For all parameters, average values are significantly better than the specifications.

# 6. Limitations

Purity of materials and environmental conditions are the main causes of limitations of the performances of the coating processes.

**Materials purity**

- quality of the bulk (blank),

*Table 1.* Summary of measurements of the parameters of VIRGO optics at LMA, compared with VIRGO specifications.

| Parameter | VIRGO requirements | LMA-VIRGO measurements |
|-----------|--------------------|------------------------|
| Average scattering | $< 5 \ 10^{-6}$ | $4 \ 10^{-6}$ over $150^2$ mm$^2$ |
| Average transmission $T$ | $10 \ 10^{-6} < T < 50 \ 10^{-6}$ | $42.9 \pm 0.2 \ 10^{-6}$ over $\varnothing 50$ mm |
| Average absorption | $< 5 \ 10^{-6}$ | $0.63 \pm 0.07 \ 10^{-6}$ over $\varnothing \ 150$ mm |
| Birefringence | $< 5 \ 10^{-4}$rad.cm$^{-1}$ | $< 5 \ 10^{-4}$rad.cm$^{-1}$ over $\varnothing 150$ mm |
| Wavefront flatness | $< 8$ nm rms over $\varnothing 150$ mm | $3.8$ nm rms over $\varnothing \ 150$ mm |
| Curvature radius | $3450 \pm 100$ m over $\varnothing 150$ mm | $3580 \pm 17$ m over $\varnothing 150$ mm |

- residuals of chemicals used for the polishing process,
- residuals of chemicals used for the cleaning process, coating,
- purity of sputtering targets,
- cleanness of the coater walls,
- purity of the components of the ion sources,
- purity of gazes used for the deposition process, ...

**Environmental conditions**

- contamination by particles,
- organic contamination,
- cleanness of boxes for transfers and transportation, ...

The control of materials purity and of environmental conditions requires to implement physico-chemical analysis tools like ESCA, RBS, AUGER, SEM, XTM, SIMS or others. The principle of SIMS (Secondary Ion Mass Spectroscopy) is shown in Fig. 31: an ion gun projects common ions (like $O^+$, $Ar^+$, $Cs^+$, $Ga^+$, ...) onto the sample to analyze. In the same time a flood gun projects an electron beam on the sample to neutralize the clusters. The sample surface ejects electrons, which are detected with a scintillator, and secondary ions which are detected by mass spectrometry with a magnetic quadrupole.

Figures 32 and 33 shows profiles obtained with a SIMS device to detect pollution by sodium, and to a less extent by potassium. Such analyzes greatly improve the efficiency of the search for the origin of unexpected problems in the deposition process.

Wavefront accuracies of $\lesssim \lambda/1000$ are a target within the 4-5 coming years, in particular for relatively gravitationally faint sources observed with VIRGO. Extremely large telescopes (see Ch. 7) are also a big challenge for researchers

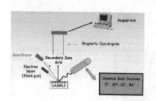

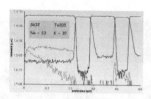

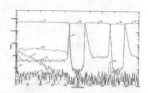

*Figure 31.* Principle of the SIMS device to detect materials polluting the deposition process.

*Figure 32.* SIMS profiles showing the growth of $^{23}Na^+$ at the transition zone where the concentration of $SiO_2$ decreases and that of $Ta_2O_5$ increases, as a function of depth in the coating.

*Figure 33.* SIMS profiles showing the growth of $Na$ at the time where $Si$ decreases and $Ta$ increases, as a function of time.

in the fields of coatings. Further progresses will aim at decreasing optical losses, and at widening their performances over a very large spectral range.

## References

Two reference books in the field of coatings are:

H. Angus Macleod, 2001, *Thin-Film Optical Filters*, Third Edition, ISBN 0 7503 0688 2, IoP, Institute of Physics Publishing, Bristol and Philadelphia

Alfred Thelen, 1988, *Design of Optical Interference Coatings*, ISBN 0-07-063786-5, McGraw-Hill Book Compagny

# CLASSICAL PROPERTIES OF
# CONTINUOUS WAVE
# OPTICAL PARAMETRIC OSCILLATORS

Thomas Coudreau[1], Thierry Debuisschert[2], Claude Fabre[1], Agnès Maître[1]

[1]*Laboratoire Kastler Brossel,*
*CNRS UMR 8552 et Université Pierre et Marie Curie*
*Case 74, UPMC, 4 Place Jussieu 75252 PARIS cedex 05*
coudreau@spectro.jussieu.fr

[2]*THALES Research and Technology - France Domaine de Corbeville, Orsay*

**Abstract**     Optical Parametric Oscillators provide a very efficient source of tunable coherent radiation. The principle of different kinds of OPOs are described. OPOs are used in astronomy for Laser Guide Star systems, and they may be used for other nonlinear optics applications in astrophysics, such as frequency conversion or parametric amplification.

**Keywords:**    optical parametric oscillators, nonlinear optics

## 1.     Introduction

The Optical Parametric Oscillator (OPO) was first predicted (Akhmanov and Khokhlov, 1962; Kingston, 1962; Kroll, 1962) and then realized at the very beginning of nonlinear optics shortly after the discovery of lasers, first in the pulsed regime (Giordmaine and Miller, 1965; Akhmanov et al., 1966) then in continuous wave (cw) (Smith et al., 1968). It generated immediately a large interest as a coherent source of broadly tunable radiation. OPOs were replaced after the 1970s by dye lasers, who possess the same advantage but are easier to use. They were then almost completely forgotten for a decade. They have just recently known a very impressive renewal of interest under the conjunction of several factors among which one can cite: the existence of all solid state lasers, both efficient and easy to use which can serve as pump beams for the OPO, the development of nonlinear crystals with low absorption and price, and

*R. Foy and F. C. Foy (eds.), Optics in Astrophysics, 343–349.*

with reproducible properties, the progress of locking techniques necessary to stabilize the system as well as the improvement in quality and flux resistance of dielectric coatings. OPOs capable of covering all the visible spectrum have thus been developed. Pulsed OPOs can now be found in many laboratories for both fundamental and applied research, and more and more in industry due to their often irreplaceable character as a compact and efficient source of tunable radiation for spectroscopy and diagnosis in the visible and infrared. cw OPOs, harder to use, have started their breakthrough more recently due to their unique character as a very broadly tunable source of highly monochromatic light. Their development has also been boosted by research made in quantum optics, since it was found that they are remarkable sources of radiation with purely quantum properties.

This chapter presents a short description of the physics of cw OPOs. After a brief presentation of the basic principles of OPOs, we will describe the properties of cw OPOs. The reader wishing to read further can read the references Byer, 1975; Brunner and Paul, 1977; Shen, 1984; JOSA, 1993; JEOS, 1997; CRAS, 2000.

## 2. Parametric interaction and amplification in nonlinear optics

If an intense electromagnetic wave, called pump wave, of frequency $\omega_0$ is sent in a crystal possessing a second order optical non linearity, of coefficient $\chi^{(2)}$, one creates in the crystal a nonlinear polarization which induces the generation of two new electromagnetic waves of frequencies $\omega_1$ and $\omega_2$ so that $\omega_0 = \omega_1 + \omega_2$. This phenomenon, one of the most important in nonlinear optics, is called "parametric generation" (Shen, 1984) and the generated waves are called respectively signal and idler. The name "signal" denotes usually the wave with the largest frequency. In this paragraph, we will describe this phenomenon in the simplest possible way so that we can extract the basic physical ideas. We suppose that all the interacting waves are plane waves propagating along the axis Oz, with wave vectors $k_i = n_i\omega_i/c$ ($i = 0, 1, 2$ respectively for pump signal and idler et $n_i$ is the linear index of refraction at frequency $\omega_i$) and we denote $E_i(z)e^{ik_iz}$ the respective complex amplitudes. Starting from the Maxwell equations, the following propagation equations can be derived for the amplitudes $E_i$ inside the crystal (Shen, 1984):

$$
\begin{aligned}
\frac{dE_0}{dz} &= i\frac{\omega_0}{2n_0c}\chi^{(2)}E_1E_2e^{-i\Delta kz} \\
\frac{dE_1}{dz} &= i\frac{\omega_1}{2n_1c}\chi^{(2)}E_0E_2^*e^{i\Delta kz} \\
\frac{dE_2}{dz} &= i\frac{\omega_2}{2n_2c}\chi^{(2)}E_0E_1^*e^{i\Delta kz}
\end{aligned}
\tag{1}
$$

where $\Delta k = k_1 + k_2 - k_0$. One can deduce immediately from Eq. 1 that the quantities :

$$\Pi_{tot} = \frac{\varepsilon_0 c}{2} \left[ n_0 |E_0|^2 + n_1 |E_1|^2 + n_2 |E_2|^2 \right] \tag{2}$$

$$\Delta N = \frac{\varepsilon_0 c}{2} \left[ \frac{n_1 |E_1|^2}{\hbar \omega_1} - \frac{n_2 |E_2|^2}{\hbar \omega_2} \right] \tag{3}$$

are constant through propagation. In particular, $\Pi_{tot}$ is the sum of the energy flux of the three waves interacting in the crystal. Its conservation shows that there is no energy storage in the crystal and that all the energy gained by the signal and idler fields through the parametric interaction is taken from the pump energy which decreases with $z$. The equations 1 can be solved analytically in the general case in terms of elliptic functions (Armstrong, 1962). One can show that the conversion efficiency is maximum when the following condition, called the phase matching condition is fulfilled:

$$\Delta k = 0 \iff n_0 \omega_0 - n_1 \omega_1 - n_2 \omega_2 = 0 \tag{4}$$

This condition ensures constructive interferences between the fields generated by parametric interaction in different points of the crystal. If this condition is not fulfilled, the amplitude of the different field oscillates as a function of $z$ on scales inversely proportional to $\Delta k$. A small phase mismatch will lower the non linearity but if it is large, the signal and idler fields are nowhere intense. A large parametric conversion efficiency can be obtained with a large phase mismatch by modifying the non linear medium, for example by a proper periodical modulation of the non linearity ("quasi phase matching").

An important configuration is the case where the input signal field on the crystal is small, the idler field zero and the pump field sufficiently large for its variation along $z$ to be negligible. The following solutions for equations 1 can be found in the case of $\Delta k = 0$:

$$E_1(z) = g E_1(0) \quad \text{and} \quad E_2(z) = i\sqrt{g^2 - 1} \sqrt{\frac{n_1 \omega_2 E_0}{n_2 \omega_1 E_0^*}} E_1^*(0) \tag{5}$$

where

$$g = \cosh\left( \frac{\chi^{(2)}}{2c} \sqrt{\frac{\omega_1 \omega_2}{n_1 n_2}} |E_0| z \right) \tag{6}$$

The complex amplitude of the signal field at the output of the crystal is amplified by a factor $g$ with respect to the input field. $g$ is always larger than one so that there is always coherent amplification of the signal field: one speaks of parametric amplification. Nonlinear optics provides an amplification different from the amplification occurring in a medium which present population inversion. For a typical value of the nonlinear coefficient of $1\ pm/V$, a pump of

1 $\mu m$ and a 1 $cm$ long crystal, the parametric gain is on the order of a few units for a pump on the order of $10^8$ $W/cm^2$. $g$ can thus be very large if pulsed pumps are used.

## 3.    Principle of Optical Parametric Oscillators

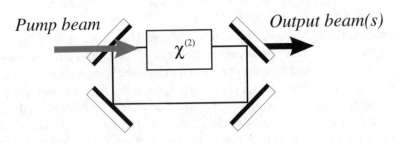

*Figure 1.*   Schematic representation of an Optical Parametric Oscillator

It is well known that by inserting an optical amplifier obtained by population inversion in an optical cavity, one can realize sources of coherent radiations, namely lasers. One can operate in the same way with parametric amplification as shown on Fig. 1. A nonlinear crystal illuminated by an input pump is inserted in an optical cavity. This cavity is represented for convenience as a ring cavity but consists usually of a linear cavity. An important difference with the laser is that there are three different fields, instead of one, which are present in the amplifying medium, all these fields being able to be recycled by the cavity mirrors. One obtain thus different types of "Optical Parametric Oscillators" or OPOs.

## 3.1    Singly resonant singly resonant OPO (SROPO)

In this case, either the signal or the idler wave amplified by the crystal is reinjected at its entrance. It is the exact analog of the laser and oscillation and emission of coherent light starts when the pump field $E_0$ reaches the value where the parametric gain, $g$, times the amplitude loss coefficient is equal to one. When the OPO is above its oscillation threshold, the energy transferred to the signal and idler waves is taken on the pump, whose amplitude decreases in the crystal which amounts to reducing the parametric gain with respect to its threshold value (of the order of a Watt). We have here a saturation of the gain which is analog to the one occurring in lasers but of a different physical origin. As for a laser, the steady state is reached when the saturated gain is equal to the losses. The output power in the signal wave increases rapidly as the pump intensity is increased. A simple model of SROPO (Brunner and Paul, 1977; Rosencher and Fabre, 2002) based on equations 1 shows that in steady state, a conversion efficiency of 100 % from pump to signal and idler can be obtained. When the cavity finesse is large (typically over 50), this steady state is predicted theoretically for a pump intensity equal to $\pi^2/4 \approx 2.4$ times the threshold intensity. Thus, despite the low value of the nonlinear effects, frequency conversion can be very significant if the signal field is recycled inside the cavity. If the pump intensity is larger than this value, reconversion through sum frequency of the signal (see Ch. 14) and idler fields occurs and the conversion efficiency is lowered. The maximal conversion obtained experimentally is 93% (Bosenberg et al., 1996).

## 3.2   Doubly resonant OPO (DROPO)

In a DROPO, the signal and idler fields are both recycled inside the cavity. Above, the oscillation threshold, both fields have a non zero value at the entrance of the crystal and solution 5 cannot be used. If the variation of the fields over the crystal length $L$ is small, equations 1 can be solved to the first order (still in the case $\Delta k = 0$):

$$E_1(L) = E_1(0) + i\frac{\omega_1}{2n_1 c}\chi^{(2)}LE_0(0)E_2^*(0) \tag{7}$$

$$E_2(L) = E_2(0) + i\frac{\omega_2}{2n_2 c}\chi^{(2)}LE_0(0)E_1^*(0) \tag{8}$$

the second term being small with respect to the first one. The variation of the signal (resp. idler) field is thus proportional to the amplitude of the idler (resp. signal) field. There is thus a crossed amplification term which is very characteristic of parametric interaction. In the case of a doubly resonant OPO in a high finesse cavity, the steady state behavior fixes a value for the pump intensity at the center of the crystal which is independent of the input pump power (Fabre et al., 1997). This is analogous to the gain clamping which occurs in laser above threshold. The energy conversion can also be very efficient in this case : four times above threshold, a 100% efficiency is predicted in the steady state regime by a simple model based on equations 1 (Byer, 1975) while an efficiency of 81% was observed experimentally (Breitenbach et al., 1995). The oscillation threshold for a DROPO (of the order of 500 mW) is significantly lowered with respect to a SROPO since both fields are enhanced by the cavity when it is close to the double resonance for both fields. With a moderate pump power, the DROPO oscillates only on restricted intervals centered on a discrete ensemble of values for the cavity length who verify the double resonance conditions (Eckart et al., 1991; Debuisschert et al., 1992). Thus, the output power is very dependent on the cavity length and a stable output is ensured only with the use of active electronics which control the cavity length.

## 3.3    Other types of OPOs

The SROPO and the DROPO are the two main categories of OPOs used today. Other configurations less common can also be useful

- the TROPO in which all three fields resonate inside the cavity. The threshold (of the order of mW) is lowered with respect to the DROPO ( Brunner and Paul, 1977; Debuisschert et al., 1992). However, the triple resonance condition for the cavity is harder to reach. In some conditions of operation, this device can be bistable, or even instable or chaotic ( Drummond et al., 1980; Lugiato et al., 1988; Pettiaux et al., 1989).
- The pump enhanced singly resonant OPO. It consists in an OPO in which the cavity is resonant for the signal and pump beams (Schiller et al., 1999). The basic properties of this device are the same as the SROPO's but with a lower threshold.
- The intracacity OPO consists in an OPO resonant for the signal beam and inserted within the pump laser cavity (Falk, 1971). This makes for a simple and compact generator of coherent light which benefits from the enhancement of the pump by the laser cavity. This cavity can be formed of totally reflecting mirrors for the pump since it is not used outside of the cavity making the cavity of very high finesse. There is however a complex nonlinear coupling between the laser and the OPO operations. The system emits not only the signal beam, with the same coherence properties as a laser beam, but also a second coherent beam, on the idler mode (which justifies its name). One can deduct from 3 that in the absence of absorption in the cavity, the output powers for these two beams are in the ratio of the frequencies $\omega_1$ and $\omega_2$.

## 4.    Conclusion

We have shown the different aspects of Optical Parametric Oscillators which explain the present interest for these sources, in fundamental as well as in applied physics. The very rapid development of compact, not power demanding sources including the pump laser and the OPO, should lead to an even wider use of such sources, in particular for industrial or medical applications.

T. Coudreau and A. Maître are also at the Pôle Matériaux et Phénomènes Quantiques FR CNRS 2437.

## References

Armstrong, J., Bloembergen, N., Ducuing, J., Pershan, P., 1962, Phys. Rev. **127**, 1918

Akhmanov, S., Khokhlov, R., 1962, Sov. Phys. Jetp **43**, 351

Akhmanov, S., Kovrigin, A., Kolosov, A., Piskarskas, A., Fadeev, V., Khokhlov, R., 1966, Sov. Phys. Jetp**3**, 241

Bosenberg, W., Drobshoff, A., J., Alexander, L., Myers, R., Byer, 1996, Opt. Lett. **21**, 1336

Breitenbach, G., Schiller, S., Mlynek, J., 1995, J. Opt. Soc. Am. **B12**, 2095

Brunner, W., Paul, H., 1977, *Theory Of Optical Parametric Amplification And Oscillation*, Progress In Optics**15**, 1

Byer, R., 1975, *Treatise In Quantum Electronics*, Academic Press, H. Rabin, C. Tang Eds, 587

Special Issue "Oscillateurs Paramétriques Optiques : Fondements et Applications", Comptes-Rendus Acad. Sci. **IV-1** Juillet 2000

Debuisschert, T., Sizmann, A., Giacobino, E., Fabre, C., 1993, J. Opt. Soc. Am. **B10**, 1668

Drummond, P., Mcneil, K., Walls, D., 1980, Optica Acta **27**, 321

Eckart, R., Nabors, C., Kosclovsky, W., Byer, R., 1991, J. Opt. Soc. Am. **B8**, 646

Fabre, C., Cohadon, P. F., Schwob, C., 1997, Quantum Semiclass. Opt. **9** 165

Falk, J. , Yarborough, J. M., Ammann, E. O., 1971, IEEE J. Quantum Electron. **QE-7**, 359

Giordmaine, J., Miller, R., 1965, Phys. Rev. Lett. **14** ,973

Special Issue of the J. Europ. Opt. Soc. **B9**, 131-295, 1997

Special Issue of J. Opt. Soc. Am. **B10**, Sept. and Nov. 1993

Kingston, R., 1962, Proc. Ire **50**, 472

Kroll, N., 1962, Phys. Rev. **127**, 1207

Rosencher, E. , Fabre, C., 2002, J. Opt. Soc Am. **19**, 1107

Lugiato, L., Oldano, C., Fabre, C., Giacobino, E., Horowicz, R., 1988, Nuovo Cimento **10** 959

Pettiaux, N., Ruo-Ding Li, Mandel, P., 1989, Optics Commun. **72**, 256

Schiller, S., Schneider, K., Mlynek, J., 1999, J. Opt. Soc. Am. **B16**, 1512

Shen, Y. R., 1984, *Nonlinear Optics*, Wiley

Smith, R., Geusic, J., Levinstein, H., Rubin, J., Singh, S., Van Uitert, L., 1968, Appl. Phys. Lett **123**, 308

# AN INTRODUCTION TO QUANTUM OPTICS

Thomas Coudreau, Claude Fabre, Serge Reynaud

*Laboratoire Kastler Brossel,*
*CNRS UMR 8552 et Université Pierre et Marie Curie*
*Case 74, UPMC, 4 Place Jussieu 75252 PARIS cedex 05*
coudreau@spectro.jussieu.fr

**Abstract**     The statistical properties of the electromagnetic field find their origin in its quantum nature. While most experiments can be interpreted relying on classical electrodynamics, in the past thirty years, many experiments need a quantum description of the electromagnetic field. This gives rises to distinct statistical properties.

**Keywords:**     quantum optics, one photon state, photon statistics

## 1.     Introduction

With the advent of extremely large aperture telescopes, there is a growing interest in the statistical properties of the field emitted by astronomical sources (Dravins, 2001). The goal is to obtain important physical information concerning the source by looking at the statistical characteristics of the light it emits. This domain was pioneered by two radio astronomers, Hanbury Brown and Twiss (1956) who measured the intensity correlation function $\langle I(t)I(t+\tau)\rangle$ of light. As this quantity is sensitive to the "granularity" of light, *i.e.* its photonic content, they opened a new and very fruitful way of investigating the quantum properties of light.

Statistical properties of light are described within the framework of quantum optics which is based on a quantized description of the electromagnetic field. In section 21.2 we will depict specific experiments which have been performed to show that a quantum description is necessary in some cases. We will describe in Section 21.3 the standard tools for the analysis of the statistical properties of light and give the results obtained for a number of sources.

*R. Foy and F. C. Foy (eds.), Optics in Astrophysics, 351–358.*

## 2.    Historical background

### 2.1    Wave approach of light

At the beginning of the 19th century, the particle description favored by Newton (1952) was replaced by a wave description : Young performed interference experiments which he interpreted as the capacity of light to be added or subtracted much like wave at the surface of water. Later on, Fresnel developed a mathematical theory of interferences and diffraction where light was seen as a scalar wave. The experiments on interferences with polarized light made by Arago made it necessary for Fresnel to revise his theory and make it a transverse vectorial wave. Maxwell showed that light was actually a wave of electric and magnetic fields in the form $\mathbf{E}\cos(\omega t - \mathbf{k}.\mathbf{r})$ with $\mathbf{E} \perp \mathbf{k}$. This electromagnetic description of light was gradually able to give a quantitative account of all the optical phenomena known at this time, with the exception of the thermal radiation and the photoelectric effect. The classical description of the radiation by a perfectly emitting body placed at thermal equilibrium led to divergences both at low and high frequency while the discharge of an electrometer occurred only with ultraviolet light, not with lower frequency light. These phenomena led to the famous introduction of the quanta by Planck in 1900 and to the advent of quantum mechanics.

### 2.2    Quantum description of light : the wave/particle duality

In 1905, Einstein wrote a very famous article (Einstein, 1965) in which he explained that light was made of unbreakable quanta of energy the basic unit being $E = h\nu$ where $E$ is the energy of one quantum, $h$ is the Planck constant and $\nu$ the light frequency. Millikan very carefully tested this experimentally and his results were in agreement with the theory. A few years later, the Compton effect was discovered. As a beam of high energy light is shone on an ensemble of electrons, diffusion might occur and light with a different wavelength is emitted. This can be seen as a collision between a quantum of light and an electron. This collision respected the conservation of the momentum. Thus, the photon, since it is its name, gained a given momentum $\hbar k$ where $\hbar = h/2\pi$ and $k$ is the wave vector. The photon now had a definite energy and momentum. Its existence was further attested by photomultipliers. When faint light is shone on a photomultiplier, the signal obtained consists in pulses well consistent with the idea of photons. But light has also a wave character which must be incorporated in the new description of light in terms of photons. In 1909, Taylor performed an interference experiment with a very faint source (equivalent to "a candle burning at a distance slightly exceeding a mile"). He recorded the interference pattern using a photographic plate. With a very short exposure time, a few bright spots were observed but with no clear behavior. As the exposure time was increased, going up to three months, the interference pattern became clearly visible. It must be noted that the source was so weak that the mean electromagnetic energy inside the experimental set-up was well below one quantum of light. This led Dirac to say that "each photon then interferes only with itself". So light seemed to have a strange double character which is usually labeled as its wave/particle duality.

In the 1920s, quantum mechanics was developed and led to a complete theory of matter including energy levels and atomic collisions. It was then possible to make the complete and *ab initio* calculation of the behavior of an atom interacting with a *classical* electric field. This so called semi classical theory explained the energy transfers by units of $h\nu$ corresponding to the separation of energy levels as well as the momentum transfer in units of $\hbar k$. The semi classical theory could explain all the previous results (black body radiation, photoeletric effect Compton effect) without referring to photons. So it seemed that the concept of photon was not so necessary : for example, the previously mentioned "observation" of the photon in the photomultiplier could also be seen as the sudden "quantum jump" of the electron of the photocathode into a continuum state induced by the classical incident wave.

In a parallel way, the ideas and methods of quantum mechanics were applied to the electromagnetic field. The electromagnetic field was shown to be quantized with properties analogous to the harmonic oscillator. Thus the energy of the field would grow in units of $h\nu$. However, in the absence of excitation, the energy is non zero and given by $E_{vacuum} = \frac{h\nu}{2}$ : even in vacuum, light has a non zero energy! This means that even if the electromagnetic field is null on average, its square is not. Through a quadratic Stark effect, vacuum can thus be coupled to atoms. This effect, known as the Lamb shift, was predicted (Bethe, 1947) and verified experimentally on the hydrogen energy levels with great precision as early as the 1950s (Lamb and Retherford, 1947). This experiment showed that indeed a quantum description of light was necessary to account for the detailed properties of the atomic levels including their lifetimes. However, all the experimental results in optics could be at that time explained using classical fields interacting with quantum atoms.

There was thus the need for optical experiments showing the flaws of classical electrodynamics. An important difference between a wave and a particle is with respect to a beam splitter : a wave can be split in two while a photon can not. An intensity correlation measurement between the two output ports of the beamsplitter is a good test as a wave would give a non zero correlation while a particle would show no correlation, the particle going either in one arm or the other. However, when one takes an attenuated source, such as the one used by Taylor, it contains single photon pulses but also a (small) fraction of two photon pulses. These two photon pulses can give correlations and thus make the experiment non conclusive.

To obtain single photon pulses, one can use the emission by a single dipole as shown below in section 21.3.1. The experiment was performed in 1977 by Kimble, Dagenais and Mandel (Kimble et al., 1977). They showed that single atoms from an atomic beam emitted light which, at small time scales, exhibited a zero correlation function. This result can not be explained through a semiclassical theory and requests a quantum description of light.

In 1986, Grangier et al. used an atomic cascade to produce a one photon state. The atomic cascade produces two photons which can be separated using wavelength filtering. The presence of one photon indicates with very high probability the presence of the second one. Thus by a trigger on one photon, one generates a true one photon state (not a mixture with a zero photon state). When this state is sent in a Mach-Zehnder interferometer, ones obtains interferences as the path difference is changed. One thus obtains interferences, sensitive to the path difference between the two arms even in situations when light has been shown to "follow" one and only one of these paths.

The previous experiments are a manifestation of the wave/particle duality of light which manifests as a particle in some experiments (the correlation after the beam splitter) and as a wave in others (the interferometer). But there exists experiments where both "classical" descriptions fail : this is the case of the experiment by Hong et al. (Hong et al., 1987). In this experiment, a non-linear material is illuminated with a laser beam and generates pair of photons which can be separated by a polarizing beamsplitter. The two photons are then recombined on a standard beamsplitter. As the position of the beamsplitter is varied, one expects for classical waves an interference pattern which is not seen experimentally. Thus the crystal is not a source of classical light. Furthermore, when the two path lengths are identical, one observes an absence of correlation between the two output ports, meaning that the two photon always exit by the same port, which is not predicted if one considers light as a classical particle.

All these experiments can be explained simply using quantum electrodynamics. There has been to date no experiment to violate this theory. We will now see what are its predictions in various cases.

## 3.    Statistical properties of the electromagnetic field

As different sources are considered, the statistical properties of the emitted field changes. A random variable $x$ is usually characterized by its probability density distribution function, $P(x)$. This function allows for the definition of the various statistical moments such as the average,

$$\bar{x} = \int_{-\infty}^{+\infty} x P(x) dx \tag{1}$$

or the variance,

$$\Delta x = \int_{-\infty}^{+\infty} (x - \bar{x})^2 P(x) dx \qquad (2)$$

Another indicator is also largely used for time varying statistical variables, the autocorrelation function

$$f(t, \tau) = \langle x(t)x(t + \tau) \rangle \qquad (3)$$

I will present here the properties of various sources when the random variable considered is the field intensity. In this case, one has access to the mean and variance via a simple photodetector. The autocorrelation function can be interpreted as the probability of detecting one photon at time $t + \tau$ when one photon has been detected at time $t$. The measurement is done using a pair of photodetectors in a start stop arrangement (Kimble et al., 1977). The system is usually considered stationary so that the autocorrelation function, which is denoted $g^{(2)}$ depends only on $\tau$ and is defined by :

$$g^{(2)}(\tau) = \frac{\langle I(t)I(t + \tau) \rangle}{\sqrt{\langle I(t) \rangle \langle I(t + \tau) \rangle}} \qquad (4)$$

where $\langle \, \rangle$ denotes a statistical average.

We provide here only a brief introduction to the statistical properties of the electromagnetic field. The interested reader will find further reading in *e.g* Mandel and Wolf (1995) and Scully and Zubairy (1997).

## 3.1 Single dipole

One considers here a single dipole (atom, ion, ...) emitting light via spontaneous emission between two energy levels. The atom is brought to the excited state by a given pumping mechanism (collision, deexcitation from an upper level ...). In this case, the mean intensity and the variance depend strongly on the pumping rate. However, the intensity autocorrelation function exhibits a striking feature. For $\tau$ going to zero, $g^{(2)}(\tau)$ also goes to zero. This can be easily explained : the dipole which emits a photon has to be reexcited by the pumping mechanism to be able to emit a second photon. The time scale of the emission is given by the linewidth of the transition, $\Gamma$. Thus between two successive emissions, a typical time $\tau = \Gamma^{-1}$ elapses and it is not possible for the source to emit simultaneously two photons.

## 3.2 Spontaneous emission by an incoherent ensemble of dipoles

An incoherent ensemble of dipoles constitutes the usual source in astrophysics (or earth). These dipoles interact weakly with each other so that no phase relation exists between the waves emitted in different regions of the source. When the medium is dilute, the probability for stimulated emission is very small and spontaneous emission dominates. This kind of radiation is called thermal or chaotic light. In this case, the probability density distribution function in terms of photon numbers can be written

$$P(m) = \frac{\bar{n}^m}{(1 + \bar{n})^{1+m}} \tag{5}$$

where the average photon number $\bar{n}$ depends on the pumping rate. The variance can be readily calculated

$$\Delta n^2 = \bar{n}^2 + \bar{n} \tag{6}$$

The characteristic property of the autocorrelation function was first demonstrated experimentally by Hanbury Brown and Twiss (1956) using two radiotelescopes. As $\tau$ goes to zero, $g^{(2)}(\tau)$ goes to two.

## 3.3     Laser field

A laser consists in a medium where stimulated emission dominates over spontaneous emission placed inside an optical cavity which recycles the optical field. Above threshold, the photon number probability density distribution is poissonian, that means that the photon arrival time are a random variable. The probability of obtaining $m$ photons during a given time interval is thus

$$P(m) = e^{-\bar{n}} \frac{\bar{n}^m}{m!} \tag{7}$$

The intensity obeys a gaussian statistics so that its variance is equal to the mean intensity

$$\Delta n^2 = \bar{n} \tag{8}$$

Finally, the field autocorrelation function is constant equal to 1.

## 3.4     Fock state

A Fock state is a state containing a fixed number of photons, $N$. These states are very hard to produce experimentally for $N \geq 2$. Their photon number probability density distribution $P(m)$ is zero everywhere except for $m = N$, their variance is equal to zero since the intensity is perfectly determined. Finally, the field autocorrelation function is constant

$$g^{(2)}(\tau) = 1 - \frac{1}{\bar{n}} \tag{9}$$

The different behaviors for the field autocorrelation function are summarized on Fig. 1.

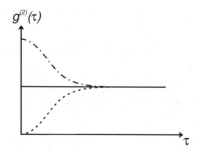

*Figure 1.* $g^{(2)}(\tau)$ for a single dipole (dashed line), thermal state (dash dotted line) and laser field (continuous line).

## 4. Conclusion

We have shown in this chapter how some experiments made it necessary in some cases to use a quantum description of light instead of the standard semi-classical theory where only the atomic part is quantized. A brief description of different fields in terms of their statistical properties was also given. This description makes it possible to discriminate between the different sources using the intensity autocorrelation function $g^{(2)}(\tau)$.

T. Coudreau is also at the Pôle Matériaux et Phénomènes Quantiques FR CNRS 2437.

## References

Bethe, H.A., 1947, *The Electromagnetic Shift of Energy Level*, Phys. Rev. **72**, 339

Dravins, D., 2001, *Quantum-Optical Signatures of Stimulated Emission* in T. Gull, S. Johansson, K. Davidson, eds., Astron. Soc. Pacific Conference series **242**, 339

Einstein, A., 1905, *Concerning an Heuristic Point of View Toward the Emission and Transformation of Light*, Ann. Phys. **17**, 132, English translation Am. J. Phys. **33**, 367 (1965)

Grangier, P., Roger, A., Aspect, A., 1986, *Experimental Evidence for a Photon Anticorrelation Effect on a Beam Splitter: a New Light on Single-Photon Interference*, Europhys. Lett. **1**, 173

R. Hanbury Brown, R., Twiss, R.Q., 1956, *Correlation between photons in two coherent beams of light*, Nature **177**, 27

Hong, C.K., Ou, Z.Y., Mandel, L., 1987, *Measurement of subpicosecond time intervals between two photons by interference*, Phys. Rev. Lett. **59**, 2044

Kimble, H.J., Dagenais, M., Mandel, L., 1977, *Photon Antibunching in Resonance Fluorescence*, Phys. Rev. Lett. **39**, 691

Lamb, W.E., Retherford, R.C., 1947, *Fine Structure of the Hydrogen Atom by a Microwave Method*, Phys. Rev. **72**, 241

Mandel, L., Wolf, P., 1995, *Optical Coherence and Quantum Optics*, Cambridge University Press

Newton, 1952, *Opticks*, Dover

Scully, M.O., Zubairy, M.S., 1997, Quantum Optics, Cambridge University Press

# ATOM INTERFEROMETRY

Arnaud Landragin, Patrick Cheinet, Florence Leduc
*BNM-SYRTE, UMR8630*
*Observatoire de Paris*
*61, avenue de l'Observatoire*
*75014 PARIS*
*FRANCE*
arnaud.landragin@obspm.fr


Philippe Bouyer
*LCFIO, UMR 8510*
*Institut d'Optique, Centre scientifique d'Orsay Bat 503*
*91403 ORSAY Cedex*
*France*

**Abstract**    The techniques of atom cooling combined with the atom interferometry make possible the realization of very sensitive and accurate inertial sensors like gyroscopes or accelerometers. Below earth-base developments, the use of these techniques in space, as proposed in the HYPER project (ass.stud), should provide extremely-high sensitivity for research in fundamental physics.

**Keywords:**    atom interferometry, laser cooling, Raman transition

## Introduction

Inertial sensors are useful devices in both science and industry. Higher precision sensors could find practical scientific applications in the areas of general relativity (Chow et al., 1985), geodesy and geology. Important applications of such devices occur also in the field of navigation, surveying and analysis of earth structures. Matter-wave interferometry has recently shown its potential to be an extremely sensitive probe for inertial forces (Clauser, 1988). First, neutron interferometers have been used to measure the Earth rotation (Colella et al., 1975) and the acceleration due to gravity (Werner et al., 1979) in the end of the seventies. In 1991, atom interference techniques have been used in

*R. Foy and F. C. Foy (eds.), Optics in Astrophysics, 359–366.*

proof-of-principle work to measure rotations (Riehle et al., 1991) and accelerations (Kasevich and Chu, 1992). In the following years, many theoretical and experimental works have been performed to investigate this new kind of inertial sensors (*Atom Interferometry*, 1997). Some of the recent works have shown very promising results leading to a sensitivity comparable to these of other kinds of sensors, as well for rotation (Gustavson et al., 2000) as for acceleration (Peters et al., 2001).

# 1. Inertial sensors based on atom interferometer: basic principle

We present here a summary of recent work with light-pulse interferometer based inertial sensors. We first outline the general principles of operation of light-pulse interferometers. This atomic interferometer (Bordé et al., 1992; Bordé et al., 1989) uses two-photon velocity selective Raman transitions (Kasevich et al., 1991), to manipulate atoms while keeping them in long-lived ground states.

## 1.1 Principle of a light pulse matter-wave interferometer

Light-pulse interferometers work on the principle that, when an atom absorbs or emits a photon, momentum must be conserved between the atom and the light field. Consequently, an atom which emits (absorbs) a photon of momentum $\hbar k_{eff}$ will receive a momentum impulse of $\delta p = -\hbar k_{eff}(+\hbar k_{eff})$. When a resonant traveling wave is used to excite the atom, the internal state of the atom becomes correlated with its momentum: an atom in its ground state $|1\rangle$ with momentum $p$ (labelled $|1, p\rangle$ ) is coupled to an excited state $|2\rangle$ of momentum $p + \hbar k_{eff}$ (labelled $|2, p + \hbar k_{eff}\rangle$) (Bordé et al., 1992). A precise control of the light-pulse duration allows a complete transfer from one state (for example $|1, p\rangle$) to the other ($|2, p + \hbar k_{eff}\rangle$) in the case of a $\pi$ pulse and a 50/50 splitting between the two states in the case of a $\pi/2$ pulse (half the duration of a $\pi$ pulse). This is analogous to, for example, a polarizing beam splitter (PBS) in optics, where each output port of the PBS (i.e. the photon momentum) is correlated to the laser polarization (i.e. the photon state). In the optical case, a precise control of the input beam polarization allows for controlling the balance between the output ports. In the case of atoms, a precise control of the light-pulse duration plays the role of the polarization control.

We use a $\pi/2 - \pi - \pi/2$ pulse sequence to coherently divide, deflect and finally recombine an atomic wavepacket. The first $\pi/2$ pulse excites an atom initially in the $|1, p\rangle$ state into a coherent superposition of states $|1, p\rangle$ and $|2, p + \hbar k_{eff}\rangle$. If state $|2\rangle$ is stable against spontaneous decay, the two parts of the wavepacket will drift apart by a distance $\hbar k T/m$ in time $T$. Each partial wavepacket is redirected by a $\pi$ pulse which induces the transitions

$|1, p\rangle \rightarrow |2, p + \hbar k_{eff}\rangle$ and $|2, p + \hbar k_{eff}\rangle \rightarrow |1, p\rangle$. After another interval $T$ the two partial wavepackets once again overlap. A final pulse causes the two wavepackets to interfere. The interference is detected, for example, by measuring the number of atoms in the $|2\rangle$ state. We obtain a large wavepacket separation by using laser cooled atoms and velocity selective stimulated Raman transitions (Kasevich et al., 1991). A very important point of these light pulse interferometers is their intrinsic accuracy thanks to the knowledge of the light frequency which defines the scaling factor of the interferometers.

## 1.2 Application to earth-based inertial sensors

Inertial forces manifest themselves by changing the relative phase of the de Broglie matter waves with respect to the phase of the driving light field, which is anchored to the local reference frame. The physical manifestation of the phase shift is a change in the probability to find the atoms in, for example, the state $|2\rangle$, after the interferometer pulse sequence as described above. A complete analytic treatment of wave packet phase shifts in the case of acceleration, gradient of acceleration and rotation together (Antoine and Bordé, 2003) can be realized with a formalism similar to ABCD matrices in light optics.

If the three light pulses of the pulse sequence are only separated in time, and not separated in space (*i.e.* if the velocity of the atoms is parallel to the laser beams), the interferometer is in a gravimeter or accelerometer configuration. In a uniformly accelerating frame with the atoms, the frequency of the driving

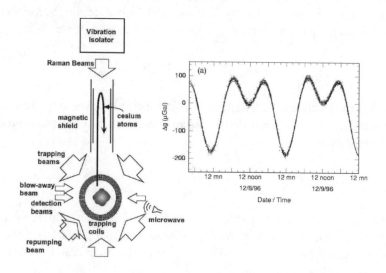

*Figure 1.*   Principle of the atom-fountain-based gravimeter achieved in S. Chu group at Stanford. The right figure shows a two days recording of the variation of gravity. The accuracy enables to resolve ocean loading effects.

laser changes linearly with time at the rate of $-k_{eff}.a$. The phase shift arises from the interaction between the light and the atoms (*Atom Interferometry*, 1997; Antoine and Bordé, 2003) and can be written:

$$\Delta\phi = \phi_1(t_1) - 2\phi_2(t_2) + \phi_3(t_3) \tag{1}$$

where $\phi_i(t_i)$ is the phase of light pulse $i$ at time $t_i$ relative to the atoms. If the laser beams are vertical, the gravitationally induced chirp can be written:

$$\Delta\phi = -k_{eff}.gT^2 \tag{2}$$

It should be noted that this phase shift does not depend on the atomic initial velocity or on the mass of the particle. Recently, an atomic gravimeter with accuracy comparable to the best corner cube has been achieved (Fig. 1, Peters et al., 2001). The main limitation of this kind of gravimeter on earth is due to spurious acceleration from the reference platform. Measuring gravity gradient may allow to overcome this problem. Indeed, using the same reference platform for two independent gravimeters enables to extract gravity fluctuations. A such apparatus (McGuirk et al., 2002), using two gravimeters as described above but sharing the same light pulses, has shown a sensitivity of $3.10^{-8}s^{-2}.Hz^{-1/2}$ and has a potential on earth up to $10^{-9}s^{-2}.Hz^{-1/2}$.

In the case of a space separation of the laser beams (i.e. if the atomic velocity is perpendicular to the direction of the laser beams), the interferometer is in a Mach-Zehnder configuration. Then, the interferometer is also sensitive to rotations, as in the Sagnac geometry (Sagnac, 1913) for light interferometers. For a Sagnac loop enclosing area $A$, a rotation $\Omega$ produces a phase shift:

$$\Delta\phi_{Sagnac} = \frac{4\pi}{\lambda v_L}\Omega.A \tag{3}$$

where $\lambda$ is the particle wavelength and $v_L$ its longitudinal velocity. The area $A$ of the interferometer depends on the distance between two pulses $L$ and on the recoil velocity $V_T = \hbar k/m$:

$$A = L^2\frac{V_T}{V_L} \tag{4}$$

Thanks to the use of massive particle, atomic interferometers can achieve a very high sensitivity. An atomic gyroscope (Gustavson et al., 2000) using thermal caesium atomic beams ($v_L \sim 300m.s^{-1}$) and with an overall interferometer length of $2m$ has demonstrated a sensitivity of $6.10^{-10}rad.s^{-1}.Hz^{-1/2}$. The apparatus consists of a double interferometer using two counter-propagating sources of atoms, sharing the same lasers. The use of the two signals enables to discriminate between rotation and acceleration (Fig. 2).

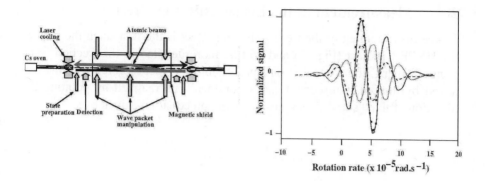

*Figure 2.* Diagram of the atomic Sagnac interferometer at Yale (Gustavson et al., 2000). Individual signals from the outputs of the two interferometers (gray lines), and difference of the two signals corresponding to a pure rotation signal (black line) versus rotation rate.

## 1.3 Cold atom gyroscope and application in space: HYPER project

Atomic Sagnac interferometers should benefit from cold atomic sources in two ways. First, reducing the velocity dispersion of the atomic sample (few millimeters per second) enables to drastically reduce the longitudinal velocity of the atoms $v_L$ (few $cm.s^{-1}$) and enhances in the same way the enclosed area and the sensitivity for a constant length. Second, the accuracy and the knowledge of the scaling factor of this gyroscope depends directly of the longitudinal velocity of the atoms and can be better controlled with cold atomic sources than with thermal beams, as it has already been demonstrated with atomic clocks ( Marion et al., 2003). Presently first prototypes based on atomic fountains of laser cooled atoms are under construction in a joint project of SYRTE and IOTA in Paris (Yver-Leduc et al., 2003) as well as at the IQO in Hannover.

Laser cooling can efficiently reduce the velocity of the atoms but cannot circumvent the acceleration due to gravity. On the ground the 1-g gravity level sets clear limitations to the ultimate sensitivities. The HYPER project (Hyper precision cold atom interferometry in space) will follow precisely this line and will benefit from the space environment, which enables very long interaction time (few seconds) and low spurious vibrational level. The sensitivity of the atomic interferometer can achieve few $10^{-12} rad.s^{-1}.Hz^{-1}$ to rotation and $10^{-12} g.Hz^{-1}$ to acceleration. This very sensitive and accurate apparatus offers the possibility of different tests of fundamental physics (ass.stud). It can realize tests of General Relativity by measuring the structure of the Lense-Thirring effect (magnitude and sign) or testing the equivalence principle on individual atoms. It can also be used to determine the fine structure constant by measuring the ratio of Planck's constant to the atomic mass.

## 2.     Measurement of the Lense-Thirring effect

The measurement of the Lense-Thirring effect is the first scientific goal of
the HYPER project (Fig. 3) and will be more detailed in this section. The
Lense-Thirring effect consists of a precession of a local reference frame (realized by inertial gyroscopes) and a non-local one realized by pointing the
direction of fixed stars. This Lense-Thirring precession is given by:

$$\Omega_{LT} = \frac{GI}{c^2} \frac{3(\omega.r)r - \omega r^2}{r^5} \tag{5}$$

The high sensitivity of atomic Sagnac interferometers to rotation rates will
enable HYPER to measure the modulation of the precession due to the Lense-
Thirring effect while the satellite orbits around the Earth. In a Sun-synchronous,
circular orbit at 700 km altitude, HYPER will detect how the direction of the
Earth's drag varies over the course of the near-polar orbit as a function of the
latitudinal position $\theta$:

$$\begin{pmatrix} \Omega_x \\ \Omega_y \end{pmatrix} \propto \frac{3}{2} \begin{pmatrix} sin(2\theta) \\ cos(2\theta) - \frac{1}{3} \end{pmatrix} \tag{6}$$

where $e_x$ and $e_y$ define the orbital plane with $\vec{e_y} \| \vec{J}$ the inertial momentum of
the Earth and $\theta \equiv arcos(r.e_x)$.

HYPER carries two atomic Sagnac interferometers, each of them is sensitive
to rotations around one particular axis, and a telescope used as highly sensitive star-tracker ($10^{-9}$ rad in the 0.3 to 3 Hz bandwidth). The two units will
measure the vector components of the gravitomagnetic rotation rate along the
two axes perpendicular to the telescope pointing direction which is directed
to a guide star. The drag variation written above describes the situation for
a telescope pointing in the direction perpendicular to the orbital plane of the
satellite. The orbit, however, changes its orientation over the course of a year
which has to be compensated by a rotation of the satellite to track continuously the guide star. Consequently the pointing of the telescope is not always
directed parallel to the normal of the orbital plane. According to the equation,
the rotation rate signal will oscillate at twice the frequency of the satellites
revolution around the Earth. The modulated signals have the same amplitude
($3.75x10^{-14}$ rad.s$^{-1}$) on the two axes but are in quadrature. The resolution
of the atomic Sagnac units (ASU) is about $3.10^{-12}$ rad.s$^{-1}$ for a drift time of
about 3s. Repeating this measurement every 3 seconds, each ASU will reach
after one orbit of 90 minutes the level of $7.10^{-14}$ rad.s$^{-1}$, in the course of one
year the level of $2.10^{-15}$ rad.s$^{-1}$, i.e. a tenth of the expected effect.

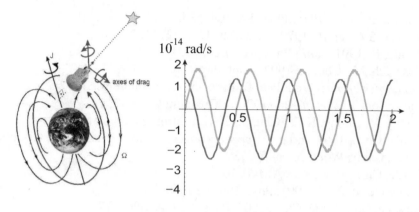

*Figure 3.* Diagram of the measurement of the Lense-Thirring effect. The black lines visualize the vector field of the Earth's drag $\Omega_{LT}$. The sensitive axes of the two ASUs are perpendicular to the pointing of the telescope. The direction of the Earth's drag varies over the course of the orbit showing the same structure as the field of a magnetic dipole. Due to this formal similarity the Lense-Thirring effect is also called gravitomagnetic effect. The modulation of the rotation rate $\Omega_{LT}$ due to Earth's gravitomagnetism as sensed by the two orthogonal ASUs in the orbit around the Earth appears at twice the orbit frequency.

## 3. Summary

Previous experiments measuring the gravitational acceleration of Earth and its gradient or rotations have been demonstrated to be very promising. Sensitivities better than 1 nrad.s$^{-1}$.Hz$^{-1/2}$ for rotation measurements and $4.10^{-9}$ g.Hz$^{-1/2}$ for a gravity measurement have already been obtained. The sensitivity of matter-wave interferometers for rotations and accelerations increases with the measurement time and can therefore be dramatically enhanced by reducing the atomic velocity. Moreover, the use of optical transitions to manipulate the atomic wave packets enables an intrinsic knowledge of the scaling factor of these inertial sensors, which is directly linked to the frequency of the transition. Therefore, combining cold atomic sources and Raman transition based atomic interferometers results in high sensitive and high accurate inertial sensors. Going to space will enhance the benefit of cold atoms by increasing the interaction time and opens up entirely new possibilities for research in fundamental physics or for inertial navigation with unprecedented precision.

## References

ESA-SCI(2000)10, July 2000.
Chow, W. *et al.*, 1985, Rev. Mod. Phys. **72**, 61
Clauser, J.F., 1988, Physica B **151**, 262

Colella, R. *et al.*, 1975, Phys. Rev. Lett. **34**, 1472

Werner, S.A., *et al.*, 1979, Phys. Rev. Lett. **42**, 1103

Riehle, F., 1991, *et al.*, Phys. Rev. Lett. **67**, 177

Kasevich, M., Chu, S., 1992, Appl. Phys. B **54**, 321

*Atom Interferometry*, ed. Paul R. Berman (Academic Press,1997).

Gustavson, T.L., *et al.*, 2000, Class. Quantum Grav. **17**, 1

Peters, A., Chung, K. Y., Chu, S., 2001, Metrologia **38**, 25

Bordé, Ch.J., 1992, in *Laser spectroscopy X*, ed. M. , Ducloy, E. , Giacobino, G. , Camy, World scientific, 239

Bordé, Ch.J., Phys. Rev. A. **140**, 10

Kasevich, M. *et al.*, 1991, Phys. Rev. Lett. **66**, 2297

Antoine, Ch., Bordé, Ch.J., 2003, Phys. Lett. A. **306**, 277

McGuirk, J.M. *et al.*, 2002, Phys. Rev. A **65**, 033605

Sagnac, M., 1913, Compt. Rend. Acad. Sci. **157**, 708

Marion, H. *et al.*, 2003, Phys. Rev. Lett. **90**, 150801

Yver-Leduc, F. *et al.*, 2003, J. Opt. B: Quantum Semiclass. Opt. **5**, 136

# AMPLITUDE AND PHASE DETECTION OF INCOHERENT SOURCE

Jean-Claude Lehureau

*Thales Research and Technology*
*92 Orsay, France*

jean-claude.lehureau@thalesgroup.com

**Abstract**     We compare homo-or heterodyne power and phase detection to the photon counting detection. The noise equivalent power of such a detector is equal to that of a
quantum limited photon counter. We show that interferometry between remote
sites can be processed by exchanging classical information. This process is well
known in radio astronomy and could apply to the optical domain. However,
vacuum fluctuation is a strong source of noise at high photon energy and limits
the performance of a non quantum interferometer compared to a conventional
optical interferometer.

**Keywords:**     coherent detection, incoherent source, thermal emission, Shottky noise, photon
counting, vacuum fluctuation, photon bunching, intensity interferometer, phase
interferometer, Hanbury Brown-Twiss

## 1.     Introduction

For a radio astronomer, a star is a source of noise; this noise can be detected and correlated from antenna to antenna in order to position precisely the
source. For optical observer, the star is a source of photons, the unique photon goes through a variety of optical paths and materialize on the focal plane
detector.

It is clear that the nature of the electromagnetic phenomena is the same for
optics and radio wave, the only experimental differences being that radiowave
photons are far below the spectral density of noise of actual detectors and that
the temperature of the source is such that each mode is statistically populated
by many photons in the radio wave domain whereas the probability of presence
of photons is very small in the optical domain.

Optical specialists used to distinguish between coherent and non coherent
waves because of the very large bandwidth of optical domain. However as

*R. Foy and F. C. Foy (eds.), Optics in Astrophysics, 367–373.*

electronic components are becoming faster, there is a growing domain of over-lap between optical and electronic bandwidth. As a matter of example, the modulation sideband of optical fiber components fills a significant part of infra red spectrum.

We investigate here whether coherent detection of optical waves issued from celestrial sources can be processed in a way similar to radio astronomy (Fig. 1).

## 2.    Heterodyne detection of optical wave

A radio wave detector is composed of :

- an antenna preamplifier

- a frequency shifter composed of local oscillator and multiplier

- an intermediate frequency amplifier (Fig. 1).

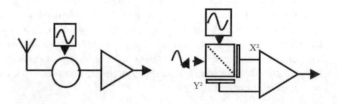

*Figure 1.*    Radiowave detector and Optical coherent detection.

In optical domain, preamplifier is no more an utopia and is in actual use in fiber communication. However quantum physics prohibits the noiseless cloning of photons: an amplifier must have a spectral density of noise greater than 1 photon/spatial mode ( a "spatial mode" corresponds to a geometrical extent of $\lambda^2/4$). Most likely, an optical heterodyne detector will be limited by the photon noise of the local oscillator and optical preamplifier will not increase the detectivity of the system.

Optical parametric oscillator (OPO, see 20) is the real equivalent to the radio frequency shifter; however OPO can be replaced by a simple addition of a local oscillator (e.g. laser) through a beam splitter. Multiplication takes place at the level of detectors. For sake of symmetry, detectors can be placed at both output of the beam splitter, the intermediate frequency is then the output of the differential amplifier.

The main source of noise of such a heterodyne detector is the photon noise that takes place at the splitting of the local oscillator. Quantum physicists see this noise as originating from vacuum fluctuation on the input arm. This gives directly the spectral density of noise at input: $h\nu/2$.

## 3. Use of heterodyne as a power detector

In term of quantum physics, the heterodyning of the signal is an amplitude and phase detector (Fig. 2). This raises a paradox similar to the EPR paradox. Because of the duality of particle and wave, detecting the electric field requires that no photon has ever been transmitted from the source to the detector (even if emission took place years ago!) and that the source shall be considered as an electromagnetic field emitter. The main practical consequence is that each sample of heterodyne signal is receiving signal and noise whereas photon detectors receive a stochastic sequence of noiseless events (photon or nil).

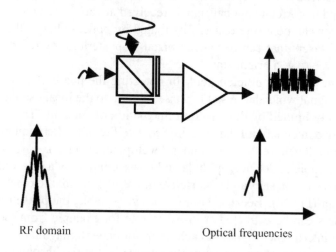

RF domain                    Optical frequencies

*Figure 2.* Mixing of a Laser oscillator with a spread spectrum Optical source

The electronic output signal is the sum of the photon noise of local oscillator (laser) and the modulation due to interference of the source with the local oscillator:

- Photon noise is stochastic phenomena and has a white spectrum over the electronic frequency domain.

- Interference corresponds to frequency shift of the optical spectrum towards the electronic frequency domain.

A noise power equivalent to one photon generates an interference signal which has an amplitude equals to twice the rms photon noise of the source. But as only the in-phase components of the source generates an interference with the local oscillator, the result is that the spectral Noise Equivalent Power of the heterodyne receiver is $h\nu$.

Finally, although heterodyne receiver is basically an amplitude-phase detector, its detectivity as a power receiver is similar to a quantum limited detector:

- The advantages of this type of detector are that even a noisy photodiode can easily be turned into a quantum limited detector and that the spectrum of the source can be analyzed precisely.

- The disadvantage is that the optical bandwidth that can be detected is equal to the output electronic bandwidth. As communication components are growing faster, this disadvantage tends to disappear.

## 4.     Use of heterodyne as a phase detector

Since a heterodyne receiver is an amplitude and phase detector, it could nicely be used to correlate optical signals received at various remote sites. The local oscillator can be a single laser distributed by optical fiber to the various sites or local lasers that can be synchronized "a posteriori" by reference to a common source (e.g. a bright star).

Heterodyne is a very efficient tool for detecting the phase of a "coherent" signal i.e. a signal which has a stable phase relation to the local oscillator. The detector is only limited by the quantum fluctuation of vacuum. This property is common use in coherent lidar. Satellite to satellite optical communications using laser as a local oscillator are under development (Fig. 3).

It is conceivable to detect amplitude and phase emitted by a celestial object at various observation sites and to correlate the results in order to create a huge interferometer (Fig. 3). Because laser can be very stable, the phase reference between lasers can be extracted at low data rate for example from the correlation of the interference signal of each laser with a high magnitude star. The main difference with communication case above is that the absolute phase of the thermal emission is meaningless; only the phase correlation from site to site can be exploited. Emission of thermal source is governed by the Planck' law. This law states that the probability of photon population of a mode is :

$$p = 1/(exp(h\nu/kT - 1))  \tag{1}$$

A mode is defined by its polarization, a geometrical extent equal to $\lambda^2/4$ and time-bandwidth equal to unity. The population of mode is $\approx 1\%$ near the emission peak:

$$\lambda T = 3000 \, \mu m.K  \tag{2}$$

In astronomical observation, the probability of population is much smaller because the angular size of the thermal source is much smaller than the Airy spot of the instrument. Therefore we will consider the correlation of signals which are much smaller than the noise. Let us normalize the pump to unity. Each output gives a signal:

$$s_1 = \sqrt{p} \times cos(\phi + \sigma_1) \qquad s_2 = \sqrt{p} \times cos(\phi + \sigma_2)  \tag{3}$$

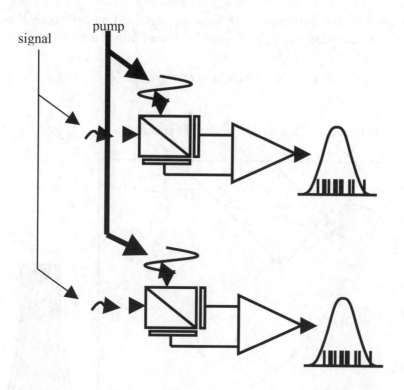

*Figure 3.* Detection of thermal source through multiple point heterodyne

where p is the population of a mode $\phi$ is the random phase of the thermal emission and $\sigma_1 - \sigma_2$ is the optical path difference between two sites. The only useful information on optical path is given by the correlation of the two outputs:

$$S = s_1 \times s_2 = p \times cos(\sigma_1 - \sigma_2) \qquad (4)$$

The noise power is normalized to unity; one needs $N \approx 1/p^2$ samples to obtain a signal equal to the noise

## 5. Exploitation of heterodyne interferences

Interference between remote sites can be processed digitally by correlating the output patterns of individual heterodyne sensors. The processing is a very simple "exclusive OR" gate and can run at teraoperations per second. The technological bottle neck is more in storing and transmitting the signals as it is today the case in radioastronomy.

A remarkable properties of very noisy channel with Gaussian noise distribution is that the channel capacity can be increased by discarding samples in the center of the distribution. This can be explained by computing the error rate when receiving a sample of value "a" issued from a modulation "±ϵ":

$$P + /P- = erfc((a - \epsilon)/\sigma)/erfc((a + \epsilon)/\sigma) \qquad (5)$$

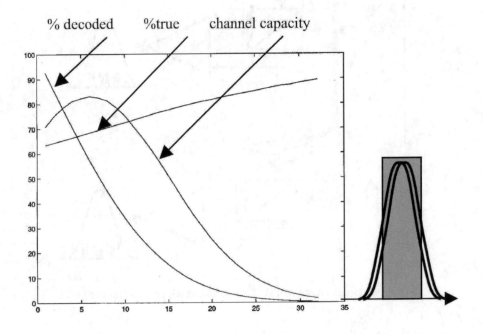

*Figure 4.*   Channel capacity of noisy channel with discarded samples ($S/N = 0.01$)

This function is decreasing towards zero. This brings the surprising result that for higher realization of the noise sample, error rate is lower. Correlating samples with high deviation from the 50/50 power division at the output of the optical beamsplitters could give interference data without overcharging the information link (Fig. 4).

## 6.    Conclusion

The beating of a faint source with a high power coherent source is a well known process to detect its phase and amplitude. The same detection equipment allows the evaluation of the power of the source with theoretical limits similar to a noiseless photon counter. Such detection apparatus are limited by the bandwidth of the electronic component; as this bandwidth is rapidly increasing, this may be a competitive solution for quantum limited detection in the far infra red. The phase of a thermal source is an useless information;

only the phase relation through two or more optical paths is of interest to the observer.

In conventional interferometry, the photon who materializes in the focal plane gives an information on the path difference between two branches of the interferometer. Every student in quantum physics has learned that the observer who detects in which branch the photon is passed has destroyed the interference pattern and can have no knowledge on path difference. However, radioastronomers who are working at multiphoton level do translate the electromagnetic phenomena into a classical observation and obtain a path relation between various observation sites. Optical heterodyne detectors obey to the same rule and can bring an interferometric information even while proceeding classical measurement. The cost of looking inside the branches of the interferometer is that information is proportional to the square of the received optical power. A naïve interpretation of quantum physics is that signal can only be measured when photons are bunched together and can be detected on both branches.

Since the absolute phase of the faint object is an useless information, the spectral quality of the local oscillator is not required. A star can be used as a local oscillator searching for interference with the faint glow of one of its planet. The observation sites must be within the coherence distance of the star while beating fringes between star and planet may be hundred to thousand times smaller. Each observation site cannot resolve the two objects; therefore only one output of the heterodyne beamsplitter can be exploited. This experiment is clearly the Hanbury Brown -Twiss. In the original experiment, bunching was observed within the coherence domain of a "round" star. More complex objects generate "bunching interferences" that could allow to estimate their shape even in presence of atmospheric perturbation.

## References

Kingston, R. H. *Detection of Optical and IR radiation*

Feyman, *Lectures on physics* Addison-Wesley, Reading (Ma)

Mandel L. and Wolf, E., 1995, *Optical Coherence and Quantum* Optics Cambridge University Press, New York

Hanbury Brown, R., Twiss, R. Q., 1954, *A new type of interferometer for use in radio astronomy*, Phil. Mag. **45**, 663

Hanbury Brown, R., Twiss, R. Q., 1956, *Correlation between photons in two coherent beams of light*, Nature **177**, 27

Quirrenbach, A., 2001, *Astronomical interferometry, from the visible to sub-mm waves*, Europhysics News **32**

# STATISTICAL WAVEFRONT SENSING

Richard G. Lane

*ARANZ - Applied Research Associates NZ Ltd*
*St Elmo Courts, 47 Hereford Str.*
*P.O. 3894, Christchurch, New Zealand*
email@aranz.com


Richard M. Clare

*Department of Electrical and Computer Engineering*
*University of Canterbury*
*New Zealand*
rmc70@@elec.canterbury.ac.nz


Marcos A. van Dam

*Lawrence Livermore National Laboratory*
*Livermore, CA 94550, USA*
vandam1@llnl.gov

**Abstract**     Wavefront sensing for adaptive optics is addressed. The most popular wavefront sensors are described. Restoring the wavefront is an inverse problem, of which the bases are explained. An estimator of the slope of the wavefront is the image centroid. The Cramèr-Rao lower bound is evaluated for several probability distribution function

**Keywords:**     wavefront sensing, wavefront sensors, image centroid, Cramèr-Rao, maximum likelihood, maximum a posteriori

## 1.     Introduction

For what is an apparently straightforward problem, wavefront sensing has produced a large number of apparently quite different solutions (for example the Shack-Hartmann (Lane and Tallon, 1992), curvature (Roddier, 1988) and pyramid (Ragazzoni, 1996) sensors). Underlying this diversity is the problem

*R. Foy and F. C. Foy (eds.), Optics in Astrophysics,* 375–395.

that phase distortions at optical frequencies cannot be measured directly, and their effects on intensity are non-linearly related to the wavefront distortion. A typical wavefront sensor manipulates the wavefront to produce intensity fluctuations that can be linearly related to the wavefront. The resulting linear equations are then solved to estimate the wavefront.

In order to accurately estimate the wavefront a number of properties are desirable in a wavefront sensor (Marcos, 2002). These are:

- Linearity: A linear relationship should exist between the wavefront and the wavefront sensing data. This ensures a unique easily obtainable solution.
- Broadband: The wavefront sensor should be able to operate across a wide range of wavelengths. Often light is limited and it is important to use every available photon to minimize the noise.
- Sensitivity: The wavefront sensing measurements need to be sensitive to changes in the wavefront. Clearly we want to make the best possible use of the available light.

Rather than attempting to focus on all aspects of wavefront sensing in this chapter we will focus on the problem of estimating parameters from the measurements. This is a problem that underlies all wavefront sensors, and indeed most instruments.

The wavefront is represented over a finite region, usually the instrument aperture, as a sum of basis functions, $\Psi(u, v)$. There is some flexibility in how we choose these functions, but essentially they should be chosen so that we can represent an arbitrary wavefront distortion, $\phi(u, v)$, by

$$\phi(u, v) = \sum_{i=1}^{\infty} a_i \Psi_i(u, v) \qquad (1)$$

where $a_i$ is the weight of the $i^{th}$ basis function.

For a circular aperture a typical set of basis functions are the Zernike polynomials (Noll, 1976), but for other geometries alternative basis functions may be more appropriate. The objective of most wavefront sensors is to produce a set of measurements, $m$, that can be related to the wavefront by a set of linear equations

$$m = \Theta a \qquad (2)$$

where $\Theta$ is the interaction matrix (Lane and Tallon, 1992) which maps the slope (or curvature) measurements to the weights of the bases.

The first problem of wavefront sensing should now be readily apparent. In practice we always have a finite number of measurements $m$, but the number of unknown parameters we need to estimate, $a$, is infinite. This problem is not unique to wavefront sensing, but typical of a general class of problem known as inverse problems. The techniques to solve this problem, discussed in Section 24.2, are common to most wavefront sensors.

The second problem is how we can obtain a linear relationship between the coefficients describing the wavefront and our measurements. It is how this linear relationship is obtained that differentiates, for example, a Shack-Hartmann and a curvature sensor. In all wavefront sensors to transform a wavefront aberration into a measurable intensity fluctuation it is necessary to propagate the wavefront. As a first approximation this propagation is described by geometric optics, and we discuss the linear relationship between the wavefront slope and the image displacement in Section 24.3.

One of the major questions that should always be asked of any estimation technique is whether is is optimal. This problem can be addressed using Cramèr-Rao bounds and we illustrate this in Section 24.4. We conclude this chapter in Section 24.5 with the application of these statistical techniques to the Shack-Hartmann and curvature sensors.

## 2. Solving the inverse problem

The problem we face is that we have to estimate a wavefront, which has an infinite number of degrees of freedom, from a finite number of measurements. At first this may seem impossible, but in reality an infinite range of possible solutions describes most practical situations, not just wavefront sensing. The key to solving the problem is that we need to make an assumption about the relative likelihood of the solutions. As an example of how this is done, consider a wavefront sensor which makes a single measurement that is sensitive to only two basis functions.

Since we have only one measurement, a conventional approach would be to estimate the coefficient of only one basis function. In this case the two factors that decide which basis function should be chosen are how sensitive the measurement is to each basis function and *a priori* how likely each basis function is.

In order to put this in the context of wavefront sensing consider estimating a wavefront aberration that we know to consist of the sum of a tilt and a coma, from the image taken at the focal plane. Section 24.3 shows that the displacement of the image is a linear function of the weightings, $a_2$ and $a_7$, of the tilt, $\Psi_2$, and coma, $\Psi_7$, terms respectively. Assuming the variance of the tilt and coma are related as in a Kolmogorov turbulence model (see Ch. 1), the expected variance of the tilt over a circular aperture, $0.455(D/r_0)^{5/3}$, is much greater than the expected variance of the coma, $0.0064(D/r_0)^{5/3}$ (Noll, 1976). This suggests that if we are going to estimate only a single coefficient we should estimate the tilt alone. However, by doing this we effectively assume that the coefficient of the coma term is zero. In practice it is unusual for the coma to be identically zero and it is in fact negatively correlated with the tilt (Noll, 1976).

A more general approach to finding a solution than simply assuming some coefficients are zero is provided by a Bayesian approach. Since the displacement $d$ is a linear function of two unknowns, $a_2$ and $a_7$, we can write

$$d = \xi a_2 + \eta a_7 + n \tag{3}$$

where $n$ is an additive noise term.

We now assume that our estimates of the two unknown modes, $\hat{a}_2$ and $\hat{a}_7$, are linear functions of the displacement, $d$,

$$\begin{aligned} \hat{a}_2 &= \alpha d \\ \hat{a}_7 &= \beta d. \end{aligned} \tag{4}$$

Our objective is to minimize the expected error, $E$, between the true coefficients and our estimates.

$$E = \left\langle (a_2 - \hat{a}_2)^2 + (a_7 - \hat{a}_7)^2 \right\rangle \tag{5}$$

where $\langle \cdot \rangle$ denotes the ensemble average. This cannot be done directly from Eq. 5 as $a_2$ and $a_7$ are unknown. However, if we substitute Eq.'s 3 and 4 into Eq. 5, and make the reasonable assumption that the measurement noise is uncorrelated with both $\hat{a}_2$ and $\hat{a}_7$, such that $\langle a_2 n \rangle$ and $\langle a_7 n \rangle$ are both equal to 0, we obtain,

$$\begin{aligned} E &= \left\langle (a_2 - \alpha(\xi a_2 + \eta a_7 + n))^2 + (y - \beta(\xi a_2 + \eta a_7 + n))^2 \right\rangle \\ &= \left\langle a_2^2 \right\rangle (1 - 2\alpha\xi + \alpha^2\xi^2 + \beta^2\xi^2) + \\ &\quad \left\langle a_7^2 \right\rangle (1 - 2\beta\eta + \alpha^2\eta^2 + \beta^2\eta^2) + \\ &\quad \left\langle a_2 a_7 \right\rangle (-2\alpha\eta - 2\beta\xi + 2\alpha^2\xi\eta + 2\beta^2\xi\eta) + \\ &\quad \left\langle n^2 \right\rangle (\alpha^2 + \beta^2). \end{aligned} \tag{6}$$

The error is a function of the covariance of the two unknown parameters, $a_2$ and $a_7$, and the noise, $n$. The minimization of the error in Eq. 6 provides a mechanism for simultaneously estimating both $a_2$ and $a_7$.

The key point is that the underdetermined system of linear equations is rendered soluble by an assumption of the prior probabilities of the unknown coefficients. It is important to realize that truncating the number of modes creates an implicit assumption on the prior probabilities of the unknown modes; we assume the higher order modes are zero.

Statistically speaking we are seeking the most likely set of basis coefficients given our observed data. Mathematically this can be written as,

$$\underset{\hat{a}}{\max} \ \{Pr\{\hat{a}|m\}\} \tag{7}$$

where $Pr$ and $|$ denote probability and conditional probability respectively. $Pr\{\hat{a}|m\}$ is difficult to express directly but is rendered more tractable by the use of Bayes theorem,

$$Pr\{\hat{a}|m\} = \frac{Pr\{m|\hat{a}\}Pr\{\hat{a}\}}{Pr\{m\}}. \tag{8}$$

This approach is equivalent to the maximum *a posteriori* (MAP) approach derived by Wallner (Wallner, 1983). The position of the maximum is unchanged by a monotonic transformation and hence further simplification can be achieved by taking the logarithm of Eq. 8

$$\log[Pr\{\hat{a}|m\}] = \log[Pr\{m|\hat{a}\}] + \log[Pr\{\hat{a}\}] - \log[Pr\{m\}]. \tag{9}$$

It is worthwhile to consider each of these terms in turn. The term $\log[Pr\{m|\hat{a}\}]$ represents how likely it would be to observe our given measurements if the parameters were equal to $a$. We can compute this by using

$$\hat{m} = \Theta\hat{a} \tag{10}$$

to estimate the measurements that we would have expected from our current wavefront estimate. The difference, $\hat{m} - m$, is then an estimate of the noise, $\hat{n}$. Assuming the noise is a zero mean Gaussian random variable we can estimate $\log[Pr\{m|\hat{a}\}]$ by

$$
\begin{aligned}
Pr\{m|\hat{a}\} &= \exp\left[-\frac{1}{2}\hat{n}^T N^{-1}\hat{n}\right] \\
&= \exp\left[-\frac{1}{2}(\hat{m} - m)^T N^{-1}(\hat{m} - m)\right]
\end{aligned} \tag{11}
$$

where $N = \langle nn^T \rangle$ is the expected noise covariance.

The $\log[Pr\{m\}]$ term of Eq. 9 can be discarded since it is not a function of $\hat{a}$, the variable we are maximizing. Put simply, however improbable our data it remains what we have measured.

The term $Pr\{\hat{a}\}$ is the key to solving an underdetermined system of equations. Given two possible solutions which would produce exactly the same measurements, we use $Pr\{\hat{a}\}$ to select the most likely. In the case of a wavefront aberration caused by atmospheric turbulence the estimated coefficients,

$\hat{a}$, are also zero mean Gaussian random variables so their probability distribution is given by:

$$Pr\{\hat{a}\} = \exp\left[-\frac{1}{2}\hat{a}^T C^{-1}\hat{a}\right]. \tag{12}$$

Here $C = \langle aa^T\rangle$ is the covariance of the basis functions used to model the turbulence. Covariance matrices are positive semi-definite by definition which implies $\hat{a}^T C^{-1}\hat{a} > 0$, and thus a defined maximum of $Pr\{\hat{a}\}$ exists.

Using Eq.'s 11 and 12 in Eq. 9 yields

$$\log[Pr\{\hat{a}|m\}] = -\frac{1}{2}\left(\hat{a}C^{-1}\hat{a} + (\hat{m}-m)^T N^{-1}(\hat{m}-m)\right). \tag{13}$$

Substituting Eq. 10 for $\hat{m}$ into Eq. 13 gives

$$\log[Pr\{\hat{a}|m\}] = -\frac{1}{2}\left(\hat{a}C^{-1}\hat{a} + (\Theta\hat{a}-m)^T N^{-1}(\Theta\hat{a}-m)\right). \tag{14}$$

Equating the partial derivative of Eq. 14 with respect to $\hat{a}$ to zero

$$\frac{\partial \log[Pr\{\hat{a}|m\}]}{\partial \hat{a}} = \hat{a}^T C^{-1} + \hat{a}^T \Theta^T N^{-1}\Theta - \Theta^T N^{-1}m = 0 \tag{15}$$

defines the reconstructor, $\Omega$, which maximize the probability,

$$\begin{aligned} \hat{a} &= (C^{-1} + \Theta^T N^{-1}\Theta)^{-1}\Theta^T N^{-1}m \tag{16}\\ &= \Omega m. \end{aligned}$$

As expected the optimal estimator is indeed a function of both the covariance of the atmospheric turbulence and the measurement noise. A matrix identity can be used to derive an equivalent form of Eq. 16 (Law and Lane, 1996)

$$\Omega = C\Theta^T(\Theta C\Theta^T + N)^{-1}. \tag{17}$$

The term inverted in Eq. 17 is the covariance of the measurements and consequently directly measurable,

$$\langle mm^T\rangle = \Theta C\Theta^T + N. \tag{18}$$

If we consider the relative merits of the two forms of the optimal reconstructor, Eq.'s 16 and 17, we note that both require a matrix inversion. Computationally, the size of the matrix inversion is important. Eq. 16 inverts an $M \times M$ (measurements) matrix and Eq. 17 a $P \times P$ (parameters) matrix. In a traditional least squares system there are fewer parameters estimated than there are measurements, ie $M > P$, indicating Eq. 16 should be used. In a Bayesian framework we are trying to reconstruct more modes than we have measurements, ie $P > M$, so Eq. 17 is more convenient.

The relationship between the noise and atmospheric covariances is also evident in Eq. 17. If the noise on the measurements is large the $N$ term dominates the inverse which means only the large eigenvalues of $C$ contribute to the inverse. Consequently only the low order modes are compensated and a smooth reconstruction results. When the data is very noisy then $\Omega$ and hence $\hat{a}$ tend to zero. If the data is very noisy, then no estimate of the basis coefficients is made.

If in the ideal case the noise on the measurements is zero, such that $N = 0$, then the measurements are explained exactly and the basis coefficients are assigned by their relative probability, which is determined by $C$.

If we decide to only estimate a finite number of basis modes we implicitly assume the coefficients of all the other modes are zero and that the covariance of the modes estimated is very large. Thus $\Theta N^{-1}\Theta$ becomes large relative to $C^{-1}$ and in this case Eq. 16 simplifies to a weighted least squares formula

$$\Omega = (\Theta^T N^{-1}\Theta)^{-1}\Theta^T N^{-1}. \tag{19}$$

Here the information on the relative probability of the basis coefficients contained in $C$ is lost.

An advantage of this approach is that we are now able to predict the form of the residual error by substituting Eq. 16 into the general form of Eq. 5,

$$E = \left\langle (a - \Omega m)(a - \Omega m)^T \right\rangle. \tag{20}$$

Then by further substituting $\Theta a + n$ for $m$ we obtain,

$$E = \left\langle \left( (I - \Omega\Theta)a - \Theta n \right)\left( (I - \Omega\Theta)a - \Theta n \right) \right\rangle \tag{21}$$

where $I$ is the identity matrix. If Eq. 21 is expanded and the noise and the signal are assumed to be uncorrelated then the residual error simplifies to,

$$E = \left\langle (I - \Omega\Theta)C(I - \Omega\Theta)^T + \Omega N\Omega^T \right\rangle. \tag{22}$$

A numerical value for the residual error can be obtained by taking the trace of each matrix in Eq. 22. Thus provided the statistics of the noise and turbulence are known, then the error in the reconstruction can be predicted.

## 2.1 Estimation example

Finally, we return to our example of a two mode turbulence, tilt and coma, with a single measurement of the centroid displacement, $d$. We assume, for simplicity, that there is no noise on the measurements, hence $N = 0$. If we assume atmospheric turbulence of severity $D/r_0 = 4$, then from Noll (Noll, 1976) we can calculate the covariance of the Zernikes,

$$C = \begin{bmatrix} 4.61 & -0.15 \\ -0.15 & 0.064 \end{bmatrix} \tag{23}$$

and the interaction matrix from (Primot et al., 1990) as $\Theta = \begin{bmatrix} 4 & 4.9 \end{bmatrix}$. If we then substitute these matrices into Eq. 17,

$$
\begin{aligned}
\begin{pmatrix} a_2 \\ a_7 \end{pmatrix} &= C\Theta^T(\Theta C\Theta^T)^{-1}d \\
&= \begin{bmatrix} 4.61 & -0.15 \\ -0.15 & 0.064 \end{bmatrix} \begin{bmatrix} 4 \\ 4.9 \end{bmatrix} \\
&\quad \left( \begin{bmatrix} 4 & 4.9 \end{bmatrix} \begin{bmatrix} 4.61 & -0.15 \\ -0.15 & 0.064 \end{bmatrix} \begin{bmatrix} 4 \\ 4.9 \end{bmatrix} \right)^{-1}d \\
&= \begin{pmatrix} 0.2551 \\ -0.0041 \end{pmatrix} d
\end{aligned}
\tag{24}
$$

we obtain the desired result of two modes being estimated from a single measurement. Note that as expected the reconstructed coefficient of tilt is much greater than that of the coma but the coma estimate is non-zero. If we now assume there is noise on the measurements we see in Fig. 1 that as $N$ increases the reconstructed coefficient of $a_2$ tends to zero as expected. The coefficient of coma, $a_7$, decreases at the same rate as that of $a_2$. It is worth noting that due to the log scale on the abscissa of Fig. 1, the solution is not unduly sensitive to the exact value of the noise covariance, $N$.

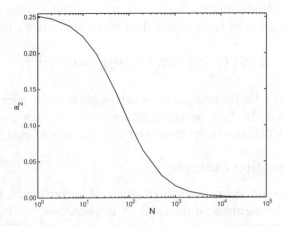

*Figure 1.*   The coefficient of tilt, $a_2$, with increasing noise covariance, $N$.

## 3. Obtaining linear measurements from a wavefront

Essentially our only tool for transforming a wavefront aberration into a measurable quantity is propagating the wavefront and observing the light intensity. This relationship between the aberrations in the wavefront at the aperture and the expected image at the detector can be modeled using Fourier analysis. We begin by defining the complex field in the aperture,

$$U_0(\mathbf{x}) = A(\mathbf{x}) \exp[i\phi(\mathbf{x})], \tag{25}$$

where $A(\mathbf{x})$ is the amplitude and $\phi(\mathbf{x})$ the phase of the wave. The Cartesian coordinates are given by $\mathbf{x} = (x, y)$ in the aperture plane and $\boldsymbol{\xi} = (\xi, \eta)$ in the measurement plane.

For a given wavelength $\lambda$ we can calculate the complex field at the measurement plane located a distance $z$ away by the Fresnel diffraction formula ( Goodman, 1996)

$$U_1(\boldsymbol{\xi}) = 1/i\lambda z \exp\left[i2\pi z/\lambda\right] \exp\left[i\pi/\lambda z\boldsymbol{\xi}^2\right] \int_{-\infty}^{\infty} U_0(\mathbf{x}) \exp\left[i\pi/\lambda z\mathbf{x}^2\right]$$
$$\exp\left[-i2\pi/\lambda z\mathbf{x} \cdot \boldsymbol{\xi}\right] d\mathbf{x}. \tag{26}$$

The intensity of the wave $I(\boldsymbol{\xi})$ is what can be measured at the detector

$$I(\boldsymbol{\xi}) = \left|1/\lambda z \int_{-\infty}^{\infty} U_0(\mathbf{x}) \exp\left[i\pi/\lambda z\mathbf{x}^2\right] \exp\left[-i2\pi/\lambda z\mathbf{x} \cdot \boldsymbol{\xi}\right] d\mathbf{x}\right|^2. \tag{27}$$

It is the shape of this distribution which imposes limits on the ability to estimate the wavefront. In the special case where the measurement plane is in fact at the focal plane, the relationship between the aperture and measurement plane simplifies to

$$I(\boldsymbol{\xi}) = 1/\lambda^2 f^2 \left|\int_{-\infty}^{\infty} \int_{-\infty}^{\infty} A(\mathbf{x}) \exp[i\phi(\mathbf{x})] \exp[-i2\pi/\lambda f\mathbf{x} \cdot \boldsymbol{\xi}] d\mathbf{x}\right|^2, \tag{28}$$

where $f$ is the focal length of the lens or mirror. Defining $\mathbf{u} = (u, v) = (\xi, \eta)/(\lambda f)$, the complex amplitudes at the aperture and the image plane form a Fourier transform pair.

Unfortunately, there is not a linear relationship between the light intensity in the measurement plane and the wavefront. This is shown in Fig. 2 which shows the intensities measured at the focal plane for a wavefront equal to pure tilt and coma terms individually. It is apparent that scaling the wavefront by $\alpha$ (in this case 5) does not result in a linear increase in the measured intensity by a factor $\alpha$. The key difference between existing existing wavefront sensors such as the Shack-Hartmann, curvature and pyramid sensors is how they transform the measured intensity data to produce a linear relationship between the measurements and the wavefront.

## 3.1     The effect of slope on the image

The simplest linear relationship is between the displacement of an image and the mean slope. If the LMS angle of arrival of the wavefront is $\boldsymbol{\theta} = (\theta_x, \theta_y)$, then the phase can be written as $\phi(\mathbf{x}) = \tilde{\phi}(\mathbf{x}) + (2\pi/\lambda)\boldsymbol{\theta} \cdot \mathbf{x}$, where $\tilde{\phi}(\mathbf{x})$ is the zero LMS slope phase. Setting $\tilde{U}_0(\mathbf{x}) = A(\mathbf{x}) \exp[i\tilde{\phi}(\mathbf{x})]$ leads to $U_0(\mathbf{x}) = \tilde{U}_0(\mathbf{x}) \exp[i(2\pi/\lambda)\boldsymbol{\theta} \cdot \mathbf{x}]$. The corresponding intensity at the detector is

$$|I(\boldsymbol{\xi})|^2 = \left| 1/\lambda z \int_{-\infty}^{\infty} \tilde{U}_0(\mathbf{x}) \exp[i\pi/\lambda z \mathbf{x}^2] \exp[-i2\pi/\lambda z \mathbf{x} \cdot (\boldsymbol{\xi} - \boldsymbol{\theta} z)] d\mathbf{x} \right|^2.$$
(29)

By comparison with Eq. 27, this shows that the effect of the wavefront slope is to displace the image by

$$\Delta \boldsymbol{\xi} = \boldsymbol{\theta} z,$$
(30)

as shown in Fig. 3.

The effect of overall slope on the wavefront can be removed by re-centering the speckle image, and in most adaptive optics systems this is performed by a planar tip/tilt mirror (Roddier and Roddier, 1993).

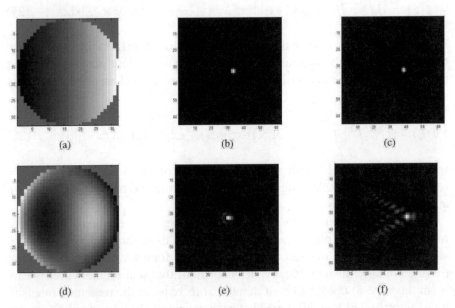

*Figure 2.*    (a) The tilt phase screen, (b) the resulting intensity distribution in the measurement plane from the tilt screen, (c) the resulting intensity distribution from the tilt screen scaled by 5, (d) the coma phase screen, (e) the resulting intensity distribution from the coma screen, and (f) the resulting intensity distribution from the coma screen scaled by 5.

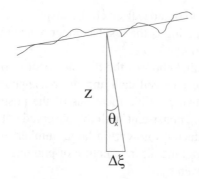

*Figure 3.* Least mean-square slope of a wavefront.

## 3.2 Estimating image displacement

Although it is apparent that the displacement of an image is related to the mean slope of the wavefront, there remains the task of estimating this displacement. The received intensity $I(\xi, \eta)$ ultimately determines how well a displacement can be estimated.

The intensity $I(\xi, \eta)$ is open to a number of different interpretations, but one of the most useful for analysis is that when $I(\xi, \eta)$ is normalized to 1, it can be thought of as a probability distribution function (pdf), $h(\xi, \eta)$, for the arrival location of a single photon. This interpretation enables statistical estimation techniques to be applied.

As an example, consider a planar wavefront from an incoherent source passing through an aberration-free circular lens. When the image is diffraction-limited, an Airy disc pattern is observed (Goodman, 1996). For an aperture of radius $1/(2\pi)$ the pdf for photon arrival is given by

$$h(\rho) = \frac{J_1(\rho)^2}{\pi \gamma \rho^2}, \tag{31}$$

where $\rho = \sqrt{u^2 + v^2}$ is the radial distance from the centre of the image and $J_1(\rho)$ is a Bessel function of order one of the first kind. The normalization constant $\gamma = \frac{4}{3\pi^2}$ ensures that $\int_0^\infty h(\rho) d\rho = 1$. To determine the location of the centre of the pdf from a speckle image, the centroid estimator is usually used (Roggemann and Welsh, 1996). In Sect. 24.3.4, it is shown that the centroid is an optimal estimator for a Gaussian PSF but not for an Airy disc.

## 3.3 Maximum-likelihood estimator

In order to discuss how we can use statistical methods to estimate the errors in measuring the displacement we use the notation $h(u|a)$. This is the conditional probability of photon arrival given the displacement $a$. In this case the parameter we are estimating is the displacement hence the conditional probability $h(u|a)$ is equal to $h(u - a)$.

Again an approach that chooses the displacement $a$ to maximize the probability of obtaining the received data appears reasonable. This is illustrated in Fig. 4, where the two possible locations of the positioning of the pdf in Fig. 4(a) is clearly more consistent with the observed data than Fig. 4(b). Let us assume an ideal detector, one with a large number of pixels and with the only noise resulting from the discrete nature of photons. The probability of D photons in a pixel is given by

$$P(D) = \frac{\lambda^D \exp[-\lambda]}{D!} \tag{32}$$

where $\lambda$ is the mean number of expected photons. If we assume that the pixels are very finely sampled, then $\lambda \to 0$, and we measure the presence of single photon locations $\mathbf{u}_i$ with a probability,

$$P(1) = \lambda \exp[-\lambda] \approx \lambda. \tag{33}$$

If we further assume the pdf is Gaussian with variance $\sigma^2$,

$$h(\mathbf{u}) = \frac{1}{2\pi\sigma^2} \exp[-\mathbf{u}^2/2\sigma^2], \tag{34}$$

the likelihood of detecting $R$ photons at positions given by $\mathbf{u}_i$ is equal to,

$$\prod_{i=0}^{R-1} h(\mathbf{u}_i|\mathbf{a}) = \prod_{i=0}^{R-1} 1/2\pi\sigma^2 \exp\left[\frac{-(u_i - \hat{a}_u)^2 - (v_i - \hat{a}_v)^2}{2\sigma^2}\right]. \tag{35}$$

Taking the natural logarithm of Eq. 35 yields

$$\sum_{i=0}^{R-1} \ln h(\mathbf{u}_i|\mathbf{a}) = R \ln 1/2\pi\sigma^2 - \sum_{i=0}^{R-1}\left[\frac{(u_i - \hat{a}_u)^2 + (v_i - \hat{a}_v)^2}{2\sigma^2}\right]. \tag{36}$$

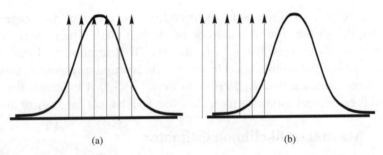

(a)                                                                 (b)

*Figure 4.*    The estimation of the pdf from the observed data where (a) the pdf is consistent with the data, and (b) the pdf is inconsistent with the data.

Differentiating with respect to $a_u$ and equating to zero to find the maximum gives

$$1/\sigma^2 \sum_{i=0}^{R-1} (u_i - \hat{a}_u) = 0, \tag{37}$$

which holds if $\hat{a}_u = 1/R \sum_{i=0}^{R-1} u_i$. Similarly, $\hat{a}_v = 1/R \sum_{i=0}^{R-1} v_i$. This is the conventional centroid estimator and its great convenience is its simple computational form. As a consequence it is widely used, regardless of whether the underlying pdf is Gaussian.

## 3.4 Variance of the estimator

Although it sounds reasonable to use the maximum likelihood to define our estimate of the displacement, there are two questions that remain. Firstly, what is the variance of the error associated with this estimate? This defines $N$ which was used in Eq. 22 to determine the error in the wavefront reconstruction. Secondly, is it possible to do better than the centroid? In other words is it optimal?

In practice the pdf of the image is sampled using a CCD array with finite pixels. Using $\Delta$ to denote the pixel width and $\delta$ to describe the offset from the origin to the first pixel the centroid can be computed from

$$\hat{a}_u = \frac{\sum_i \sum_j (i\Delta - \delta)\tilde{h}(i\Delta, j\Delta)}{\sum_i \sum_j \tilde{h}(i\Delta, j\Delta)} \tag{38}$$

where $\tilde{h}$ is used to denote the actual measured intensity. The variance of this estimate depends strongly on the type of noise that predominates in the CCD. If it is read noise then each pixel can be modeled as having a noise variance equal to $\eta^2$. The overall noise variance is equal to,

$$E\left[(\hat{a}_u - a_u)^2\right] = \eta^2 \sum_i \sum_j (i\Delta - \delta)^2 \tag{39}$$

This variance grows without bound with the size of the CCD and dictates that the smallest practical number of pixels should be used. A $2 \times 2$ array (or quad-cell) with $\delta = \frac{1}{2}$, is thus the configuration that dominates many existing slope sensors (Tyler and Fried, 1982).

In future improvements in technology may mean that that read noise no longer is the dominant noise source, and Poisson noise arising from the quantum nature of light is in fact the limiting factor. In this case the variance of the centroid noise is equal to,

$$E\left[(\hat{a}_u - a_u)^2\right] = \frac{1}{R} \sum_i \sum_j (i\Delta - \delta)^2 h(i\Delta, j\Delta) \tag{40}$$

which can be approximated by a continuous integral

$$E\left[(\hat{a}_u - a_u)^2\right] = \frac{1}{R} \int_{-\infty}^{\infty} \int_{-\infty}^{\infty} (\hat{a}_u - a_u)^2 h(u|a) du dv. \tag{41}$$

For the Gaussian pdf in Eq. 34

$$\int_{-\infty}^{\infty} \int_{-\infty}^{\infty} \frac{u^2}{2\pi\sigma^2} \exp[-\frac{u^2 + v^2}{2\sigma^2}] du dv = \sigma^2. \tag{42}$$

The traditional analysis has then used a Gaussian to approximate both the Airy disk and sinc$^2$ distribution as in Fig. 6 (Primot et al., 1990; Welsh and Gardner, 1989). The value of $\sigma$ in Eq. 41 is determined by matching the Gaussian approximation and the actual intensity at both the peak and the $\exp[-1]$ points (Welsh and Gardner, 1989). When this is done we find

$$E\left[(\hat{a}_u - a_u)^2\right] = \frac{(1.22\lambda f)^2}{(2\sqrt{2}d)^2} \tag{43}$$

for a circular lenslet of diameter $d$ and focal length $f$, and

$$E\left[(\hat{a}_u - a_u)^2\right] = \frac{(1.05\lambda f)^2}{(2\sqrt{2}d)^2} \tag{44}$$

for a square lenslet with a dimension $d$.

An interesting result occurs when instead of using the Gaussian approximation the Airy disk is used directly in the calculations. In this case the variance goes to infinity,

$$\begin{aligned} E[(\hat{\mathbf{a}} - \mathbf{a})^2] &= \int_{-\infty}^{\infty} \int_{-\infty}^{\infty} (\mathbf{u} - \mathbf{a})^2 \frac{J_1(\mathbf{u} - \mathbf{a})^2}{\pi(\mathbf{u} - \mathbf{a})^2} du dv \\ &= 2\pi \int_0^{\infty} \frac{J_1(\rho)^2}{\pi} \rho d\rho \\ &= \infty \end{aligned} \tag{45}$$

Physically, the reason for this is that the probability of detecting photons very far away from the centre of the Airy disc does not decay quickly enough (Irwan and Lane, 1999). The second moment of the PSF corresponding to any discontinuous aperture measured on any infinite plane, not just on the focal plane, is also infinite (Teague, 1982). The reason this happens is that the Fourier transform of any discontinuous function decays as $|\mathbf{x}|^{-1}$ far away from the origin. Hence, the PSF, which is the modulus squared, decays as $|\mathbf{x}|^{-2}$. The area of integration increases as $|\mathbf{x}|^2$, so the integral does not converge. Although there are very few photons far from the centre, a centroid calculation weights them

(a)            (b)

*Figure 5.* Simulation of photon arrival for 100 photons for (a) the Airy disk and (b) the Gaussian approximation.

too highly. Figure 5 shows a simulation of the photon arrival assuming an Airy disk and a Gaussian approximation for 100 photons in each case. The presence of outlying photons from the Airy disk is readily apparent.

Truncating the plane constrains the centroid estimate to a certain region, making the variance finite. Since the truncated plane is placed where the centre is expected to be we are implicitly adding prior information (van Dam and Lane, 2000). The smaller the plane, the more the centroid is effectively localized and the more prior information is assumed. Therefore, by adding prior information, truncating the plane can improve the centroid estimate, even though some photons are lost. The optimal solution is to maximize the likelihood directly,

$$\max_{\mathbf{a}} \left\{ \prod_{i=0}^{N-1} h(\sqrt{(u_i - a_u)^2 + (v_i - a_v)^2}) \right\}, \tag{46}$$

but this is more computationally intensive (van Dam and Lane, 2000).

## 4. Cramèr-Rao Lower Bounds

Finally we need to compare the variance of our estimator with the best attainable. It can be shown that The Cramèr-Rao lower bound (CRLB) is a lower bound on the variance of an unbiased estimator (Kay, 1993). The quantities estimated can be fixed parameters with unknown values, random variables or a signal and essentially we are finding the best estimate we can possibly make.

The bound can be calculated provided that $\frac{\partial h(u|a)}{\partial a}$ and $\frac{\partial^2 h(u|a)}{\partial a^2}$ exist and are absolutely integrable, which is the case for all the PSFs of interest (van Trees, 1968). The CRLB can be extended to any number of measurements and is

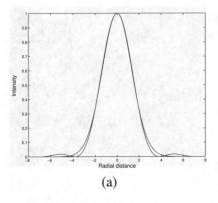
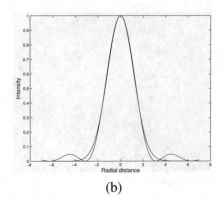

(a)                                                                              (b)

*Figure 6.*   Comparison of the (a) Airy disc (oscillating) and Gaussian (monotonically decreasing) distributions, and (b) sinc$^2$ (oscillating) and Gaussian (monotonically decreasing) distributions.

formally stated as follows: If $\hat{a}$ is an unbiased estimator of $a$,

$$E\left[(\hat{a}_u - a_u)^2\right] \geq \left(E\left\{\left[\frac{\partial \ln h(u|a)}{\partial a}\right]^2\right\}\right)^{-1} \tag{47}$$

or equivalently

$$E\left[(\hat{a}_u - a_u)^2\right] \geq \left(-E\left[\frac{\partial^2 \ln h(u|a)}{\partial a^2}\right]\right)^{-1}. \tag{48}$$

Note that by unbiased we mean that when the noise tends to zero we expect $\hat{a} \to a$. The left-hand side can be written simply as $\mathrm{Var}(\hat{a})$, the variance of the estimate, since the estimator is unbiased. The CRLB for some PDFs of interest is now computed.

## 4.1    CRLB for a Gaussian distribution

A Gaussian PDF in one dimension centred at $u = a$ with variance $\sigma^2$ is defined as

$$h(u|a) = 1/\sqrt{2\pi\sigma^2} \exp\left[-(u-a)^2/2\sigma^2\right]. \tag{49}$$

A direct evaluation of Eq. 49 gives

$$
\begin{aligned}
E\left[(\hat{a}_u - a_u)^2\right] &\geq \left(-E\left[\frac{\partial^2 \ln h(u|a)}{\partial a^2}\right]\right)^{-1} \\
&= \left(E\left[\frac{\partial^2}{\partial a^2} \frac{(u-a)^2}{2\sigma^2}\right]\right)^{-1} \\
&= \left(E\left[1/\sigma^2\right]\right)^{-1} \\
&= \sigma^2.
\end{aligned}
\tag{50}
$$

This demonstrates that provided the pdf of photon arrival is Gaussian the centroid is indeed optimal.

Figure 7 shows that for the maximum likelihood estimator the variance in the slope estimate decreases as the telescope aperture size increases. For the centroid estimator the variance of the slope estimate also decreases with increasing aperture size when the telescope aperture is less than the Fried parameter, $r_0$ (Fried, 1966), but saturates when the aperture size is greater than this value.

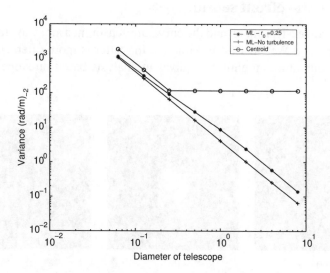

*Figure 7.* The variance in the slope estimate versus aperture size for the centroid and maximum likelihood estimators for turbulence defined by $r_0 = 0.25$.

## 4.2     CRLB for square and round apertures

For a square aperture of length $L$, the CRLB for the best variance is (van Dam and Lane, 2000)

$$E\left[(\hat{a}_u - a_u)^2\right] \geq 3/4 \left(\lambda f/\pi L\right)^2 \tag{51}$$

and for an aperture of diameter $d$

$$E\left[(\hat{a}_u - a_u)^2\right] \geq \left(\lambda f/\pi d\right)^2. \tag{52}$$

## 4.3     CRLB for a speckle pattern

It is worth noting that it is the average curvature of the pdf that determines our ability to estimate a parameter. As an example consider estimating the displacement of a known speckle pattern as in Fig. 8(a). If we use a centroid estimator for the displacement of the speckle the displacement accuracy approaches that of a Gaussian with a dimension equal to the width of the speckle image, Fig. 8(b). In contrast, if we use a maximum likelihood estimator which matches the detail in the speckle we obtain a performance approaching that of a Gaussian with a width equal to a single speckle, Fig.8(c). This is because when we look at Eq. 48 it is the average curvature of the intensity and not its size that defines our ability to estimate displacement.

## 5.     The wavefront sensors

Both the Shack-Hartmann and the curvature sensor measure wavefront slope (Roggemann and Welsh, 1996). Although the latter is more often considered to measure curvature an entirely equivalent analysis based on slope is possi-

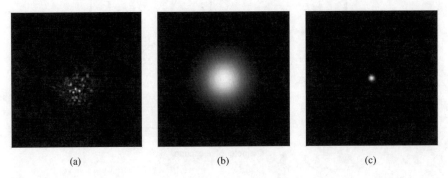

      (a)                              (b)                             (c)

*Figure 8.*     (a) The speckle image at the focal plane, (b) centroid estimator of the displacement, and (c) maximum likelihood estimator of the displacement.

ble (Marcos, 2002). We conclude with a brief discussion of how statistical techniques can be applied to these sensors.

## 5.1    The Shack-Hartmann Approach

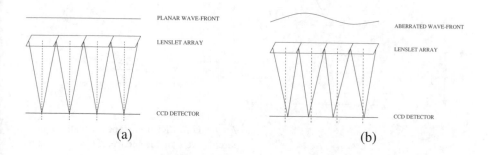

*Figure 9.* The Shack-Hartmann sensor with (a) a planar wavefront and (b) an aberrated wavefront. The dashed lines are the perpendicular bisectors of the lenslets.

One obvious way to increase the available measurements is to subdivide the aperture and measure the mean slope over each portion. This is shown in Figure 9. Clearly if we subdivide the aperture into $N$ subapertures and measure the mean slope across each subaperture we obtain $2N$ measurements.

As a simple rule of thumb if a simple least squares estimate is employed the number of modes estimated should be half the number of measurements. If a Bayesian approach is employed the number of modes estimated should be at least the number of measurements.

The analysis of this sensor is straightforward. We estimate the noise covariance using the techniques outlined in Section 24.3.4 and the use the general Bayesian formulae to calculate the reconstructor and predict performance.

## 5.2    Curvature sensor

The curvature sensor first proposed by Roddier (Roddier, 1988), does not make measurements at the focal plane. Instead measurements are taken at two planes symmetrically distributed around the focal plane as shown in Fig. 10. These measurements are best thought of as blurred images of the aperture and consequently our ability to measure the tilt is affected since as we move from the focal plane the light is spread over a wider angle. Consequently we would expect the tilt signal to degrade as we move away from the aperture. This can be formalized by the CRLB and Fig. 11 shows the best attainable tilt performance (Marcos, 2002).

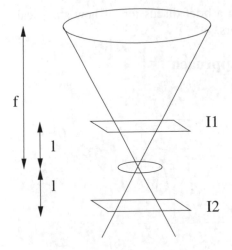

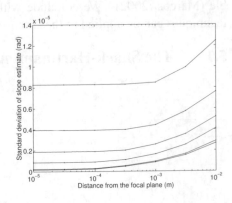

*Figure 11.* CRLB of slope estimate for different levels of turbulence, $D = 1$ m and 1 photon. The curves represent, from top to bottom, $D/r_0 = 40, 20, 10, 5, 1$ and 0.

*Figure 10.* A schematic diagram of the wavefront curvature sensor set-up.

## 6.    Conclusion

In this paper we have endeavored to show some of the statistical principles underlying wavefront sensing. It is important to realize that the problem is inherently one of parameter estimation; because as such it is part of a significant literature beyond traditional wavefront sensing.

## Acknowledgments

R.M. Clare wishes to acknowledge the financial assistance provided by the Foundation of Research, Science and Technology, New Zealand in the form of a Bright Future Scholarship.

## References

van Dam, M.A., Lane, R.G., 2000, *Wave-front slope sensing*, JOSA. A **17**, 1319

Fried, D.L., 1966, *Optical Resolution through a randomly inhomogeneous medium for very long and very short exposures*, JOSA. A **56**, 1376

Goodman, J., 1996, *Introduction to Fourier optics* (McGraw-Hill, New York), pp. 63-75.

Irwan, R., Lane, R.G., 1999, *Analysis of optimal centroid estimation applied to Shack-Hartmann sensing*, Applied Optics **38**, 6737

Kay, S.M., 1993, *Fundamentals of statistical signal processing estimation theory*, Prentice-Hall, New Jersey, pp 27-81

Lane, R.G., Tallon, M., 1992, *Wave-front reconstruction using a Shack-Hartmann sensor*, JOSA.A **31**, 6902

Law, N.F., Lane, R.G., 1996, *Wavefront estimation at low light levels*, Opt. Comms. **126**, 19

Van Dam, M.A., Lane, R.G., 2002, *Wave-front sensing from defocused images by use of wave-front slopes*, Appl. Opt. **26**, 5497

Noll, R., 1976, *Zernike Polynomials and atmospheric turbulence*, JOSA.A **56**, 207

Primot, J., Rousset, G., Fontanella, J.-C., 1990, *Deconvolution from wave-front sensing: a new technique for compensating turbulence-degraded images*, JOSA.A **7**,1598

Ragazzoni, R., 1996, *Pupil plane wavefront sensing with an oscillating prism*, J. Mod. Opt. **43** 289

Roddier, F., 1988, *Curvature sensing and compensation: a new concept in adaptive optics*, Applied Opt. **27**, 1223

Roddier, C., Roddier, F., 1993, *Wave-front reconstruction from defocused images and the testing of ground-based optical telescopes*, JOSA.A **10**, 2277 (1993).

M. C. Roggemann, M.C., Welsh, B., 1996, *Imaging through turbulence*, CRC Press, Boca Raton, Fla.

Teague, M.R., 1982, *Irradiance moments: their propagation and use for unique retrieval of phase*, JOSA **72**, 1199

van Trees, H.L., 1968, *Detection, estimation and modulation theory*, Part 1, pp 66-85, Wiley, NY

Tyler, G.A., Fried, D.L., 1982, *Image-position error associated with a quadrant detector*, JOSA **72**, 804

Wallner, E.P., 1983, *Optimal wave-front correction using slope measurements* JOSA, **73**, 1771

Welsh, B.M., Gardner, C.S., 1989, *Performance analysis of adaptive-optics systems using laser guide stars and slope sensors*, JOSA.A **6**, 1913

# INTRODUCTION TO IMAGE RECONSTRUCTION AND INVERSE PROBLEMS

Eric Thiébaut

*Observatoire de Lyon (France)*
thiebaut@obs.univ-lyon1.fr

**Abstract**     Either because observed images are blurred by the instrument and transfer medium or because the collected data (e.g. in radio astronomy) are not in the form of an image, image reconstruction is a key problem in observational astronomy. Understanding the fundamental problems underlying the deconvolution (noise amplification) and the way to solve for them (regularization) is the prototype to cope with other kind of inverse problems.

**Keywords:**     image reconstruction, deconvolution, inverse problems, regularization

## 1.     Image formation

### 1.1     Standard image formation equation

For an incoherent source, the observed image $y(\mathbf{s})$ in a direction $\mathbf{s}$ is given by the standard image formation equation:

$$y(\mathbf{s}) = \int h(\mathbf{s}|\mathbf{s}')\, x(\mathbf{s}')\, \mathrm{d}\mathbf{s}' + n(\mathbf{s}) \tag{1}$$

where $x(\mathbf{s}')$ is the object brightness distribution, $h(\mathbf{s}|\mathbf{s}')$ is the instrumental point spread function (PSF), and $n(\mathbf{s})$ accounts for the noise (source and detector). The PSF $h(\mathbf{s}|\mathbf{s}')$ is the observed brightness distribution in the direction $\mathbf{s}$ for a point source located in the direction $\mathbf{s}'$.

Figure 1 shows the simulation of a galaxy observed with a PSF which is typical of an adaptive optics system. The noisy blurred image in Fig. 1 will be used to compare various image reconstruction methods described in this course.

*R. Foy and F. C. Foy (eds.), Optics in Astrophysics*, 397–421.

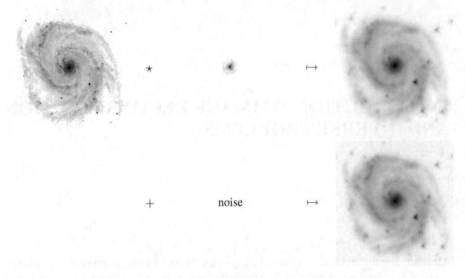

*Figure 1.* Simulation of the image of a galaxy. From top-left to bottom-right: true object brightness distribution, PSF (instrument + transfer medium), noiseless blurred image, noisy blurred image. The PSF image was kindly provided by F. and C. Roddier.

## 1.2    Discrete form of image formation equation

For discretized data (e.g. pixels), Eq. (1) can be written in matrix form:

$$\mathbf{y} = \mathbf{H} \cdot \mathbf{x} + \mathbf{n} \tag{2}$$

where $\mathbf{H}$ is the response matrix, $\mathbf{y}$, $\mathbf{x}$ and $\mathbf{n}$ are the data, object brightness distribution and noise vectors. These vectors have the same layout; for instance, the data vector corresponding to a $N \times N$ image writes:

$$\mathbf{y} = \left[ y(1,1), y(2,1), \dots, y(N,1), y(1,2), \dots, y(N,2), \dots, y(N,N) \right]^{\mathrm{T}}.$$

## 1.3    Shift invariant PSF

Within the isoplanatic patch of the instrument and transfer medium, the PSF can be assumed to be shift invariant:

$$h(\mathbf{s}|\mathbf{s}') = h(\mathbf{s} - \mathbf{s}'), \tag{3}$$

for $\|\mathbf{s} - \mathbf{s}'\|$ small enough. In this case, the image formation equation involves a convolution product:

$$y(\mathbf{s}) = \int h(\mathbf{s} - \mathbf{s}') \, x(\mathbf{s}') \, d\mathbf{s}' + n(\mathbf{s}) \tag{4a}$$

$$\xrightarrow{\text{FT}} \quad \hat{y}(\mathbf{u}) = \hat{h}(\mathbf{u}) \, \hat{x}(\mathbf{u}) + \hat{n}(\mathbf{u}) \tag{4b}$$

where the hats denote Fourier transformed distributions and $\mathbf{u}$ is the spatial frequency. The Fourier transform $\hat{h}(\mathbf{u})$ of the PSF is called the modulation transfer function (MTF).

In the discrete case, the convolution by the PSF is diagonalized by using the discrete Fourier transform (DFT):

$$\hat{y}_u = \hat{h}_u \, \hat{x}_u + \hat{n}_u \tag{5}$$

where index $u$ means $u$-th spatial frequency $\mathbf{u}$ of the discrete Fourier transformed array.

## 1.4 Discrete Fourier transform

The 1-D discrete Fourier transform (DFT) of a vector $\mathbf{x}$ is defined by:

$$\hat{x}_u = \sum_{k=0}^{N-1} x_k \, e^{-2i\pi uk/N} \quad \overset{\text{FT}}{\longleftarrow} \quad x_k = \frac{1}{N} \sum_{u=0}^{N-1} \hat{x}_u \, e^{+2i\pi uk/N}$$

where $i \equiv \sqrt{-1}$ and $N$ is the number of elements in $\mathbf{x}$. The $N \times N$ discrete Fourier transform (2-D DFT) of $\mathbf{x}$ writes:

$$\hat{x}_{u,v} = \sum_{k,l} x_{k,l} \, e^{-2i\pi(uk+vl)/N} \quad \overset{\text{FT}}{\longleftarrow} \quad x_{k,l} = \frac{1}{N^2} \sum_{u,v} \hat{x}_u \, e^{+2i\pi(uk+vl)/N}$$

where $N$ is the number of elements along each dimension. Using matrix notation:

$$\hat{\mathbf{x}} = \mathbf{F} \cdot \mathbf{x} \quad \overset{\text{FT}}{\longleftarrow} \quad \mathbf{x} = \mathbf{F}^{-1} \cdot \hat{\mathbf{x}} = \frac{1}{N_{\text{pixel}}} \mathbf{F}^{H} \cdot \hat{\mathbf{x}}$$

where $N_{\text{pixel}}$ is the number of elements in $\mathbf{x}$, and $\mathbf{F}^{H}$ is the conjugate transpose of $\mathbf{F}$ given by (1-D case):

$$F_{u,k} \equiv \exp(-2i\pi uk/N).$$

In practice, DFT's can be efficiently computed by means of fast Fourier transforms (FFT's).

## 2. Direct inversion

Considering the diagonalized form (5) of the image formation equation, a very tempting solution is to perform straightforward direct inversion in the Fourier space and then Fourier transform back to get the deconvolved image.

In other words:

$$\mathbf{x}^{(\text{direct})} = \text{FFT}^{-1} \left( \frac{\text{FFT}\left( \includegraphics{galaxy} \right)}{\text{FFT}\left( \quad \cdot \quad \right)} \right) = \includegraphics{noise}$$

The result is rather disappointing: the deconvolved image is even *worse* than the observed one! It is easy to understand what happens if we consider the Fourier transform of the direct solution:

$$\hat{x}_u^{(\text{direct})} = \hat{y}_u/\hat{h}_u = \hat{x}_u + \hat{n}_u/\hat{h}_u \tag{6}$$

which shows that, in the direct inversion, the perfect (but unknown) solution $\hat{x}_u$ get corrupted by a term $\hat{n}_u/\hat{h}_u$ due to noise. In this latter term, division by small values of the MTF $\hat{h}_u$, for instance at high frequencies where the noise dominates the signal (see Fig. 2a), yields very large distortions. Such noise amplification produces the high frequency artifacts displayed by the direct solution.

Instrumental transmission (convolution by the PSF) is always a smoothing process whereas noise is usually non-negligible at high frequencies, the noise amplification problem therefore *always* arises in deconvolution. This is termed as *ill-conditioning* in inverse problem theory.

## 3.     Truncated frequencies

A crude way to eliminate noise amplification is to use a suitable cutoff frequency $u_{\text{cutoff}}$ and to write the solution as:

$$\hat{x}_u^{(\text{cutoff})} = \begin{cases} \dfrac{\hat{y}_u}{\hat{h}_u} & \text{for } |u| < u_{\text{cutoff}} \\[2mm] 0 & \text{for } |u| \geq u_{\text{cutoff}} \end{cases} \tag{7}$$

In our example, taking $u_{\text{cutoff}} = 80$ frequels guarantees that the noise remains well below the signal level (see Fig. 2a); the resulting image is shown in Fig. 3a. The improvement in resolution and quality is clear but the stiff truncature of frequencies is responsible of ripples which can be seen around bright sources in Fig. 3a. A somewhat smoother effect can be obtained by using other kind of inverse filters like the Wiener filter described in the next section.

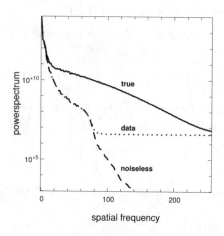

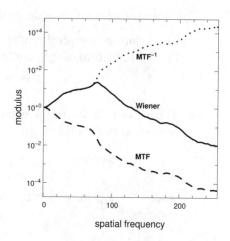

*Figure 2a.* Profiles of the power-spectra of the true brightness distribution, the noiseless blurred image, and the actual data (noisy and blurred image). Clearly the noise dominates after frequency $\simeq$ 80 frequels.

*Figure 2b.* Profiles of the modulation transfer function (MTF), its inverse and Wiener inverse-filter.

*Figure 3a.* Restored image using a simple cutoff frequency ($u_{\text{cutoff}} = 80$) in the deconvolution.

*Figure 3b.* Image obtained by using Wiener inverse-filter.

## 4.    Wiener inverse-filter

The Wiener inverse-filter is derived from the following two criteria:

- The solution is given by applying a linear filter $\mathbf{f}$ to the data and the Fourier transform of the solution writes:

$$\hat{x}_u^{(\text{Wiener})} \equiv \hat{f}_u^{(\text{Wiener})} \hat{y}_u \,. \tag{8}$$

- The expected value of the quadratic error with respect to the true object brightness distribution must be as small as possible:

$$\mathbf{f}^{(\text{Wiener})} = \arg\min_{\mathbf{f}} \mathrm{E}\left\{ \left\| \mathbf{x}^{(\text{Wiener})} - \mathbf{x}^{(\text{true})} \right\|^2 \right\} \tag{9}$$

where $\mathrm{E}\{x\}$ is the expected value of $x$. For those not familiar with this notation,

$$\arg\min_{\mathbf{f}} \phi(\mathbf{f})$$

is the element $\mathbf{f}$ which minimizes the value of the expression $\phi(\mathbf{f})$.

To summarize, Wiener inverse-filter is the linear filter which insures that the result is as close as possible, on average and in the least squares sense, to the true object brightness distribution.

In order to derive the expression of the filter, we write the expected value $\epsilon$ of the quadratic error:

$$\epsilon \equiv \mathrm{E}\left\{ \left\| \mathbf{x}^{(\text{Wiener})} - \mathbf{x}^{(\text{true})} \right\|^2 \right\} = \mathrm{E}\left\{ \sum_k \left( x_k^{(\text{Wiener})} - x_k^{(\text{true})} \right)^2 \right\}$$

by Parseval's theorem (in its discrete form):

$$\epsilon = \frac{1}{N_{\text{pixel}}} \mathrm{E}\left\{ \sum_u \left| \hat{x}_u^{(\text{Wiener})} - \hat{x}_u^{(\text{true})} \right|^2 \right\} = \frac{1}{N_{\text{pixel}}} \mathrm{E}\left\{ \sum_u \left| \hat{f}_u \hat{y}_u - \hat{x}_u^{(\text{true})} \right|^2 \right\}.$$

The extremum of $\epsilon$ (actually a minimum) is reached for $\mathbf{f}$ such that:

$$\frac{\partial \epsilon}{\partial \mathrm{Re}\{\hat{f}_u\}} = 0 \quad \text{and} \quad \frac{\partial \epsilon}{\partial \mathrm{Im}\{\hat{f}_u\}} = 0, \quad \forall u \,.$$

Since the two partial derivatives of $\epsilon$ have real values, the minimum of $\epsilon$ obeys:

$$\frac{\partial \epsilon}{\partial \mathrm{Re}\{\hat{f}_u\}} + \mathrm{i}\, \frac{\partial \epsilon}{\partial \mathrm{Im}\{\hat{f}_u\}} = 0, \quad \forall u$$

$$\Longleftrightarrow \mathrm{E}\left\{ \hat{y}_u^\star \left( \hat{f}_u \hat{y}_u - \hat{x}_u \right) \right\} = 0, \quad \forall u$$

$$\Longleftrightarrow \mathrm{E}\left\{ \left( \hat{h}_u \hat{x}_u + \hat{n}_u \right)^\star \left( \hat{f}_u \left( \hat{h}_u \hat{x}_u + \hat{n}_u \right) - \hat{x}_u \right) \right\} = 0, \quad \forall u$$

$$\Longleftrightarrow \left( |\hat{h}_u|^2 \hat{f}_u - \hat{h}_u^\star \right) \mathrm{E}\{ |\hat{x}_u|^2 \} + \hat{f}_u \, \mathrm{E}\{ |\hat{n}_u|^2 \} = 0, \quad \forall u$$

where $z^\star$ denotes the complex conjugate to $z$. The last expression has been obtained after some simplifications which apply when the signal and the noise are *uncorrelated* and when the noise is centered: $E\{\hat{n}_u\} = 0$. The final relation can be solved to obtain the expression of the Wiener inverse-filter:

$$\hat{f}_u^{(\text{Wiener})} = \frac{\hat{h}_u^\star}{|\hat{h}_u|^2 + \dfrac{E\{|\hat{n}_u|^2\}}{E\{|\hat{x}_u|^2\}}}. \tag{10}$$

Figure 2b and Eq. (10) show that the Wiener inverse-filter is close to the direct inverse-filter for frequencies of high signal-to-noise ratio (SNR), but is strongly attenuated where the SNR is poor:

$$\hat{f}_u^{(\text{Wiener})} \simeq \begin{cases} 1/\hat{h}_u & \text{for SNR}_u \gg 1 \\[2mm] 0 & \text{for SNR}_u \ll 1 \end{cases} \quad \text{with SNR}_u \equiv \sqrt{\frac{|\hat{h}_u|^2 \, E\{|\hat{x}_u|^2\}}{E\{|\hat{n}_u|^2\}}}.$$

The Wiener filter therefore avoids noise amplification and provides the best solution according to some *quality criterion*. We will see that these features are common to all other methods which correctly solve the deconvolution inverse problem. The result of applying Wiener inverse-filter to the simulated image is shown in Fig. 3b.

Wiener inverse-filter however yields, possibly, unphysical solution with *negative values* and *ripples* around sharp features (e.g. bright stars) as can be seen in Fig. 3b. Another drawback of Wiener inverse-filter is that spectral densities of noise and signal are usually *unknown* and must be guessed from the data. For instance, for white noise and assuming that the spectral density of object brightness distribution follows a simple parametric law, e.g. a power law, then:

$$\begin{cases} E\{|\hat{n}_u|^2\} = \text{constant} \\[2mm] E\{|\hat{x}_u|^2\} \propto \|u\|^{-\beta} \end{cases} \implies \hat{f}_u = \frac{\hat{h}_u^\star}{|\hat{h}_u|^2 + \alpha \, \|u\|^\beta}, \tag{11}$$

where $\|u\|$ is the length of the $u$-th spatial frequency. We are left with the problem of determining the best values for $\alpha$ and $\beta$; this can be done by means of Generalized Cross-Validation (GCV, see related section).

## 5. Maximum likelihood

Maximum likelihood methods are commonly used to estimate parameters from noisy data. Such methods can be applied to image restoration, possibly with additional constraints (e.g., positivity). Maximum likelihood methods are however not appropriate for solving ill-conditioned inverse problems as will be shown in this section.

## 5.1      Unconstrained maximum likelihood

The maximum likelihood (ML) solution is the one which maximizes the probability of the data $\mathbf{y}$ given the model among all possible $\mathbf{x}$:

$$\mathbf{x}^{(\text{ML})} = \arg \max_{\mathbf{x}} \Pr\{\mathbf{y}|\mathbf{x}\} = \arg \min_{\mathbf{x}} \phi_{\text{ML}}(\mathbf{x}) \qquad (12a)$$

where

$$\phi_{\text{ML}}(\mathbf{x}) \propto -\log(\Pr\{\mathbf{y}|\mathbf{x}\}) + \text{constant} \qquad (13b)$$

is the likelihood penalty related to the log-likelihood up to an additive constant (and also up to a multiplicative strictly positive factor).

### 5.1.1      Unconstrained ML for Gaussian noise.      In the case of Gaussian noise, the log-likelihood reads:

$$-\log(\Pr\{\mathbf{y}|\mathbf{x}\}) = \frac{1}{2}(\mathbf{H}\cdot\mathbf{x} - \mathbf{y})^{\text{T}} \cdot \mathbf{W} \cdot (\mathbf{H}\cdot\mathbf{x} - \mathbf{y}) + \text{constant}$$

where $\mathbf{H}\cdot\mathbf{x}$ is the model of the data (see Eq. (2)) and the weighting matrix $\mathbf{W}$ is the inverse of the covariance matrix of the data:

$$\mathbf{W} = \text{Cov}(\mathbf{y})^{-1}. \qquad (14b)$$

Taking $\eta = 2$ and dropping any additive constant, the likelihood penalty $\phi_{\text{Gauss}}(\mathbf{x})$ writes:

$$\phi_{\text{Gauss}}(\mathbf{x}) \equiv (\mathbf{H}\cdot\mathbf{x} - \mathbf{y})^{\text{T}} \cdot \mathbf{W} \cdot (\mathbf{H}\cdot\mathbf{x} - \mathbf{y}). \qquad (15b)$$

The minimum of $\phi_{\text{Gauss}}(\mathbf{x})$ obeys:

$$\left.\frac{\partial \phi_{\text{Gauss}}(\mathbf{x})}{\partial x_k}\right|_{\mathbf{x}^{(\text{ML})}} = 0, \quad \forall k.$$

The gradient of the penalty is:

$$\nabla \phi_{\text{Gauss}}(\mathbf{x}) = 2\,\mathbf{H}^{\text{T}} \cdot \mathbf{W} \cdot (\mathbf{H}\cdot\mathbf{x} - \mathbf{y}), \qquad (16b)$$

then $\mathbf{x}^{(\text{ML})}$ solves of the so-called *normal equations*:

$$\mathbf{H}^{\text{T}} \cdot \mathbf{W} \cdot (\mathbf{H}\cdot\mathbf{x}^{(\text{ML})} - \mathbf{y}) = 0 \qquad (17b)$$

which yields a unique minimum provided that $\mathbf{H}^{\text{T}} \cdot \mathbf{W} \cdot \mathbf{H}$ be positive definite (by construction this matrix is positive but may not be definite). The solution of the normal equations is:

$$\mathbf{x}^{(\text{ML})} = (\mathbf{H}^{\text{T}} \cdot \mathbf{W} \cdot \mathbf{H})^{-1} \cdot \mathbf{H}^{\text{T}} \cdot \mathbf{W} \cdot \mathbf{y} \qquad (18b)$$

which is the well known solution of a weighted linear least squares problem[1]. However we have not yet proven that the ML solution is able to smooth out the noise. We will see that this is not the case in what follows.

**5.1.2  Unconstrained ML for Gaussian white noise.**  For Gaussian stationary noise, the covariance matrix is diagonal and proportional to the identity matrix:

$$\text{Cov}(\mathbf{y}) = \text{diag}(\sigma^2) = \sigma^2\, \mathbf{I} \iff \mathbf{W} \equiv \text{Cov}(\mathbf{y})^{-1} = \sigma^{-2}\, \mathbf{I}.$$

Then the ML solution reduces to:

$$\mathbf{x}^{(\text{ML})} = (\mathbf{H}^{\text{T}} \cdot \mathbf{H})^{-1} \cdot \mathbf{H}^{\text{T}} \cdot \mathbf{y}.$$

Taking the discrete Fourier transform of $\mathbf{x}^{(\text{ML})}$ yields:

$$\hat{x}_u^{(\text{ML})} = \frac{\hat{h}_u^{\star}\, \hat{y}_u}{|\hat{h}_u|^2} = \frac{\hat{y}_u}{\hat{h}_u} \tag{19b}$$

which is exactly the solution obtained by the direct inversion of the diagonalized image formation equation. As we have seen before, this is a *bad solution* into which the noise is amplified largely beyond any acceptable level. The reader must not be fooled by the particular case considered here for sake of simplicity (Gaussian stationary noise and DFT to approximate Fourier transforms), this disappointing result is very general: *maximum likelihood only is unable to properly solve ill-conditioned inverse problems.*

## 5.2  Constrained maximum likelihood

In the hope that additional constraints such as *positivity* (which must hold for the restored brightness distribution) may avoid noise amplification, we can seek for the *constrained maximum likelihood* (CML) solution:

$$\mathbf{x}^{(\text{CML})} = \arg\min_{\mathbf{x}} \phi_{\text{ML}}(\mathbf{x}) \quad \text{subject to} \quad x_j \geq 0, \quad \forall j. \tag{20b}$$

Owing to the constraints, no direct solution exists and we must use *iterative methods* to obtain the solution. It is possible to use bound constrained version of optimization algorithms such as conjugate gradients or limited memory variable metric methods (Schwartz and Polak, 1997; Thiébaut, 2002) but multiplicative methods have also been derived to enforce non-negativity and deserve particular mention because they are widely used: *RLA* (Richardson, 1972; Lucy, 1974) for Poissonian noise; and *ISRA* (Daube-Witherspoon and Muehllehner, 1986) for Gaussian noise.

---

[1]For general linear least squares problems, rather than directly use Eq. (18b) to find $\mathbf{x}^{(\text{ML})}$, it is generally faster to solve the normal equations by Cholesky factorization of $\mathbf{H}^{\text{T}} \cdot \mathbf{W} \cdot \mathbf{H}$ or to compute the least squares solution from QR or LQ factorizations (see e.g., Press et al., 1992).

**5.2.1    General non-linear optimization methods.**    Since the penalty may be non-quadratic, a non-linear multi-variable optimization algorithm must be used to minimize the penalty function $\phi_{\mathrm{ML}}(\mathbf{x})$. Owing to the large number of parameters involved in image restoration (but this is also true for most inverse problems), algorithms requiring a limited amount of memory must be chosen. Finally, in case non-negativity is required, the optimization method must be able to deal with bound constraints for the sought parameters. Non-linear conjugate gradients (CG) and limited-memory variable metric methods (VMLM or L-BFGS) meet these requirements and can be modified (using gradient projection methods) to account for bounds (Schwartz and Polak, 1997; Thiébaut, 2002). Conjugate gradients and variable metric methods only require a user supplied code to compute the penalty function and its gradient with respect to the sought parameters. All the images in this course where restored using VMLM-B algorithm a limited-memory variable metric method with bound constraints (Thiébaut, 2002).

**5.2.2    Richardson-Lucy algorithm.**    RLA has been obtained independently by Richardson (1972) and Lucy (1974) on the basis of probabilistics considerations. Exactly the same algorithm has also been derived by others authors using the *expectation-maximization* (EM) procedure (Dempster et al., 1977). RLA is mostly known by astronomers whereas EM is mostly used in medical imaging; but, again, they are just different names for exactly the same algorithm. RLA yields the constrained maximum likelihood solution for Poissonian noise:

$$\mathbf{x}^{(\mathrm{RLA})} = \arg\min_{\mathbf{x}} \phi_{\mathrm{Poisson}}(\mathbf{x}) \quad \text{subject to} \quad x_j \geq 0, \quad \forall j \qquad (21a)$$

where:

$$\phi_{\mathrm{Poisson}}(\mathbf{x}) = \sum_j \left[\tilde{y}_j - y_j \log \tilde{y}_j\right] \qquad (22b)$$

and $\tilde{\mathbf{y}} \equiv \mathbf{H}\cdot\mathbf{x}$ is the model of the observed image. Starting with an initial guess $\mathbf{x}^{(0)}$ (for instance a uniform image), RLA improves the solution by using the recursion:

$$x_j^{(k+1)} = \left(\mathbf{H}^{\mathrm{T}} \cdot \mathbf{q}^{(k)}\right)_j x_j^{(k)} \quad \text{with} \quad q_j^{(k)} = y_j/\tilde{y}_j^{(k)} \qquad (23b)$$

where $\mathbf{q}^{(k)}$ is obtained by element-wise division of the data $\mathbf{y}$ by the model $\tilde{\mathbf{y}}^{(k)} \equiv \mathbf{H} \cdot \mathbf{x}^{(k)}$ for the restored image $\mathbf{x}^{(k)}$ at $k$-th iteration. The following pseudo-code implements RLA given the data $\mathbf{y}$, the PSF $\mathbf{h}$, a starting solution $\mathbf{x}^{(0)}$ and a number of iterations $n$:

    *RLA*($\mathbf{y}$, $\mathbf{h}$, $\mathbf{x}^{(0)}$, $n$)

      $\hat{\mathbf{h}} := \mathrm{FFT}(\mathbf{h})$                           (store MTF)

$$\text{for } k = 0, ..., n - 1$$

$$\tilde{\mathbf{y}}^{(k)} := \text{FFT}^{-1}\left(\hat{\mathbf{h}} \times \text{FFT}(\mathbf{x}^{(k)})\right) \qquad (k\text{-th model})$$

$$\mathbf{x}^{(k+1)} := \mathbf{x}^{(k)} \times \text{FFT}^{-1}\left(\hat{\mathbf{h}}^{\star} \times \text{FFT}(\mathbf{y}/\tilde{\mathbf{y}}^{(k)})\right) \qquad (\text{new estimate})$$

$$\text{return } \mathbf{x}^{(n)}$$

where multiplications $\times$ and divisions $/$ are performed element-wise and $\hat{\mathbf{h}}^{\star}$ denotes the complex conjugate of the MTF $\hat{\mathbf{h}}$. This pseudo-code shows that each RLA iteration involves 4 FFT's (plus a single FFT in the initialization to compute the MTF).

### 5.2.3    Image Space Reconstruction Algorithm.

ISRA (Daube-Witherspoon and Muehllehner, 1986) is a multiplicative and iterative method which yields the constrained maximum likelihood in the case of Gaussian noise. The ISRA solution is obtained using the recursion:

$$x_j^{(k+1)} = x_j^{(k)} \frac{(\mathbf{H}^{\mathrm{T}} \cdot \mathbf{W} \cdot \mathbf{y})_j}{(\mathbf{H}^{\mathrm{T}} \cdot \mathbf{W} \cdot \mathbf{H} \cdot \mathbf{x}^{(k)})_j} \qquad (24b)$$

A straightforward implementation of ISRA is:

$\text{ISRA}(\mathbf{y}, \mathbf{h}, \mathbf{W}, \mathbf{x}^{(0)}, n)$

$\quad \hat{\mathbf{h}} := \text{FFT}(\mathbf{h})$              (store MTF)

$\quad \mathbf{r} := \text{FFT}^{-1}\left(\hat{\mathbf{h}}^{\star} \times \text{FFT}(\mathbf{W} \cdot \mathbf{y})\right)$      (store numerator)

$\quad \text{for } k = 0, ..., n - 1$

$\quad\quad \tilde{\mathbf{y}}^{(k)} := \text{FFT}^{-1}\left(\hat{\mathbf{h}} \times \text{FFT}(\mathbf{x}^{(k)})\right)$     ($k$-th model)

$\quad\quad \mathbf{s}^{(k)} := \text{FFT}^{-1}\left(\hat{\mathbf{h}}^{\star} \times \text{FFT}(\mathbf{W} \cdot \tilde{\mathbf{y}}^{(k)})\right)$   ($k$-th denominator)

$\quad\quad \mathbf{x}^{(k+1)} := \mathbf{x}^{(k)} \times \mathbf{r}/\mathbf{s}^{(k)}$        (new estimate)

$\quad \text{return } \mathbf{x}^{(n)}$

which, like RLA, involves 4 FFT's per iteration. In the case of stationary noise ($\mathbf{W} = \sigma^{-2}\,\mathbf{I}$), ISRA can be improved to use only 2 FFT's per iteration:

$\text{ISRA}(\mathbf{y}, \mathbf{h}, \mathbf{x}^{(0)}, n)$

$\quad \hat{\mathbf{h}} := \text{FFT}(\mathbf{h})$            (compute MTF)

$\quad \mathbf{r} := \text{FFT}^{-1}\left(\hat{\mathbf{h}}^{\star} \times \text{FFT}(\mathbf{y})\right)$       (store numerator)

$\quad \text{for } k = 0, ..., n - 1$

$\quad\quad \mathbf{s}^{(k)} := \text{FFT}^{-1}\left(|\hat{\mathbf{h}}|^2 \times \text{FFT}(\mathbf{x}^{(k)})\right)$    ($k$-th denominator)

$\quad\quad \mathbf{x}^{(k+1)} := \mathbf{x}^{(k)} \times \mathbf{r}/\mathbf{s}^{(k)}$        (new estimate)

$\quad \text{return } \mathbf{x}^{(n)}$

Multiplicative algorithms (ISRA, RLA and EM) are very popular (mostly RLA in astronomy) because they are very simple to implement and their very first iterations are very efficient. Otherwise their convergence is much slower than other optimization algorithms such as constrained conjugate gradients or variable metric. For instance, the result shown in Fig. 4c was obtained in 30 minutes of CPU time on a 1 GHz Pentium III laptop by VMLM-B method, against more than 10 hours by Richardson-Lucy algorithm.

Multiplicative methods are only useful to find a non-negative solution and, owing to the multiplication in the recursion, they leave unchanged any pixel that happens to take zero value[2].

At the cost of deriving specialized algorithms, multiplicative methods can be generalized to other expressions of the penalty to account for different noise statistics[3] (Lantéri et al., 2001) and can even be used to explicitly account for regularization (Lantéri et al., 2002).

## 5.3    Maximum likelihood summary

Constraints such as positivity may help to improve the sought solution (at least any un-physical solution is avoided) because it plays the role of a floating support (thus limiting the effective number of significant pixels). This kind of constraints are however only effective if there are enough *background pixels* and the improvement is mostly located near the edges of the support. In fact, there may be good reasons to *not* use non-negativity: e.g. because there is no significant background, or because a background bias exists in the raw image and has not been properly removed.

Figures 4b and 4c show that neither unconstrained nor non-negative maximum likelihood approaches are able to recover a usable image. Deconvolution by unconstrained/constrained maximum likelihood yields noise amplification — in other words, the maximum likelihood solution remains ill-conditioned (i.e. a *small* change in the data due to noise can produce arbitrarily *large* changes in the solution): *regularization* is needed.

When started with a *smooth* image, iterative maximum likelihood algorithms can achieve some level of regularization by early stopping of the iterations *before* convergence (see e.g. Lantéri et al., 1999). In this case, the regularized solution is not the maximum likelihood one and it also depends on the initial solution and the number of performed iterations. A better solution is to explicitly account for additional regularization constraints in the penalty criterion. This is explained in the next section.

---

[2]This *feature* is however useful to implement support constraints when the support is known in advance.
[3]If the exact statistics of the noise is unknown, assuming stationary Gaussian noise is however more robust than Poissonian (Lane, 1996).

# 6. Maximum *a posteriori* (MAP)

## 6.1 Bayesian approach

We have seen that the maximum likelihood solution:

$$\mathbf{x}^{(\mathrm{ML})} = \arg \max_{\mathbf{x}} \Pr\{\mathbf{y}|\mathbf{x}\},$$

which maximizes the probability of the data given the model, is unable to cope with noisy data. Intuitively, what we rather want is to find the solution which has maximum probability given the data $\mathbf{y}$. Such a solution is called the maximum *a posteriori* (MAP) solution and reads:

$$\mathbf{x}^{(\mathrm{MAP})} = \arg \max_{\mathbf{x}} \Pr\{\mathbf{x}|\mathbf{y}\}. \tag{25b}$$

From Baye's theorem:

$$\Pr\{\mathbf{x}|\mathbf{y}\} = \frac{\Pr\{\mathbf{y}|\mathbf{x}\} \Pr\{\mathbf{x}\}}{\Pr\{\mathbf{y}\}},$$

and since the probability of the data $\mathbf{y}$ alone does not depend on the unknowns $\mathbf{x}$, we can write:

$$\mathbf{x}^{(\mathrm{MAP})} = \arg \max_{\mathbf{x}} \Pr\{\mathbf{y}|\mathbf{x}\} \Pr\{\mathbf{x}\}.$$

Taking the log-probabilities:

$$\mathbf{x}^{(\mathrm{MAP})} = \arg \min_{\mathbf{x}} \Big[ -\log\big(\Pr\{\mathbf{y}|\mathbf{x}\}\big) - \log\big(\Pr\{\mathbf{x}\}\big) \Big].$$

Finally, the maximum *a posteriori* solves:

$$\mathbf{x}^{(\mathrm{MAP})} = \arg \min_{\mathbf{x}} \phi_{\mathrm{MAP}}(\mathbf{x}) \tag{26a}$$

where the *a posteriori* penalty reads:

$$\phi_{\mathrm{MAP}}(\mathbf{x}) = \phi_{\mathrm{ML}}(\mathbf{x}) + \phi_{\mathrm{prior}}(\mathbf{x}) \tag{27b}$$

with:

$$\phi_{\mathrm{ML}}(\mathbf{x}) = -\eta \log \Pr\{\mathbf{y}|\mathbf{x}\} + \text{constant} \tag{28c}$$
$$\phi_{\mathrm{prior}}(\mathbf{x}) = -\eta \log \Pr\{\mathbf{x}\} + \text{constant} \tag{28d}$$

where $\eta > 0$. $\phi_{\mathrm{ML}}(\mathbf{x})$ is the *likelihood* penalty and $\phi_{\mathrm{prior}}(\mathbf{x})$ is the so-called *a priori* penalty. These terms are detailed in what follows. Nevertheless, we can already see that the only difference with the maximum likelihood approach, is that the MAP solution must minimize an additional term $\phi_{\mathrm{prior}}(\mathbf{x})$ which will be used to account for regularization.

## 6.2    MAP: the likelihood penalty term

We have seen that minimizing the likelihood penalty $\phi_{\mathrm{ML}}(\mathbf{x})$ enforces *agreement with the data*. Exact expression of $\phi_{\mathrm{ML}}(\mathbf{x})$ should depends on the known statistics of the noise. However, if the statistics of the noise is not known, using a least-squares penalty is more robust (Lane, 1996). In the following, and for sake of simplicity, we will assume Gaussian stationary noise:

$$\phi_{\mathrm{ML}}(\mathbf{x}) = (\mathbf{H} \cdot \mathbf{x} - \mathbf{y})^{\mathrm{T}} \cdot \mathbf{W} \cdot (\mathbf{H} \cdot \mathbf{x} - \mathbf{y})$$

$$= \frac{1}{\sigma^2} \sum_k \left( (\mathbf{H} \cdot \mathbf{x})_k - y_k \right)^2 \tag{29a}$$

$$= \frac{1}{N_{\mathrm{pixel}} \, \sigma^2} \sum_u \left| \hat{h}_u \, \hat{x}_u - \hat{y}_u \right|^2 . \tag{29b}$$

where the latter expression has been obtained by (discrete) Parseval's theorem.

## 6.3    MAP: the *a priori* penalty term

The a priori penalty $\phi_{\mathrm{prior}}(\mathbf{x}) \propto -\log \mathrm{Pr}\{\mathbf{x}\}$ allows us to account for additional constraints not carried out by the data alone (i.e. by the likelihood term). For instance, the prior can enforce agreement with some preferred (e.g. smoothness) and/or exact (e.g. non-negativity) properties of the solution. At least, the prior penalty is responsible of regularizing the inverse problem. This implies that the prior must provide information where the data alone fail to do so (in particular in regions where the noise dominates the signal or where data are missing). Not all prior constraints have such properties and the enforced *a priori* must be chosen with care. Taking into account additional a priori constraints has also some drawbacks: it must be realized that the solution will be biased toward the prior.

## 6.4    Needs of a tunable *a priori*

Usually the functional form of the *a priori* penalty $\phi_{\mathrm{prior}}(\mathbf{x})$ is chosen from *qualitative* (non-deterministic) arguments, for instance:

> "*Since we know that noise mostly contaminates the higher frequencies, we should favor the smoothest solution among all the solutions in agreement with the data.*"

Being qualitative, the relative strength of the *a priori* with respect to the likelihood must therefore be adjustable. This can be achieved thanks to an *hyperparameter* $\mu$:

$$\phi_{\mathrm{MAP}}(\mathbf{x}) = \phi_\mu(\mathbf{x}) \equiv \phi_{\mathrm{ML}}(\mathbf{x}) + \mu \, \phi_{\mathrm{prior}}(\mathbf{x}) . \tag{30}$$

Alternatively $\mu$ can be seen as a Lagrange multiplier introduced to solve the constrained problem: minimize $\phi_{\mathrm{prior}}(\mathbf{x})$ subject to $\phi_{\mathrm{ML}}(\mathbf{x})$ be equal to some value (see Gull's method below). Also note that there can be more than one hyperparameter. For instance, $\alpha$ and $\beta$ in our parameterized Wiener filter in Eq. (11) can be seen as such hyperparameters.

## 6.5 Smoothness prior

Because the noise usually contaminates the high frequencies, smoothness is a very common regularization constraint. Smoothness can be enforced if $\phi_{\text{prior}}(\mathbf{x})$ is some measure of the roughness of the sought distribution $\mathbf{x}$, for instance (in 1-D):

$$\phi_{\text{roughness}}(\mathbf{x}) = \sum_j [x_{j+1} - x_j]^2. \tag{31}$$

The roughness can also be measured from the Fourier transform $\hat{\mathbf{x}}$ of the distribution $\mathbf{x}$:

$$\phi_{\text{prior}}(\mathbf{x}) = \sum_u w_u \, |\hat{x}_u|^2$$

with spectral weights $w_u \geq 0$ and being a non-decreasing function of the spatial frequency, e.g.:

$$w_u = |\mathbf{u}|^\beta \quad \text{with} \quad \beta \geq 0.$$

The regularized solution is easy to obtain in the case of *Gaussian white noise* if we choose a *smoothness* prior measured in the Fourier space. In this case, the MAP penalty writes:

$$
\begin{aligned}
\phi_\mu(\mathbf{x}) &= \phi_{\text{Gauss}}(\mathbf{x}) + \mu \, \phi_{\text{prior}}(\mathbf{x}) \\
&= (\mathbf{H} \cdot \mathbf{x} - \mathbf{y})^{\text{T}} \cdot \mathbf{W} \cdot (\mathbf{H} \cdot \mathbf{x} - \mathbf{y}) + \mu \sum_u w_u \, |\hat{x}_u|^2 \\
&= \frac{1}{\sigma^2} \sum_j \left( (\mathbf{H} \cdot \mathbf{x})_j - y_j \right)^2 + \mu \sum_u w_u \, |\hat{x}_u|^2 \\
&= \frac{1}{N_{\text{pixel}} \, \sigma^2} \sum_u \left| \hat{h}_u \, \hat{x}_u - \hat{y}_u \right|^2 + \mu \sum_u w_u \, |\hat{x}_u|^2
\end{aligned}
$$

of which the *complex* gradient is:

$$\frac{\partial \phi_\mu(\mathbf{x})}{\partial \text{Re}(\hat{x}_u)} + i \, \frac{\partial \phi_\mu(\mathbf{x})}{\partial \text{Im}(\hat{x}_u)} = \frac{2}{N_{\text{pixel}} \, \sigma^2} \, \hat{h}_u^\star \, (\hat{h}_u \, \hat{x}_u - \hat{y}_u) + 2 \, \mu \, w_u \, \hat{x}_u.$$

The root of this expression is the MAP solution:

$$\hat{x}_u^{[\mu]} \equiv \frac{\hat{h}_u^\star \, \hat{y}_u}{\left| \hat{h}_u \right|^2 + \mu \, N_{\text{pixel}} \, \sigma^2 \, w_u} \tag{32}$$

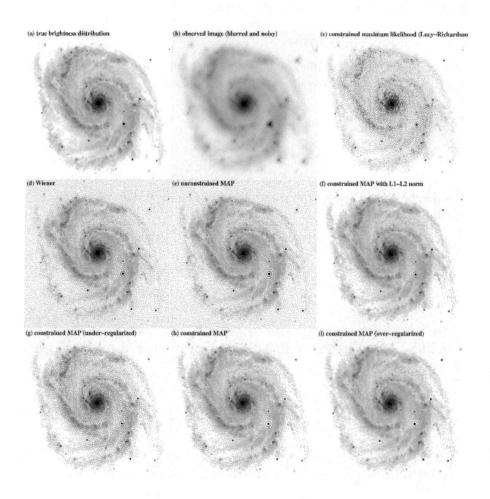

*Figure 4.*    Comparison of deconvolution results by various methods and regularization levels. From top-left to bottom-right: (a) true object, (b) observed image, (c) non-negative maximum likelihood (ML) solution, (d) solution by Wiener inverse-filter, (e) unconstrained maximum *a posteriori* (MAP) solution with quadratic smoothness, (f) non-negative MAP solution with $\ell_1 - \ell_2$ norm, (g) under-regularized non-negative MAP solution with quadratic smoothness, (h) non-negative MAP solution with quadratic smoothness, (i) over-regularized non-negative MAP solution with quadratic smoothness. The regularization level $\mu_{\mathrm{GCV}}$ has been obtained from generalized cross-validation (GCV).

Taking $\mu N_{\text{pixel}} \sigma^2 w_u = \text{E}\{|\hat{n}_u|^2\}/\text{E}\{|\hat{x}_u|^2\}$ or $w_u = \|u\|^\beta$ and $\alpha = \mu N_{\text{pixel}} \sigma^2$, this solution is identical to the one given by Wiener inverse-filter in Eq. (11). This shows that Wiener approach is a particular case in MAP framework.

## 6.6    Other kind of regularization terms

There exist many different kind of regularization which enforce different constraints or similar constraints but in a different way.

For instance, the smoothness regularization in Eq. (31) is quadratic with respect to the sought parameters and is a particular case of Tikhonov regularization:

$$\phi_{\text{Tikhonov}}(\mathbf{x}) = (\mathbf{D} \cdot \mathbf{x})^{\text{T}} \cdot (\mathbf{D} \cdot \mathbf{x})$$

where $\mathbf{D}$ is either the identity matrix or some differential operator. Quadratic regularization may also be used in the case of a Gaussian prior or to find a solution satisfying some known correlation matrix (Tarantola and Valette, 1982):

$$\phi_{\text{GP}}(\mathbf{x}) = (\mathbf{x} - \mathbf{p})^{\text{T}} \cdot \mathbf{C}_{\mathbf{x}}^{-1} \cdot (\mathbf{x} - \mathbf{p})$$

where $\mathbf{C}_{\mathbf{x}}$ is the assumed correlation matrix (when $\mathbf{p} = 0$) or covariance matrix with respect to the prior $\mathbf{p}$ which is also the default solution when there are no data.

The well-known maximum entropy method (MEM) can be implemented thanks to a non-quadratic regularization term which is the so-called negentropy:

$$\phi_{\text{MEM}}(\mathbf{x}) = \sum_j \left[ p_j - x_j + x_j \log\left(\frac{x_j}{p_j}\right) \right]$$

where $\mathbf{p}$ is some prior distribution.

Non-quadratic $\ell_1 - \ell_2$ norms applied to the spatial gradient may be used to enforce a smoothness constraints but avoid ripples around point-like sources or sharp edges. In effect, the $\ell_1 - \ell_2$ norm will prevent small differences of intensity between neighbor pixels but put less constraints for large differences. The effectiveness of using such a norm to measure the roughness of the sought image is shown in Fig. 4f which no longer exhibits ripples around the bright stars.

## 7.    Choosing the hyperparameter(s)

In order to finally solve our inverse problem, we have to choose an adequate level of regularization. This section presents a few methods to select the value of $\mu$. Titterington et al. (1985) have made a comparison of the results obtained from different methods for choosing the value of the hyperparameters.

## 7.1     Gull's approach

For Gaussian noise, the MAP solution is given by minimizing:

$$\phi_\mu(\mathbf{x}) = \chi^2(\mathbf{x}) + \mu \, \phi_{\text{prior}}(\mathbf{x})$$

$$\text{where:} \quad \chi^2 \equiv [\mathbf{m}(\mathbf{x}) - \mathbf{y}]^{\text{T}} \cdot \mathbf{W} \cdot [\mathbf{m}(\mathbf{x}) - \mathbf{y}] \,,$$

where $\mathbf{m}(\mathbf{x})$ is the model. For a perfect model, the expected value of $\chi^2$ is equal to the number of measurements: $E\{\chi^2\} = N_{\text{data}}$. If the regularization level is too small, the MAP model will tend to overfit the data resulting in a value of $\chi^2$ smaller than its expected value. On the contrary, if the regularization level is too high, the MAP model will be too biased by the a priori and $\chi^2$ will be larger than its expected value. These considerations suggest to choose the weight $\mu$ of the regularization such that:

$$\chi^2(\mathbf{x}^{[\mu]}) = E\{\chi^2\} = N_{\text{data}}$$

where $\mathbf{x}^{[\mu]}$ is the MAP solution obtained for a given $\mu$. In practice, this choice tends to *oversmooth* the solution. In fact, the model being derived from the data, it is always biased toward the data and, even for a correct level of regularization, the expected value of $\chi^2$ must be less than $N_{\text{data}}$. For a parametric model with $M$ free parameters adjusted so as to minimize $\chi^2$, the correct formula is:

$$E\{\chi^2\} = N_{\text{data}} - M \,,$$

the difference $N_{\text{data}} - M$ is the so-called number of degrees of freedom. This formula cannot be directly applied to our case because the parameters of our model are correlated by the regularization and $M \ll N_{\text{param}}$. Gull (1989), considering that the prior penalty is used to control the effective number of free parameters, stated that $M \simeq \mu \, \phi_{\text{prior}}(\mathbf{x})$. Following Gull's approach, the hyperparameter $\mu$ and the MAP solution are obtained by:

$$\mathbf{x}^{(\text{Gull})} = \arg \min_{\mathbf{x}} \phi_\mu(\mathbf{x}) \quad \text{subject to:} \quad \phi_\mu(\mathbf{x}) = N_{\text{data}} \qquad (33)$$

with $\phi_\mu(\mathbf{x}) = \chi^2(\mathbf{x}) + \mu \, \phi_{\text{prior}}(\mathbf{x})$ and also possibly subject to $x_j \geq 0, \forall j$. This method is rarely used because it requires to have a very good estimation of the absolute noise level.

## 7.2     Cross-validation

Cross-Validation methods make use of the fact that it is possible to estimate missing measurements when the solution of an inverse problem is obtained.

### 7.2.1     Ordinary cross-validation.     Let:

$$\mathbf{x}^{[\mu,k]} = \arg \min \phi_\mu(\mathbf{x}|\mathbf{y}^{[k]}) \quad \text{where} \quad \mathbf{y}^{[k]} \equiv \{y_j : j \neq k\} \qquad (34)$$

be the regularized solution obtained from the *incomplete data set* where the $k$-th measurement is missing; then:

$$\tilde{y}^{[\mu,k]} \equiv \left( \mathbf{H} \cdot \mathbf{x}^{[\mu,k]} \right)_k \tag{35}$$

is the predicted value of the missing data $y_k$. The *cross-validation* penalty is the weighted sum of the quadratic difference between the predicted value and the real measurement:

$$\mathrm{CV}(\mu) \equiv \sum_k \frac{(\tilde{y}^{[\mu,k]} - y_k)^2}{\sigma_k^2} \tag{36}$$

where $\sigma_k^2$ is the variance of the $k$-th measurement (noise is assumed to be uncorrelated). For a given value of the hyperparameter, $\mathrm{CV}(\mu)$ measures the statistical ability of the inversion to predict the value of missing data. A good choice for the hyperparameter is the one that minimizes $\mathrm{CV}(\mu)$ since it would achieve the best prediction capability. $\mathrm{CV}(\mu)$ can be re-written into a more workable expression:

$$\mathrm{CV}(\mu) = \sum_k \frac{(\tilde{y}_k^{[\mu]} - y_k)^2}{\sigma_k^2 \, (1 - a_{k,k}^{[\mu]})^2} \tag{37}$$

where:

$$\tilde{\mathbf{y}}^{[\mu]} = \mathbf{H} \cdot \mathbf{x}^{[\mu]} \tag{38}$$

is the model for a given $\mu$ and $\mathbf{a}^{[\mu]}$ is the so-called *influence matrix*:

$$a_{k,l}^{[\mu]} = \frac{\partial \tilde{y}_k^{[\mu]}}{\partial y_l} . \tag{39}$$

### 7.2.2 Generalized cross-validation.

To overcome some problems with ordinary cross-validation, Golub et al. (1979) have proposed the generalized cross-validation (GCV) which is a weighted version of CV:

$$\mathrm{GCV}(\mu) = \sum_k w_k^{[\mu]} \frac{(\tilde{y}^{[\mu,k]} - y_k)^2}{\sigma_k^2} = \frac{\sum_k (\tilde{y}_k^{[\mu]} - y_k)^2 / \sigma_k^2}{\left[ 1 - \frac{1}{N_{\text{pixel}}} \sum_k a_{k,k}^{[\mu]} \right]^2} . \tag{40}$$

### 7.2.3 (G)CV in the case of deconvolution.

In our case, i.e. *gaussian white noise* and *smoothness* prior, the MAP solution is:

$$\hat{x}_u^{[\mu]} \equiv \frac{\hat{h}_u^{\star} \, \hat{y}_u}{\left| \hat{h}_u \right|^2 + \mu \, r_u} \quad \text{with} \quad r_u = N_{\text{pixel}} \, \sigma^2 \, w_u.$$

the Fourier transform of the corresponding model is:

$$\hat{\tilde{y}}_u^{[\mu]} \equiv \hat{h}_u \, \hat{\tilde{x}}_u^{[\mu]} = \frac{\left|\hat{h}_u\right|^2 \hat{y}_u}{\left|\hat{h}_u\right|^2 + \mu \, r_u} = q_u^{[\mu]} \, \hat{y}_u \quad \text{with} \quad q_u^{[\mu]} = \frac{\left|\hat{h}_u\right|^2}{\left|\hat{h}_u\right|^2 + \mu \, r_u}$$

and all the diagonal terms of the influence matrix are *identical*:

$$a_{k,k}^{[\mu]} = \frac{\partial \tilde{y}_k^{[\mu]}}{\partial y_k} = \frac{1}{N_{\text{pixel}}} \sum_u q_u^{[\mu]}.$$

In our case, i.e. gaussian white noise and smoothness prior, CV and GCV have the same expression:

$$\mathrm{CV}(\mu) = \mathrm{GCV}(\mu) = \frac{\sum_u \left(1 - q_u^{[\mu]}\right)^2 |\hat{y}_u|^2}{\left[1 - \frac{1}{N_{\text{pixel}}} \sum_u q_u^{[\mu]}\right]^2}$$

which can be evaluated for different values of $\mu$ in order to find its minimum.

## 8.    Myopic and blind deconvolution

So far, we considered image deconvolution assuming that the PSF was *perfectly known*. In practice, this is rarely the case. For instance, when the PSF is measured by a calibration procedure, it is corrupted by some level of *noise*. Moreover, if the observing conditions change, the calibrated PSF can mismatch the actual PSF. It may even be the case that the PSF cannot be properly calibrated at all, because it is varying too rapidly, or because there is no time or no means to do such a calibration. What can we do to cope with that?

In this case, since the unknown are the object brightness distribution $\mathbf{x}$ and the actual PSF $\mathbf{h}$, the MAP problem has to be restated as:

$$\{\mathbf{x}, \mathbf{h}\}^{(\mathrm{MAP})} = \arg\max_{\{\mathbf{x},\mathbf{h}\}} \Pr\{\mathbf{x}, \mathbf{h}|\mathbf{y}\} \tag{41}$$

where the data are $\mathbf{y} = \{\mathbf{y}_{\text{obj}}, \mathbf{y}_{\text{PSF}}\}$, $\mathbf{y}_{\text{obj}}$ being the observed image of the object and $\mathbf{y}_{\text{PSF}}$ being the calibration data. Expanding the previous equation:

$$
\begin{aligned}
\{\mathbf{x}, \mathbf{h}\}^{(\mathrm{MAP})} &= \arg\max_{\{\mathbf{x},\mathbf{h}\}} \Pr\{\mathbf{x}, \mathbf{h}|\mathbf{y}\} \\
&= \arg\max_{\{\mathbf{x},\mathbf{h}\}} \frac{\Pr\{\mathbf{y}|\mathbf{x}, \mathbf{h}\} \, \Pr\{\mathbf{x}, \mathbf{h}\}}{\Pr\{\mathbf{y}\}} \\
&= \arg\max_{\{\mathbf{x},\mathbf{h}\}} \frac{\Pr\{\mathbf{y}|\mathbf{x}, \mathbf{h}\} \, \Pr\{\mathbf{x}\} \, \Pr\{\mathbf{h}\}}{\Pr\{\mathbf{y}\}} \\
&= \arg\max_{\{\mathbf{x},\mathbf{h}\}} \Pr\{\mathbf{y}|\mathbf{x}, \mathbf{h}\} \, \Pr\{\mathbf{x}\} \, \Pr\{\mathbf{h}\} \\
&= \arg\min_{\{\mathbf{x},\mathbf{h}\}} \left( -\log \Pr\{\mathbf{y}|\mathbf{x}, \mathbf{h}\} - \log \Pr\{\mathbf{x}\} - \log \Pr\{\mathbf{h}\} \right)
\end{aligned}
$$

we find that the sought PSF and object brightness distribution are a minimum of:

$$\phi_{\text{myopic}}(\mathbf{x}, \mathbf{h}) = \phi_{\text{ML}}(\mathbf{y}|\mathbf{x}, \mathbf{h}) + \mu_{\text{obj}}\, \phi_{\text{obj}}(\mathbf{x}) + \mu_{\text{PSF}}\, \phi_{\text{PSF}}(\mathbf{h}) \qquad (42)$$

where $\phi_{\text{ML}}(\mathbf{y}|\mathbf{x}, \mathbf{h}) \propto -\log \Pr\{\mathbf{y}|\mathbf{x}, \mathbf{h}\}$ is the likelihood penalty and where $\phi_{\text{obj}}(\mathbf{x}) \propto -\log \Pr\{\mathbf{x}\}$ and $\phi_{\text{psf}}(\mathbf{h}) \propto -\log \Pr\{\mathbf{h}\}$ are regularization terms enforcing the a priori constraints for the sought distributions.

Assuming Gaussian noise and if the calibration data is given by an image of a point-like source, the MAP criterion writes:

$$\begin{aligned}
\phi_{\text{myopic}}(\mathbf{x}, \mathbf{h}) = \quad & (\mathbf{h} \odot \mathbf{x} - \mathbf{y}_{\text{obj}})^{\text{T}} \cdot \mathbf{W}_{\text{obj}} \cdot (\mathbf{h} \odot \mathbf{x} - \mathbf{y}_{\text{obj}}) \\
& + (\mathbf{h} - \mathbf{y}_{\text{PSF}})^{\text{T}} \cdot \mathbf{W}_{\text{PSF}} \cdot (\mathbf{h} - \mathbf{y}_{\text{PSF}}) \\
& + \mu_{\text{obj}}\, \phi_{\text{obj}}(\mathbf{x}) + \mu_{\text{PSF}}\, \phi_{\text{PSF}}(\mathbf{h}) \qquad (43)
\end{aligned}$$

where $\odot$ denotes convolution and where $\mathbf{W}_{\text{obj}}$ and $\mathbf{W}_{\text{PSF}}$ are the weighting matrices for the object and PSF images respectively.

Solving such a *myopic* deconvolution problem is much more difficult because its solution is highly non-linear with respect to the data. In effect, whatever are the expressions of the regularization terms, the criterion to minimize is no longer quadratic with respect to the parameters (due to the first likelihood term). Nevertheless, a much more important point to care of is that unless enough constraints are set by the regularization terms, the problem may not have a unique solution.

A possible algorithm for finding the solution of the myopic problem is to proceed by successive regularized deconvolutions. At every stage, a new estimate of the object is obtained by a first regularized deconvolution given the data, the constraints and an estimate of the PSF, then another regularized deconvolution is used to obtain a new estimate of the PSF given the constraints, the data and the previous estimate of the object brightness distribution:

$$\begin{aligned}
\mathbf{x}^{(k+1)} = \quad & \arg \min_{\mathbf{x}} \Big[ (\mathbf{h}^{(k)} \odot \mathbf{x} - \mathbf{y}_{\text{obj}})^{\text{T}} \cdot \mathbf{W}_{\text{obj}} \cdot (\mathbf{h}^{(k)} \odot \mathbf{x} - \mathbf{y}_{\text{obj}}) \\
& + \mu_{\text{obj}}\, \phi_{\text{obj}}(\mathbf{x}) \Big]
\end{aligned}$$

$$\begin{aligned}
\mathbf{h}^{(k+1)} = \quad & \arg \min_{\mathbf{h}} \Big[ (\mathbf{h} \odot \mathbf{x}^{(k+1)} - \mathbf{y}_{\text{obj}})^{\text{T}} \cdot \mathbf{W}_{\text{obj}} \cdot (\mathbf{h} \odot \mathbf{x}^{(k+1)} - \mathbf{y}_{\text{obj}}) \\
& + (\mathbf{h} - \mathbf{y}_{\text{PSF}})^{\text{T}} \cdot \mathbf{W}_{\text{PSF}} \cdot (\mathbf{h} - \mathbf{y}_{\text{PSF}}) \\
& + \mu_{\text{PSF}}\, \phi_{\text{PSF}}(\mathbf{h}) \Big]
\end{aligned}$$

where $\mathbf{x}^{(k)}$ and $\mathbf{h}^{(k)}$ are the sought distributions at $k$-th iteration. Such an iterative algorithm does reduce the global criterion $\phi_{\text{MAP}}(\mathbf{x}, \mathbf{h})$ but the final

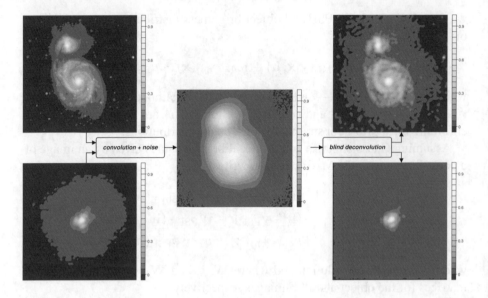

*Figure 5.* Example of blind deconvolution in action. Left: true object brightness and PSF. Middle: simulation of corresponding observed image. Right: the two components found by blind deconvolution. Source: Thiébaut, 2002.

solution depends on the initial guess $\mathbf{x}^{(0)}$ or $\mathbf{h}^{(0)}$ unless the regularization terms warrant unicity.

Myopic deconvolution has enough flexibility to account for different cases depending on the signal-to-noise ratio of the measurements:

- In the limit $\mathbf{W}_{\mathrm{PSF}} \rightarrow +\infty$, the PSF is perfectly characterized by the calibration data (i.e. $\mathbf{h} \leftarrow \mathbf{y}_{\mathrm{PSF}}$) and myopic deconvolution becomes identical to **conventional deconvolution**.

- In the limit $\mathbf{W}_{\mathrm{PSF}} \rightarrow 0$ or if no calibration data are available, myopic deconvolution becomes identical to **blind deconvolution** which involves to find the PSF and the brightness distribution of the object from only an image of the object.

Stated like this, conventional and blind deconvolution appear to be just two extreme cases of the more general myopic deconvolution problem. We however have seen that conventional deconvolution is easier to perform than myopic deconvolution and we can anticipate that blind deconvolution must be far more difficult.

Nevertheless a number of blind deconvolution algorithms have been devised which are able to notably improve the quality of real (i.e. noisy) astronomical images (e.g. Ayers and Dainty, 1988; Lane, 1992; Thiébaut and Conan, 1995).

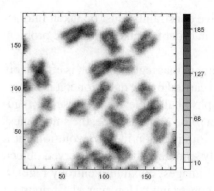

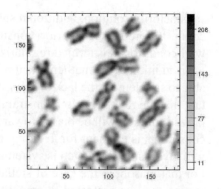

*Figure 6a.* Microscopic image of chromosomes (courtesy Jean-Claude Bernengo from the Centre Commun de Quantimétrie, Université Claude Bernard, Lyon, France).

*Figure 6b.* Microscopic image of chromosomes improved by blind deconvolution.

For instance and following the MAP approach, blind deconvolution involves the minimization of the join criterion (Thiébaut and Conan, 1995; Thiébaut, 2002):

$$\phi_{\text{blind}}(\mathbf{x}, \mathbf{h}) = (\mathbf{h} \odot \mathbf{x} - \mathbf{y}_{\text{obj}})^{\text{T}} \cdot \mathbf{W}_{\text{obj}} \cdot (\mathbf{h} \odot \mathbf{x} - \mathbf{y}_{\text{obj}})$$
$$+ \mu_{\text{obj}}\,\phi_{\text{obj}}(\mathbf{x}) + \mu_{\text{PSF}}\,\phi_{\text{PSF}}(\mathbf{h})\,. \tag{44}$$

Figure 5 shows an example of blind deconvolution by the resulting algorithm applied to simulated data. Of course the interest of blind deconvolution is not restricted to astronomy and it can be applied to other cases for which the instrumental response cannot be properly calibrated for instance in medical imaging (see Fig. 6a and Fig. 6b).

## 9. Concluding remarks

We have seen how to properly solve for the inverse problem of image deconvolution. But all the problems and solutions discussed in this course are not specific to image restoration and apply for other problems.

Inverse problems are very common in experimental and observational sciences. Typically, they are encountered when a large number of parameters (as many as or more than measurements) are to be retrieved from measured data assuming a model of the data – also called the direct model. Such problems are ill-conditioned in the sense that a simple inversion of the direct model applied directly to the data yields a solution which exhibits significant, or even dominant, features which are completely different for a small change of the input data (for instance due to a different realization of the noise). Since the objective constraints set by the data alone are not sufficient to provide a unique and satisfactorily solution, additional subjective constraints must be taken into account. Enforcing such a priori constraints in order to make the inverse problem well-conditioned is termed regularization.

In addition to the mathematical requirement that regularization is effectively able to supplement the lack of information carried by the data alone, it is important that the regularization constraints be physically relevant because the regularized solution will be biased toward the a priori. For instance, the smoothness constraints is efficient for avoiding noise amplification in deconvolution but it will give a solution which is systematically smoother than the observed object. For that reason, in the case of stellar field images, maximum entropy regularization may be preferred to smoothness constraints. However, the important point is more what type of constraints is set by the regularization rather than how exactly this is implemented. There are many different ways to measure the roughness (first or second derivatives, ...) but they shall give solutions which differ only for details.

The level of regularization must be tuned with care: too high and the result will be excessively biased toward the a priori which means that not all the informational contents of data is extracted; too low and the solution will be corrupted by artifacts due to the amplification of the noise. A number of methods have been devised to objectively find the good level of regularization. Among others, generalized cross validation (GCV) chooses the level of regularization for which the solution of the inverse problem has the best capability to predict missing measurements. Generally, solving inverse problems can be stated as constrained optimization of some criterion, the so-called penalty function. There is therefore a strong link between inverse problems and non-linear constrained optimization methods.

## Acknowledgments

All the results shown in this chapter were processed with *Yorick* which is a free data processing software available for Unix, MacOS/X and MS-Windows. The yorick home site is `ftp://ftp-icf.llnl.gov/pub/Yorick`.

The book *"Inverse Problem Theory"* (Tarantola, 1987) is a very good introduction to the subject (the first part of the book, *"Discrete Inverse Problem"*, is freely downloadable at: `http://www.ipgp.jussieu.fr/~tarant`).

## References

Ayers, G.R., Dainty, J.C., 1988, *Iterative blind deconvolution and its applications*, Opt. Lett. **13**, 547

Daube-Witherspoon, M.E., Muehllehner, G., 1986, *An iterative image space reconstruction algorithm suitable for volume etc.*, IEEE Trans. Med. Imaging, **5**, 61

Dempster, A.P., Laird, N.M., and Rubin, D.B., 1977, *Maximum likelihood from incomplete data via the em algorithm*, J. R. Stat. Soc. B **39**, 1

Golub, G.H., Heath, M., and Wahba, G., 1979, *Generalized cross-validation as a method for choosing a good ridge parameter*, Technometrics **21**,215

Gull, S.F., 1989, *Maximum Entropy and Bayesian Methods*, Kluwer Academic, Chapter *Developments in maximum entropy data analysis*, pp 53–72

Lane, R.G., 1992, *Blind deconvolution of speckle images*, JOSA.A **9**, 1508

Lane, R.G., 1996, *Methods for maximum-likelihood deconvolution*, JOSA.A **13**, 1992

Lantéri, H., Soummer, R., Aime, C., 1999, *Comparison between ISRA and RLA algorithms. Use of a Wiener Filter based stopping criterion*, A&AS, **140**, 235

Lantéri, H., Roche, M., and Aime, C., 2002, *Penalized maximum likelihood image restoration with positivity constraints: multiplicative algorithms*, Inverse Problems **18**, 1397

Lantéri, H., Roche, M., Cuevas, O., Aime, C., 2001, *A general method to devise maximum-likelihood signal restoration multiplicative algorithms with non-negativity constraints*, Signal Processing **81**, 945

Lucy, L.B., 1974, *An iterative technique for the rectification of observed distributions*, ApJ **79**, 745

Press, W.H., Teukolsky, S.A., Vetterling, W.T., Flannery, B.P., 1992, *Numerical Recipes in C*, Cambridge University Press, 2nd edition

Richardson, W.H., 1972, *Bayesian-based iterative method of image restoration*, JOSA **62**, 55

Schwartz, A., Polak, E., 1997, *Family of projected descent methods for optimization problems with simple bounds*, Journal of Optimization Theory and Applications **92**, 1

Tarantola, A., 1987, *Inverse Problem Theory*, Elsevier

Tarantola, A., Valette, B., 1982, *Generalized nonlinear inverse problems solved using the least squares criterion*, Reviews of Geophysics and Space Physics **20**, 219

Thiébaut, E., Conan, J.-M., 1995, *Strict a priori constraints for maximum likelihood blind deconvolution*, JOSA.A, **12**, 485

Thiébaut, É., 2002, *Optimization issues in blind deconvolution algorithms*, SPIE **4847**, 174

Titterington, D.M., 1985, *General structure of regularization procedures in image reconstruction*, Astron. Astrophys. **144**, 381

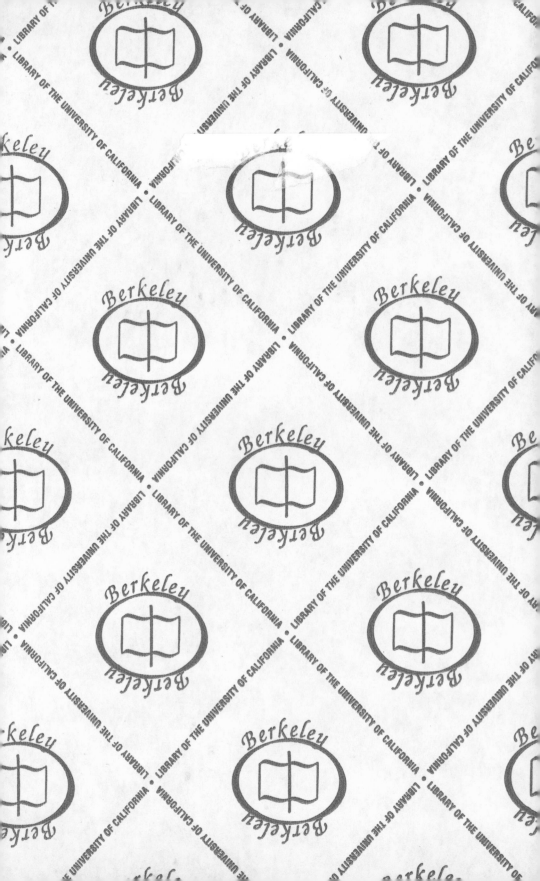